99-ECO-3

EVALUATING ENDOCRINE DISRUPTION IN RECEIVING WATERS: SCREENING FOR BIOMARKERS

by:

Patricia V. Cline, Ph.D.
Golder Associates Inc.

2002

Water Environment Research Foundation

The Water Environment Research Foundation, a not-for profit organization, funds and manages water quality research for its subscribers through a diverse public-private partnership between municipal utilities, corporations, academia, industry, and the federal government. WERF subscribers include municipal and regional water and wastewater utilities, industrial corporations, environmental engineering firms, and others that share a commitment to cost-effective water quality solutions. WERF is dedicated to advancing science and technology addressing water quality issues as they impact water resources, the atmosphere, the lands, and quality of life.

For more information, contact:
Water Environment Research Foundation
601 Wythe Street
Alexandria, VA 22314-1994
Tel: (703) 684-2470
Fax: (703) 299-0742
www.werf.org
werf@werf.org

This report was co-published by the following organizations. For non-subscriber sales information, contact:

Water Environment Federation
601 Wythe Street
Alexandria, VA 22314-1994
Tel: (800) 666-0206
Tel: (703) 684-2452
Fax: (703) 684-2492
www.wef.org
pubs@wef.org

IWA Publishing
Alliance House
12 Caxton Street, London SW1H 0QS,
United Kingdom
Tel: +44 (0)20 7654 5500
Fax: +44 (0)20 7654 5555
www.iwapublishing.com
publications@iwap.co.uk

Library of Congress Catalog Card Number: 2002112605

WERF ISBN 978-1-893664-58-6 IWAP ISBN: 1-84339-650-6 WEF ISBN: 978-1-57278-209-9
ISBN-13 978-1-843-39650-5

Golder Associates Inc.

The research on which this report is based was funded, in part, by the United States Environmental Protection Agency through Cooperative Agreement No. CR-827345-01 with the Water Environment Research Foundation (WERF). Unless an EPA logo appears on the cover, this report is a publication of WERF, not EPA. Funds awarded under the Cooperative Agreement cited above were not used for editorial services, reproduction, printing, or distribution.

ACKNOWLEDGMENTS

The project team acknowledges and thanks Dr. David L. Sedlak and Edward Kolodziej (University of California — Berkeley), Byron Freeman, Ph.D., (Institute of Ecology, University of Georgia) and John Higman (St. Johns River Water Management District) for their contributions to this project.

Report Preparation

Principal Investigator:
Patricia V. Cline, Ph.D.
Golder Associates Inc.

Project Team:
Nancy Denslow, Ph.D.
Protein Chemistry Core Lab, University of Florida

Peter Meyer
Hydrosphere

Stephen Goudey, Ph.D.
HydroQual

Aylin Lewellen
Golder Associates Inc.

Isabel Johnson
Golder Associates Inc.

Project Subcommittee

Philip B. Dorn, Ph.D., Chair, *Shell Global Solutions*

Paul F. Gagliardo, P.E., *The City of San Diego*

James R. Rolston, *Pinellas County Utilities Operations Department*

David L. Sedlak, Ph.D., *University of California — Berkeley*

Ralph G. Stahl, Jr., Ph.D., D.A.B.T., *DuPont Company*

Octavia Conerly, *U.S. Environmental Protection Agency*

Water Environment Research Foundation Staff

Senior Project Manager: Margaret Stewart

ABSTRACT AND BENEFITS

Abstract:

Low concentrations of certain chemicals may alter normal endocrine system functions in various animals, and adversely affect their growth, reproduction, and/or development. Chemical analyses, biomarkers, and in vitro assays of domestic wastewater discharges to surface water bodies suggest that effluent may contribute to this endocrine disruption.

Low concentrations of hormones and chemicals that mimic estrogen have been detected in some domestic wastewater effluents and associated watersheds (e.g. U.S. Geological Survey Reconnaissance, Kolpin et al., 2002). Furthermore, several studies have demonstrated that physiological changes in fish, including the presence of vitellogenin protein in male fish and/or intersex conditions, may be attributed to exposure to estrogenic compounds.

The project team explored approaches to evaluate the potential for biomarker formation as a result of effluent exposures, and the subsequent relevance of emerging assays and physiological measurements on potential adverse impacts to individuals or fish populations in receiving streams.

The results point to a threshold estrogen concentration that must be exceeded to produce biomarkers in fish; exceeding this level even intermittently may produce detectable changes in vitellogenin levels. Reproductive success was impaired significantly in the presence of an anti-estrogen, with less impact for exposure only to estrogen.

Benefits:

- Provides wastewater managers with a framework to address the effects of endocrine disrupting compounds, and estrogenic constituents in particular, on receiving streams.
- Presents laboratory data on biomarker formation and implications for reproductive success.
- Illustrates options for biomarker field study design, implementation, and interpretation.
- Explains the benefits and limitations of biomarker measurements.

Key Words — Effluent, estrogen, hormone, vitellogenin, fathead minnow, bass

TABLE OF CONTENTS

LIST OF TABLES

LIST OF FIGURES

LIST OF ACRONYMS

11-KT	11-ketotestosterone
BOD	biological oxygen demand
bp	base pair
BSA	bovine serum albium
cDNA	copy DNA
cfs	cubic foot per second
CPRG	chlorophenyl-red-D-galactopyranoside
E2	17β-estradiol
EC_{20}	effective concentration to 20%
EDC	endocrine disrupting chemical
EDSTAC	Endocrine Disruptor Screening and Advising Committee
EE2	ethynylestradiol
ELISA	enzyme-linked immunosorbent assay
ER	estrogen receptor
ERα	estrogen receptor alpha
ERE	estrogen response elements
FHM	fathead minnow
GaEPD	Georgia Environmental Protection Division
GSI	gonadosomatic index
H&E	hematoxylin and eosin
HAAS	hormonally active agents
HDPE	high density polyethylene
HSI	hepatosomatic index
HYP	hypertrophied
ICI	7a-[9-[(4,4,5,5,5,-Pentafluoropentyl)sulphinyl]nonyl]-estra-1,3,5(10) -triene-3,17b-diol (anti-estrogen)
kDa	kilodalton
LAS	land application system
LC_{50}	lethal concentration to 50%
Mab-3G2	vitellogenin monoclonal antibody
MeOH	methane
MFO	mixed-function oxygenase
mgd	million gallons per day
mRNA	messenger RNA
NAWQA	National Water Quality Assessment Program
ng	nanogram
nM	nanomolar
NOEC	no observed effect concentration
NPDES	national pollutant discharge elimination system
NRC	National Research Council
PCR	polymerase chain reaction
pg	picogram
pM	picomolar
PnPP	P nitrophenol phosphate

ppb	parts per billion
ppm	parts per million
RIA	radio-immunoassay
rRNA	ribosomal RNA
RT–PCR	reverse transcription-polymerase chain reaction
S	Siemens
SOP	standard operating procedures
SPC	enlarged spaces
SPE	solid phase extraction
SWAP	source water assessment plan
T	testosterone
TBSTZ	tris-buffered saline containing tween-20 and azide
TEG	tri-ethylene glycol
TIE	toxicity identification evaluation
TMB	tetramethylbenzidine
TRC	total residual chlorine
U	undifferentiated
USEPA	United States Environmental Protection Agency
USGS	United States Geological Survey
UV	ultraviolet
Vtg	vitellogenin
WERF	Water Environment Research Foundation
WWTP	wastewater treatment plant
YES	yeast estrogen screen

EXECUTIVE SUMMARY

ES.1 Introduction

In recent years, researchers have voiced concerns about endocrine disruption — when chemicals cause adverse effects by interfering with natural hormonal functions in receptors. This project focused on identifying the potential for estrogenic constituents in domestic wastewater effluents to induce changes in physiological responses (biomarkers) in fish, and relating emerging assays for endocrine disrupting compounds (EDCs) to adverse effects on individuals or populations.

The purpose of this project was to provide environmental managers with a framework to evaluate the potential for endocrine disruption in wild fish exposed to effluents, and to educate them on the uses and limitations of biomarker measurements. The project team conducted laboratory studies and field tests to evaluate biomarker formation and its relationship to significant adverse effects caused by endocrine disruption. Specifically, the team developed and validated in vivo and in vitro assays and developing laboratory exposure protocols to evaluate biomarker formation. The researchers screened effluents and performed a short-term fathead minnow reproductive assay to link biomarker formation and reproductive success. Biomarkers were measured in wild and caged fish in watershed locations with diverse land uses. The findings were compared with emerging literature on this topic.

ES.2 Research Approach

Research for this project was conducted in two phases: laboratory research and a field-based study. Phase 1, the laboratory research, was devoted to validating and applying widely used laboratory-based assays for measuring potential biomarker formation and effects. Laboratory studies included effluent and sediment screening and a 21-day reproductive study to relate biomarker formation with reproductive success. In addition to measuring responses in fish following exposures, the project team evaluated the yeast estrogen screen (YES) assay as a more rapid/sensitive in vitro assay to detect estrogenicity of effluent or surface water samples.

In Phase 2 — a field study conducted in the two North Georgia watersheds — wild fish (largemouth bass, spotted bass, and redeye bass) were collected from the Chattahoochee and Conasauga rivers in the spring and fall of 2001. Caged fathead minnows also were placed at the three Conasauga River sampling stations to compare to the laboratory studies and observations in bass. Overall, 124 fathead minnows and 266 bass were evaluated during the two sampling events. Several biomarkers, considered indicators of endocrine homeostasis and health, also were measured in these fish. Water samples were collected for general water quality tests, the YES assay, and, in the fall, chemical tests for steroids.

The project team's initial research concern was to evaluate the occurrence and ecological significance of EDC biomarkers in watersheds, particularly as they may relate to domestic wastewater effluent discharges. Figure 2-1 illustrates a conceptual model that provides a framework to understand issues in potential biomarker formation and adverse effects.

Biomarkers can be useful as sensitive integrators for exposure to EDCs, when the individual causative agents in effluents may not be chemically identified or may differ among effluent types. Biomarkers can reflect initial biochemical response to contaminants; however, researchers interpreting these results must consider a range of factors (habitat, diet, sex, and age) and whether the response would result in irreversible adverse effects on the individual or population.

Attendees of public meetings may ask such general questions as, "What about endocrine disruptors in your effluent?" Wastewater treatment plant officials' responses should focus the concerns to specific effects or compound types (e.g. estrogenic responses). At that time, they can identify potential assays and propose their approach to interpret and communicate the results.

Wastewater treatment plants have the potential to release compounds, including normal human steroid hormones and synthetic mimics, into the environment. Studies have shown that it is possible to measure the following compounds in sewage outflows: 17 β-estradiol (E2, normal female steroid hormone), ethynylestradiol (the birth control pill), and surfactants such as p-nonylphenol and nonylphenol ethoxylates. Therefore, currently there is an emphasis on determining the estrogenic potential of domestic wastewaters, providing a focus for tools and assays to be applied. However, continued research may indicate that other measurements may be relevant.

A plant manager seeking to answer the question, "What is the likelihood that biomarkers would be detected in the receiving stream?" must consider several issues. Facilities vary in influent characteristics, treatment efficiencies, attenuation potential in the receiving stream, and receptor sensitivity/exposure patterns. Therefore, potential adverse impacts and biomarker formation/interpretation should be placed in the context of site-specific conditions.

To evaluate potential impacts of EDCs in effluent, decisions must be made on project objectives, study protocols and measurements, and interpretation. The following factors, which were included as critical components of this study, may guide future studies:

- Biomarker selection,
- Test species,
- Exposure (concentration/duration) and response,
- Spatial/temporal issues,
- Artifacts and background, and
- Biomarkers and reproductive success.

ES.3 Biomarkers and Assays

ES.3.1 In Vivo Assays

A number of important reproductive biomarkers have emerged to evaluate the effects of EDCs on fish (see Figure ES-1). A full lifecycle test can provide the most comprehensive data set, but can be cost prohibitive. Biomarker measurements can be sensitive, show rapid response times, and provide insight into potential adverse physiological responses. However, there are limitations associated with interpreting these measurements.

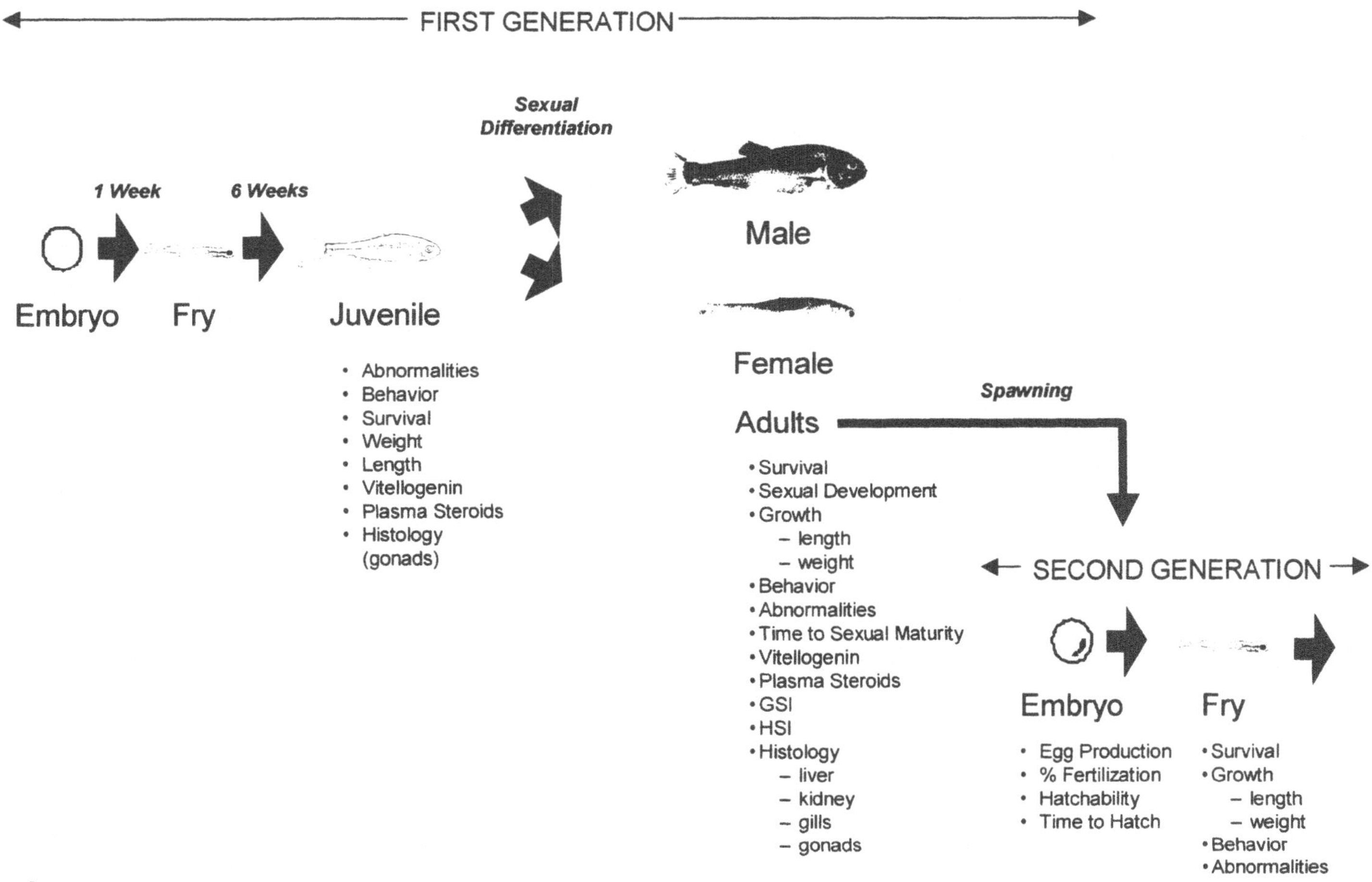

Figure ES-1. Evaluation of Biomarkers and Reproductive Success.

The most commonly reported biomarker associated with potential estrogenic effects of effluents is elevated levels of vitellogenin — the normally female-specific egg-yolk precursor protein — in male plasma. The relevance of this and other biomarkers depends on exposure factors (concentration and duration), and the age, size, and sex of fish. For example, vitellogenin is a primary biomarker in adult fathead minnows. However, studies on medaka exposed to estrogenic compounds 0 to 100 days post-hatch show intersex conditions (ova–testis) as a sensitive endpoint. Therefore, the biomarker selection is linked to the test species and exposure regime. Before selecting biomarkers, the availability of validated assays for the test species of interest is needed, and preliminary assays are recommended before effluent exposures.

Over the course of this 2-year project, the project team used several biomarkers to measure potential endocrine modulating effects of wastewaters on fish, but vitellogenin was the primary biomarker used. Various studies have included additional measurements as applicable — such as gonadosomatic index (GSI), hepatosomatic index (HSI), histopathology, and plasma hormone levels. In addition, two molecular-based assays to measure the induction of genes for fathead minnow vitellogenin and estrogen receptor were measured in one laboratory study and also in the caged fathead minnows from the field study. Fecundity (egg production) was measured in a 21-day flow-through study.

Observations from the adult fathead minnow studies suggest the following.

- Vitellogenin provides a more significant response than GSI with short exposure durations (e.g. 7 days) at environmentally relevant concentrations.
- Changes in secondary sex characteristics (feminization of male fathead minnows) were observed with exposures to relatively high concentrations (1 nM) of E2. These were not
- observed at lower concentrations (0.1 nM) and shorter durations (7 to 10 days), when vitellogenin levels were significantly above controls.
- Additional research is needed; however, the project team's research suggests that mRNA measurements may be more sensitive than protein measurements to estrogen exposure.
- Adult fathead minnows were obtained from various suppliers. Unexpectedly, male control fish showed variable detectable levels of vitellogenin, generally with lower levels in the fall. The source of test specimens and acclimation are critical components in study design.
- It is difficult to obtain sufficient fathead minnow plasma to analyze vitellogenin and hormones without further refinement of protocols. Vitellogenin levels in plasma are more stable (they fluctuate less on a daily cycle) than hormones.
- The EDC screening program indicates (anti)-estrogenic or (anti)-androgenic effects in adult fathead minnows can be detected in the 21-day reproductive study. However, reduction in vitellogenin in females would not necessarily be observed in shorter-term effluent exposures because of the longer half-life of vitellogenin in plasma, inherent variability in protein levels in females, and likely low concentrations of these types of EDCs in effluents.
- Fecundity provides a more direct measure of an ecologically relevant effect than protein biomarkers. Reproductive tests are more costly, require conditioning to obtain

comparable egg production in test vessels before exposures, require longer exposures, and are potentially more variable than vitellogenin measurements.

Because many assays are not standardized, there is likely to be variability in results from different studies. Differences in vitellogenin assay protocols (e.g. antibody type, sample, and standard curve preparation) can lead to differences in detection limit and/or reported values. However, relative responses may be consistent.

Levels of plasma steroid hormones (testosterone, 11-ketotestosterone, and estradiol) have been measured in several studies; altered ratios of 11-ketotestosterone and estradiol may indicate exposures to EDCs. Additional information on baseline levels and circadian or seasonal patterns may be needed to interpret results appropriately.

New assays are likely to evolve, and current assays will become more standardized. As more studies are performed, a better foundation will emerge for interpreting results as natural patterns of physiological responses are documented and relationships to whole-organism responses are clarified.

ES.3.2 In Vitro Assays

Several assays are being developed that can be used to evaluate the estrogenicity of chemicals and water samples. The YES assay has been used to evaluate the estrogenicity of treatment plant influent, effluents, biosolids, and surface water samples. The recombinant yeast contains the human estrogen receptor, which is activated in the presence of bioavailable estrogen or estrogen mimics. This activates a reporter gene, inducing a color change from yellow to red.

The assay can be performed quickly (approximate incubation time of 4 days) on several samples simultaneously (96-well plate). During incubation, the yeast grows and color develops. The plate is read at two wavelengths — one to detect the development of the red color, the other to measure turbidity in the well associated with the yeast.

When samples are concentrated sufficiently, the YES assay can detect estrogenicity that would be below thresholds for induction of vitellogenin in fish. This assay may be best used in addition to in vivo assays to evaluate significance. However, it also can guide understanding of temporal variability of effluents, removal efficiencies, and surface water spatial patterns.

The YES assay was used during this 2-year project to evaluate stability of exposure solutions and to screen effluents and surface water samples; over time, protocols evolved. The assay was highly responsive using control solutions of E2; application to environmental samples was more complex. The project team noted the following about the assay.

- False positive responses from bacterial contamination of the yeast occurred during some initial tests.
- Reduced response of E2 spikes in surface water samples compared to control water suggested that components in environmental samples can inhibit response.

- Visual observations of color development did not always correlate with absorbance readings.

Continued application of this assay to environmental samples is needed to standardize protocols and improve data interpretation. If this assay is to be used as a component of an effluent study, the project team recommends that all raw data (absorbance readings) for the samples and the standard curve be provided with a picture of the developed plate. Protocols should include the procedures for standards, as well as samples.

ES.4 Test Species

The selection of test species for site-specific laboratory or field studies may be based on relevance to the watershed, assay development, baseline of biomarker data, ease of handling and/or suppliers, and species sensitivity. Test species most frequently reported for laboratory studies to screen EDCs include fathead minnows, medaka, juvenile trout, and zebrafish. A range of species, including carp, bass, flounder, and roach, has been evaluated for wild fish studies.

Fathead minnows, which are the staple of aquatic toxicology and the Environmental Protection Agency's (EPA's) preferred test species, were selected as the primary species for laboratory tests in this project. Fathead minnows begin their annual reproductive cycle in mid-spring and finish in the late fall. They are fractional spawners, whose cycles are driven by water temperature and photoperiod. Because they are small fish, their blood volume is limited. During the fathead minnow reproductive period, females are able to spawn eggs every 4 to 7 days. Because they are not synchronized with each other, it is difficult to measure changes in plasma levels of hormones and vitellogenin in females. But because considerable information exists regarding this universal toxicological model and because the fish reach maturity relatively quickly, it is possible to conduct full lifecycle studies to measure important reproductive endpoints.

Although fathead minnows were the dominant laboratory species, one study compared the responses of fathead minnows and juvenile rainbow trout. Data suggest that plasma vitellogenin levels will increase in trout at concentrations below the threshold for fathead minnows.

The North Georgia field study targeted largemouth bass, which have been described as being sensitive to endocrine disruption. Reproductive cycles of bass differ from fathead minnows. Largemouth bass are relatively well synchronized in their reproductive cycles and typically spawn only once in the late spring. They have sufficient blood volume to allow collection of enough plasma to test for vitellogenin and sex steroid hormone levels. Females' gonads mature in unison, making it easier to stage their maturity level. Some studies have been completed on their annual reproductive cycle, which allowed the project team to interpret its data.

ES.5 Exposure (Concentration/Duration) and Biomarker Response

ES.5.1 Thresholds/Endpoints/Species Sensitivity

Changes in physiological responses (biomarkers) as a result of exposure to effluent constituents are a function of concentration (E2 equivalents), exposure duration, and species sensitivity. Research studies vary in design of the in vivo tests. However, the maximum acceptable toxicant concentration (MATC) derived from current published studies illustrates ranked responses as a function of E2 concentrations. These suggest that below concentrations where vitellogenin induction would be detected in fathead minnows (~ 0.1 nM), ova–testis would develop in medaka (post-hatch exposure) and vitellogenin would be measurable in juvenile trout. The rankings on this table are likely to become refined over time as additional studies are published.

ES.5.2 Positive E2 Controls and Effluent Screening for Vitellogenin Induction in Male Fish

To address the question, "What is the likelihood that a detectable EDC biomarker would result from effluent exposure?" the project team's study focused primarily on vitellogenin induction in adult male fathead minnows. Male vitellogenin levels for various E2 exposure regimes are illustrated in Figure 4-3. Based on these tests, researchers determined that a 7-day static renewal exposure provided sufficient contact with 0.1 nM of E2 to show a significant increase in vitellogenin, while limiting the volume of effluent required to perform a laboratory screening assay. At this lower E2 concentration, which may be more representative of effluents, significant changes to GSI were not observed from a 7-day exposure. This suggests that vitellogenin is the more sensitive biomarker for estrogenic response.

Of the 15 effluents screened for fathead minnow vitellogenin induction using the static renewal protocol, the initial seven samples are shown in Figure 5-1; one of these effluents shows significant vitellogenin induction, and two others show slight elevation. The screening was done on grab samples that were not optimized to maximize potential estrogenicity.

In addition to the static renewals, support was provided for a longer (~21 day) effluent flow-through study to compare responses in fathead minnows and juvenile trout. Fathead minnows did not respond, suggesting that the concentration of estrogen-like substances was below their required threshold. However, with the same effluent, juvenile rainbow trout responded by inducing vitellogenin to levels that were significantly higher than background.

The project team drew the following conclusions from these studies.

- Short-term male fathead minnow exposure to effluent grab samples can produce increased vitellogenin levels. However, only one sample showed a response greater than the 0.1-nM E2 (27-ng/L) positive control.
- These preliminary screening assays illustrate a rapid response for this biomarker. However, the grab sample should reflect a time of likely greatest response. Additional renewals at facility could better capture periods of peak activity.
- The comparison of juvenile trout and fathead minnows demonstrates that trout are more sensitive to vitellogenin induction.

- The fact that vitellogenin was detected in the negative controls of the juvenile trout may reflect some prior exposure to estrogenic compounds that increased sensitivity in this exposure.

ES.5.3 Changes in Vitellogenin Protein and mRNA in a 7-day Static Exposure

The duration of exposure to produce a measurable biomarker response was evaluated further, comparing patterns of vitellogenin protein and mRNA. The underlying principles are that following exposure to estrogenic compounds, mRNA concentrations would increase and ultimately result in accumulation of vitellogenin in plasma. When no further exposure to E2 occurs, mRNA concentrations would decrease over a few days, while vitellogenin would remain elevated for weeks.

The project team performed a 7-day static exposure of female fathead minnows to 1-nM concentrations of E2, ICI 182,780 (a potent anti-estrogen), 5-androstan-17 methyl-17-ol-3-one (an androgen), and flutamide (an anti-androgen). This time period was sufficient to induce plasma vitellogenin to 30 mg/ml in response to 1 nM of E2 by 24 hours. The level of vitellogenin was induced even more, to 140 mg/ml, after 7 days. In this same timeframe, only flutamide appeared to also induce plasma vitellogenin to 20 mg/ml after 7 days.

In this same 7-day timeframe, vitellogenin mRNA was induced for both E2 and flutamide exposures, in concordance with the plasma vitellogenin measurements. The mRNA assay not only is more sensitive than the plasma vitellogenin assay, but also more direct since mRNAs have short half-lives and their presence is related directly to the action of a hormone at the gene level. The surprising result from this experiment was that the anti-estrogen, ICI, did not decrease vitellogenin mRNA in females below controls as anticipated.

In general, plasma vitellogenin is a long-term biomarker. It is present in males for up to a month after an exposure, depending on the concentration of contaminant used. Vitellogenin mRNA, on the other hand, is an immediate-exposure biomarker, since the half-life of vitellogenin mRNA is short — about 5 days. Thus, the two biomarkers track different types of events. Plasma vitellogenin would track any exposure in the last 30 days, while vitellogenin mRNA would track any exposure in the last 5 days.

ES.6 Study Implementation Issues

During the course of the project, researchers conducted experiments to evaluate sample-storage issues and potential laboratory contributions to biomarker formation. For example, the project team set up a series of tanks that contained various plastics or laboratory cleaning agents. Several of these materials produce an estrogenic response. To avoid biasing the findings, these compounds were not used in the experiments.

The YES assay was used to measure potential estrogenicity of water stored in plastic and glass containers, and potential decreases in response of E2 control solutions resulting from sorption.

ES.7 Background

Sediments have the potential to sorb chemicals from water and reduce their potential to impact organisms. Conversely, sediments with high levels of these retained chemicals may be an ongoing source of impact to the water column. The project team conducted a sediment study to evaluate each of these factors (see Fig 6-9). In the first experiment, control sediment was added to one of two tanks containing 1 nM of E2 solution. This demonstrated that adding sediments to the tanks lowered the effective E2 concentration as measured by the vitellogenin assay. This suggests that sediments play a large role in sequestering xenobiotics out of the water column and/or increasing degradation rates. The project team measured the potential for a number of sediments to release hormonally active agents into the water column. Team members measured the induction of plasma vitellogenin as the hormonally active agent was released from the sediment. The results of this study suggest that sediments are a possible reservoir for hormonally active agents in the environment, and underscore their potential as a source of estrogen-like contaminants.

As shown in figure ES-2, vitellogenin concentrations in bass are quite different in the two seasons and between males and females (see Figure 6-4). The levels for both males and females in the spring were much higher than in the fall, reflecting anticipated seasonal patterns for this biomarker. Similar to vitellogenin, the project team saw a much greater differential in GSI measurements among the sites in the spring than in the fall. Determining differences among stations during each season for these biomarkers must distinguish changes from the normal cycles. No significant differences for these biomarkers in the field study.

ES.8 Biomarkers and Reproductive Success

Like vitellogenin levels, physiological indicators in male plasma can be difficult to interpret because of limitations in understanding acclimation, normal levels/variability, and impacts on reproductive success. Reproductive success is commonly considered to be a more ecologically relevant endpoint for EDCs. To evaluate the relationship between vitellogenin and reproductive success in fathead minnows, the project team performed 21-day flow-through exposures. Fathead minnow responses were compared in a 21-day flow-through experiment for exposure to estrogen (E2), anti-estrogen (ICI), a mixture of both, and the negative control. Concentrations of control solutions were 0.5 nM, significantly above the threshold for vitellogenin induction and higher than would be expected for most effluent samples. Results of male vitellogenin levels are shown in Figure ES-3. Plasma vitellogenin levels in both males and females increased significantly with exposure to E2, alone or in combination with ICI. GSI decreased significantly, compared to controls, only with exposure to the ICI–E2 mixture.

The project team measured reproductive success by noting the cumulative number of eggs spawned. Significant decreases in egg production were observed for the ICI and the ICI–E2 mixture, and a lesser decrease was observed for the E2 exposure. Therefore, the presence of the anti-estrogen had a significant impact on egg production, but did not contribute to vitellogenin increases.

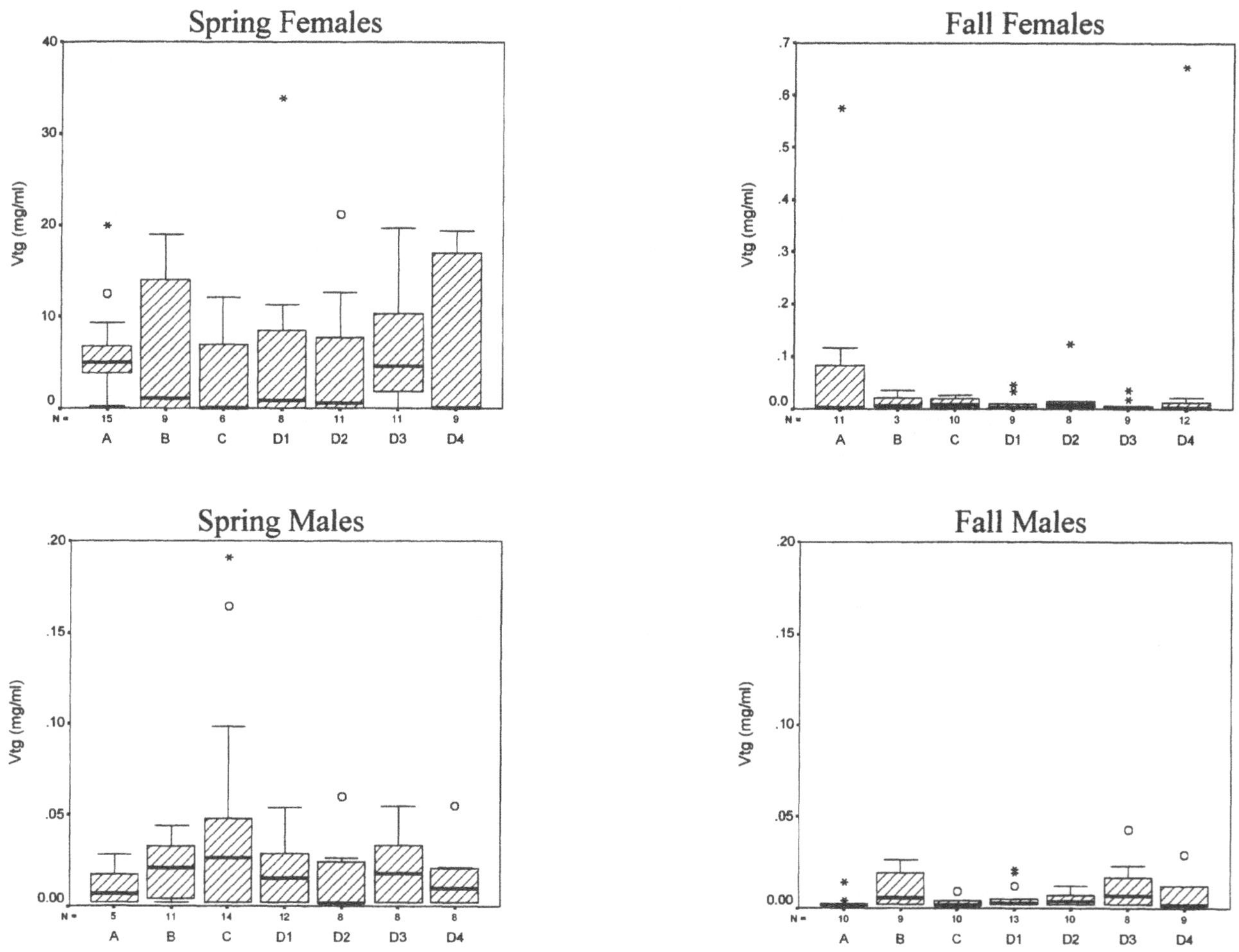

Note:
A = Forested; B = Agricultural; C = Urban; (Conausaga Stations shown on Figure C-3)
D1 = Station #1; D2 = Station #2; D3 = Station #3; D4 = Station #4 (Chattahoochee Stations shown on Figure C-2)
Station A predominantly redeye.

Figure ES-2. Comparison of Plasma Vitellogenin Levers for Bass Collected at Sampling Locations on the Conasauga and Chattahoochee Rivers (Spring and Fall 2001).

These two compounds may have physiological effects in fish that do not occur through the estrogen receptor. This underscores the necessity to look at multiple biomarkers of reproduction when assessing effects and to consider the complexity of potential interferences in reproductive function.

ES.9 Recommendations

Basic research in several areas is needed to improve the use and interpretation of biomarkers and the relevance in prioritizing actions to protect aquatic resources. For example, in regard to analytical techniques, the project team identified the need to:

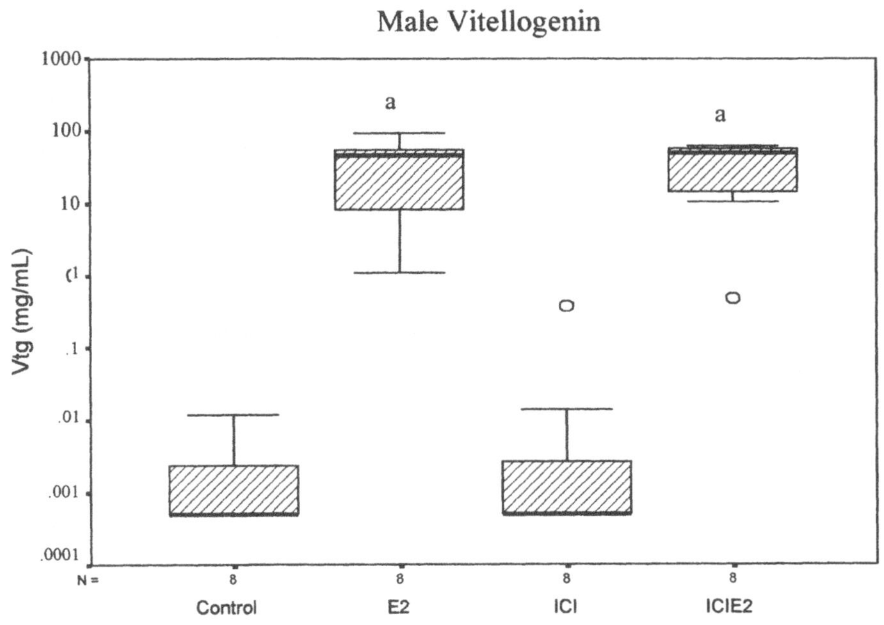

Control = Triethylene glycol solvent control; E_2 = 17B-estradiol (0.5nM);
ICI = ICI #182,780 (anti-estrogen) (0.5 nM); $ICIE_2$ = Mixture 0.5nM ICI + 0.5nM

[a] Statistically different from control.

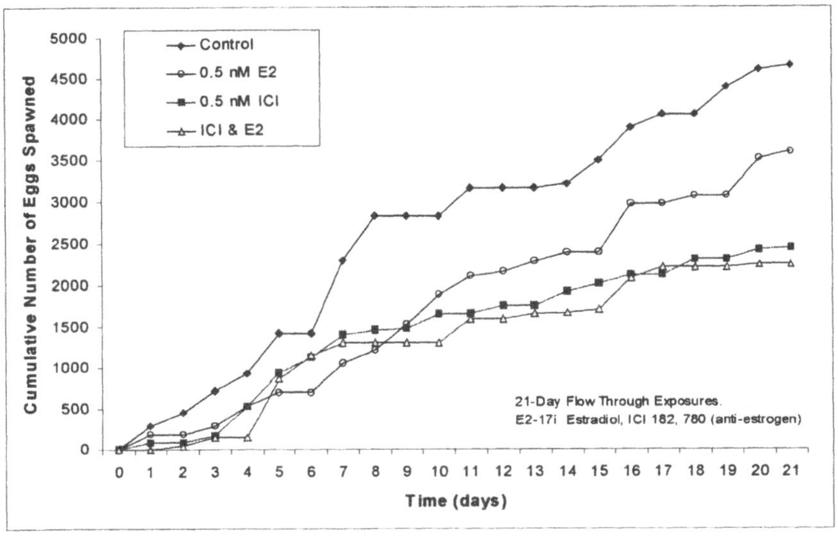

Figure ES-3. Short-term Fathead Minnow Reproductive Study: Plasma Vitellogenin and Fecundity Response to Estrogen and Anti-Estrogen.

- Increase standardization and validation of protocols, including biases and interferences.
- Develop additional assays to capture the wider range of critical effects.
- Explore the use of emerging technologies, including a cDNA chip being developed to measure changes in multiple genes at one time, to provide a more comprehensive fingerprint of the biochemical status of a receptor.
- Further develop in vitro assays of environmental media to document correlations with in vivo responses.

The project team also identified the need to improve understanding of EDCs in effluents and surface water bodies, including spatial and temporal patterns, by:

- Providing additional data on environmental levels of surface waters and/or sediments.
- Using in vitro assays and/or screening-level in vivo assays to evaluate the range of effluent responses.
- Evaluating potential analysis of sediments downstream from a discharge point to determine if these reflect a more cumulative response.

In terms of biological response, the project team concluded that:

- Data suggest possible increased vitellogenin sensitivity during periods of active spawning when natural levels tend to be elevated, and inhibited response in the fall. This should be verified further, and the implications for mechanisms for controlling the formation of this biomarker should be clarified.
- A seasonal study with fathead minnows should be performed to determine if periods of reduced vitellogenin response also correlate with reduction in potential adverse impacts on developing oocytes.

While research in these areas progresses, a facility may have preliminary data that suggest biomarkers may be detected associated with estrogenic compounds in its effluent. Based on current knowledge, the facility may consider the following options.

- Place this information in the context of existing understanding of effluent receiving stream characteristics and studies.
- If action is required, use in vitro assays to characterize temporal patterns in effluent characteristics.
- Carefully consider seasonal and background considerations in evaluating biomarker data.
- Critically evaluate and consider uncertainties in all data collected.
- Although it is more expensive, consider following the EPA protocol that includes evaluating reproductive success in addition to biomarker formation.

CHAPTER 1.0

INTRODUCTION

1.1 Project Overview

An increasing body of knowledge is being accumulated on the presence of endocrine disruptor biomarkers in aquatic systems, particularly as related to estrogenic compounds in discharges from domestic wastewater treatment plants (WWTPs). This is a challenging technical issue because of the complexity of the endocrine system, and the use of an array of new endpoints and techniques that are not completely understood in terms of predicting a detrimental effect at the population level.

This is the final report of a 2-year project, "An Assessment of the Occurrence and Ecological Significance of Endocrine Disrupting Chemicals in Watersheds." The project included a range of laboratory and field studies that interface with current surveys being performed in complex watersheds. These provide a foundation to critically evaluate current studies, and a framework to implement site-specific studies. One goal of the project team was to address such questions of potential broad interest to domestic WWTP managers as:

- What is the likelihood that my effluent would produce a detectable biomarker?
- What other sources (natural or anthropogenic) may be present in this watershed that may also produce these effects?
- What is the potential area affected by my facility's discharge?
- How long would a fish be exposed before a response could be observed? Is the response reversible?
- Were the fish collected at an appropriate time in their lifecycle? How does this influence the observed measurements? What is the geographic range of the fish? Which species are most sensitive?
- What is the possibility that the observed histology is an artifact (i.e., occurs at some frequency in natural populations)?
- How accurate or precise are the assays? What common problems are encountered?
- If a biomarker is detected, will reproduction be impaired? What is an appropriate endpoint?

The subject of "ecological significance of endocrine disruption biomarkers in watersheds" is broad and complex. This report seeks to address this issue as it relates primarily to domestic wastewater discharges to receiving streams, and the nature and magnitude of potential impacts resulting from the presence of estrogenic compounds. In addition, while studies have been published on the impact of endocrine disrupting chemicals (EDCs) on invertebrates, the focus of this study remained on potential impacts on fish. This report is intended to consolidate core issues and techniques to address the broader concerns and endpoints beyond the traditional effluent acute and chronic toxicity testing that may be important in protecting aquatic resources.

The project team sought to meet the project objectives (Table 1-1) by conducting critical evaluations of assays, focused laboratory experiments to address selected questions, and ongoing literature review to compare results from this study and expand the foundation for summarizing issues. The implementation process included applying a series of biomarker measurements, with protocols refined and/or modified as the study continued. An overview of the various tests/studies performed during this project is shown in Figure 1-1.

Table 1-1. Endocrine Disruptor Biomarkers in Watersheds: Project Objectives

- Define uses, applicability, and limitations of biomarker measurements.
- Clarify current fundamental understanding of biomarkers of endocrine disruption in watersheds including background conditions, natural fluctuations in hormone levels seasonally, and relationship of biomarkers to impairment of reproduction.
- Screen selected effluents for nature and magnitude of biomarker response.
- Extrapolate laboratory studies to complexities of field investigations linking potential for measurable impacts when discharged into receiving waters of varying characteristics.
- Document boundary conditions to guide hypotheses for potential watershed impacts including range of species sensitivity, dilution effects, and other contributing factors (e.g., contaminated sediment).
- Implement a field study to test hypotheses developed from the literature and laboratory studies, to include background sites, seasonal sampling, agricultural areas and/or contaminated sediment areas.
- Provide guidance for managers on issues related to endocrine-modulating chemicals in watersheds based on the current study.

1.2 Key Concepts

For many involved in managing and operating WWTPs, there already is an integration of a number of technical backgrounds to address the traditional regulatory requirements for monitoring effluents (e.g., chemical testing and bioassays), conducting watershed assessments, and evaluating business and engineering alternatives associated with treatment. Many of the traditional toxicity testing programs already have been designed to evaluate growth, reproduction, and development in vertebrates. These programs capture many of the critical endpoints required for identifying endocrine disruption and identifying chemicals that alter

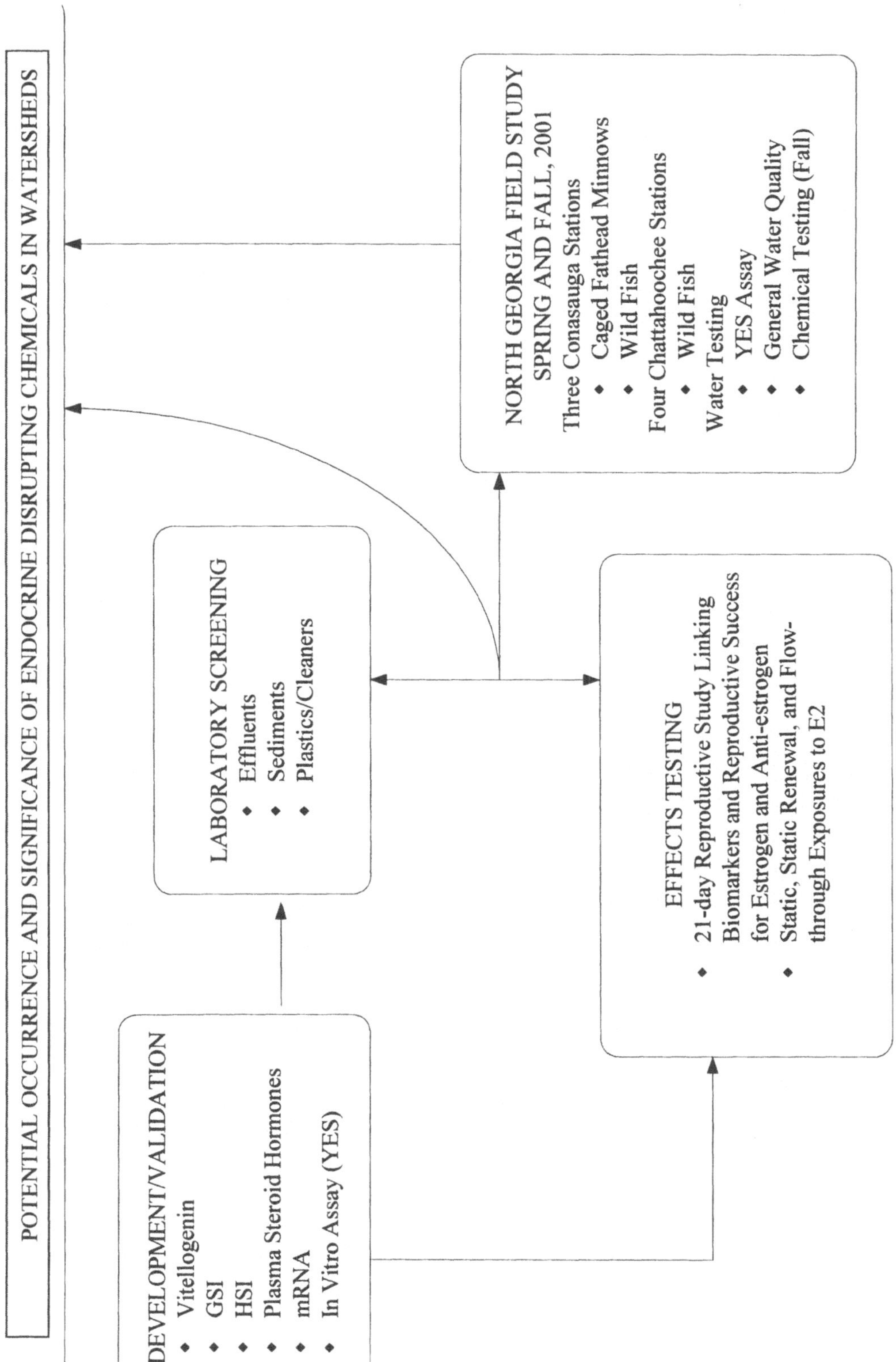

Figure 1-1. Project Overview

development, reduce fertility, produce abnormal behaviors, etc. A traditional wastewater management framework can be applied to address the importance of potential adverse impacts from endocrine disruption. New techniques and endpoints, however, also may be applied to detect this type of impairment. Because these new methods have not undergone rigorous validation, it is important to continue to pay close attention to the information gathered from measuring changes in biomarkers, and to determine whether the information really points to significant alterations in wildlife endocrine systems.

A number of books and articles can provide a background on endocrine disruptors, in vitro assays, biomarker measurement techniques, and the regulatory challenge. Van Der Kraak et al. (2001) summarize why this issue is so complicated. For example:

- Many chemicals, including natural and synthetic hormones and plant constituents as well as selected pesticides, surfactants, and plasticizers, are hormonally active. They represent a broad range of chemical characteristics making the exposure/response relationships difficult to establish.
- Exposure to an EDC does not necessarily result in an adverse outcome and may depend on the level, duration, and timing of exposure.
- The endocrine system is very complex. Multiple hormones work through vastly different mechanisms to produce the same effect on an individual. In addition, hormones may cause different effects depending on the individual's nutritional status, age, gender, and species.
- Adverse effects may occur during a critical period of development, but not be manifested until later in the individual's lifetime.
- Exposure to certain chemicals may interfere with the biosynthesis, transport, availability, or metabolism of hormones or proteins, as well as interact directly with receptors and signaling processes.
- Defining the association between chemical exposure and physiological dysfunction is particularly challenging because a number of normal physiological factors also can impact growth, reproduction, and survival.
- Physiological endpoints (e.g., hormone levels, histological changes, and hormonally regulated proteins) may not relate directly to the performance of the individual or population fitness. Furthermore, these static measurements may not capture the important dynamic nature of an individual's response mechanism.

Adding to the complexities inherent in understanding the endocrinology of a range of aquatic species, new techniques are being developed to provide additional tools to clarify mechanisms and responses. As biotechnology expands, tools are being developed to measure target proteins and identify DNA or messenger RNA sequences. As these tools are applied to studies of ecological effects, their uses and limitations must be defined. The range and patterns of biomarker variability present under normal, unimpacted conditions must be determined to provide an adequate baseline and reference point for interpretation. The methods used in this study are discussed with the intent of broadening understanding of the application and data interpretation.

1.2.1 Endocrine Disruptor

The term endocrine disruptors gained popularity when Colborn and Clement (1992) used it to refer to potential adverse effects of environmental contaminants via hormonal mechanisms. The U.S. Environmental Protection Agency (USEPA) formed the Endocrine Disruptor Screening and Advising Committee (EDSTAC) in 1996 to develop a process for testing and classifying chemicals that may have the potential to adversely impact normal endocrine-related functions. The EDSTAC (1998) has defined an endocrine disruptor as "an exogenous chemical substance or mixture that alters the structure or functions(s) of the endocrine system and causes adverse effects at the level of the organism, its progeny, populations, or subpopulations of organisms, based on scientific principles, data, weight of evidence, and the precautionary principle." In the context of this report, constituents identified as potential "endocrine-disrupting chemicals" are designated as EDCs.

The National Research Council (NRC; 1999) conducted an independent study of endocrine disruption and chose the term hormonally active agents, or HAAs, without regard to the specific mode or mechanism of action of the chemical. This was agreed upon because few studies have shown conclusively that exposure actually leads to endocrine system dysfunction. The NRC also broadened the issue because it became apparent that the initial focus only on estrogen-like compounds was too narrow. Other compounds needed to be studied as well, including compounds that behaved as anti-estrogens, androgens and anti-androgens, or compounds that affected thyroid function.

The new term hormonally active agents, however, has not displaced the use of endocrine disruption in the public's view or in published literature. The NRC report "Hormonally Active Agents in the Environment," however, does provide insights into the challenges, complexities, and controversies that continue to surround this subject, including disagreements on the interpretation of such basic terms as disruption and adverse.

While it is important to be aware of these issues, the focus of this study was to provide, to the extent available, specific information on the nature of responses and uncertainties as they relate to research on effluents and their underlying constituents.

1.2.2 Biomarkers/Bioindicators

Biomarkers are molecules, biochemical pathways, or cellular processes in experimental animals that change in response to contaminated habitats and are indicative of the exposure. Normally, these are quantifiable and the magnitude of the response is proportional to the extent of contamination.

NRC (1987) defined biomarkers as "[i]ndicators of events in biological systems or samples" and (1989) as "[t]ools that can be used to clarify the relationship, if any, between exposure to a xenobiotic substance and disease." Biomarkers generally are classified into one of following three categories, based on their relation to the exposure–disease continuum.

- *Biomarkers of exposure* are defined as the identification of an exogenous substance within the biologic system, the interactive product between a xenobiotic compound and the endogenous components, or other events in the biologic system related to exposure.

- *Biomarkers of effect* are defined as any changes that are qualitatively or quantitatively predictive of health impairment or potential impairment resulting from exposure.
- *Biomarkers of susceptibility* are defined as indicators that the health of an organism is especially sensitive to the challenge of exposure to a xenobiotic compound.

There are several advantages to using biomarkers to evaluate effects of EDCs. Biomarkers can show rapid and sensitive physiological response, represent a spatial and temporal integration of the measure of bioavailable chemicals more effectively than chemical analyses of the media. Biomarkers indicate evidence of exposure to chemicals that metabolize rapidly, integrate pharmacodynamic and toxicological interactions from potentially additive influences of chemicals in mixtures, and indicate potential causes of population responses.

However, concerns have been raised regarding potential misuse of physiological measurements (Munkittrick and Van Der Kraak, 1999). These include the need to develop a baseline of data on normal levels and ranges, to understand factors other than EDCs that can influence biomarker response, and to clarify the relationship between the presence of biomarkers to adverse effects on individuals or populations.

Munkittrick and Van Der Kraak (1999) emphasize that study objectives and uses and limitations of biomarker data should be clarified in the study design.

1.3 Effluents

Evidence of estrogenic activity in domestic WWTP effluents has been reported in a number of studies over the past several years. The potential endocrine disrupting effects were reported near discharges based on observations of vitellogenin induction and intersex conditions in gonads. Studies showed vitellogenin increases in caged rainbow trout (Harries et al., 1997), intersex gonad conditions in roach fishes (Jobling et al., 1998), and vitellogenin and intersex conditions in flounder (Allen et al., 1999).

The underlying causes for estrogenic affects have been attributed to a number of natural and synthetic hormones and alkylphenols and alkylphenolethoxolates. The analytical protocols for detecting low levels of these compounds in effluents and surface waters continue to be refined, and the range of constituents that may show interferences in hormonal function will continue to be refined.

CHAPTER 2.0

PROBLEM FORMULATION

2.1 Framing a Site-specific Conceptual Model

The traditional hazard-assessment approach combines effluent chemical testing and/or acute and chronic effluent bioassay data to predict concentrations and effects in receiving streams. Effluents are regulated based on the assumption that if receiving water concentrations are below the effect level of concern, no adverse effects would be observed. In some cases, a mixing zone is allowed to combine the effluent with the receiving waters. As a third control measure, biological criteria/bioassessments provide numerical values or narrative statements that describe the integrity of communities.

This overall approach is consistent with evaluating EDCs. However, a baseline of methods, effects, and data analysis has yet to be developed to evaluate mechanism-related variability for biomarkers.

Formulating a conceptual model to evaluate the development of potential biomarkers and their implications to receiving streams provides a framework for integrating ongoing research findings to a specific site situation. Figure 2-1 illustrates key components that guide the analysis of potential stream impacts.

If a facility manager is asked, "What about endocrine disruptors?" the question must be focused and alternatives to address this issue addressed in an iterative and tiered approach.

This problem formulation provides a structure clarifying components for evaluating the potential ecological significance of estrogenic constituents in receiving streams, which is discussed in detail in subsequent chapters. There are three main aspects for defining the potential for EDC-related impacts in a receiving stream:

1. Defining the source term,
2. Clarifying dose–response patterns for the chemicals/receptors of interest, and
3. Characterizing the receiving stream's potential for attenuation or formation of biomarkers.

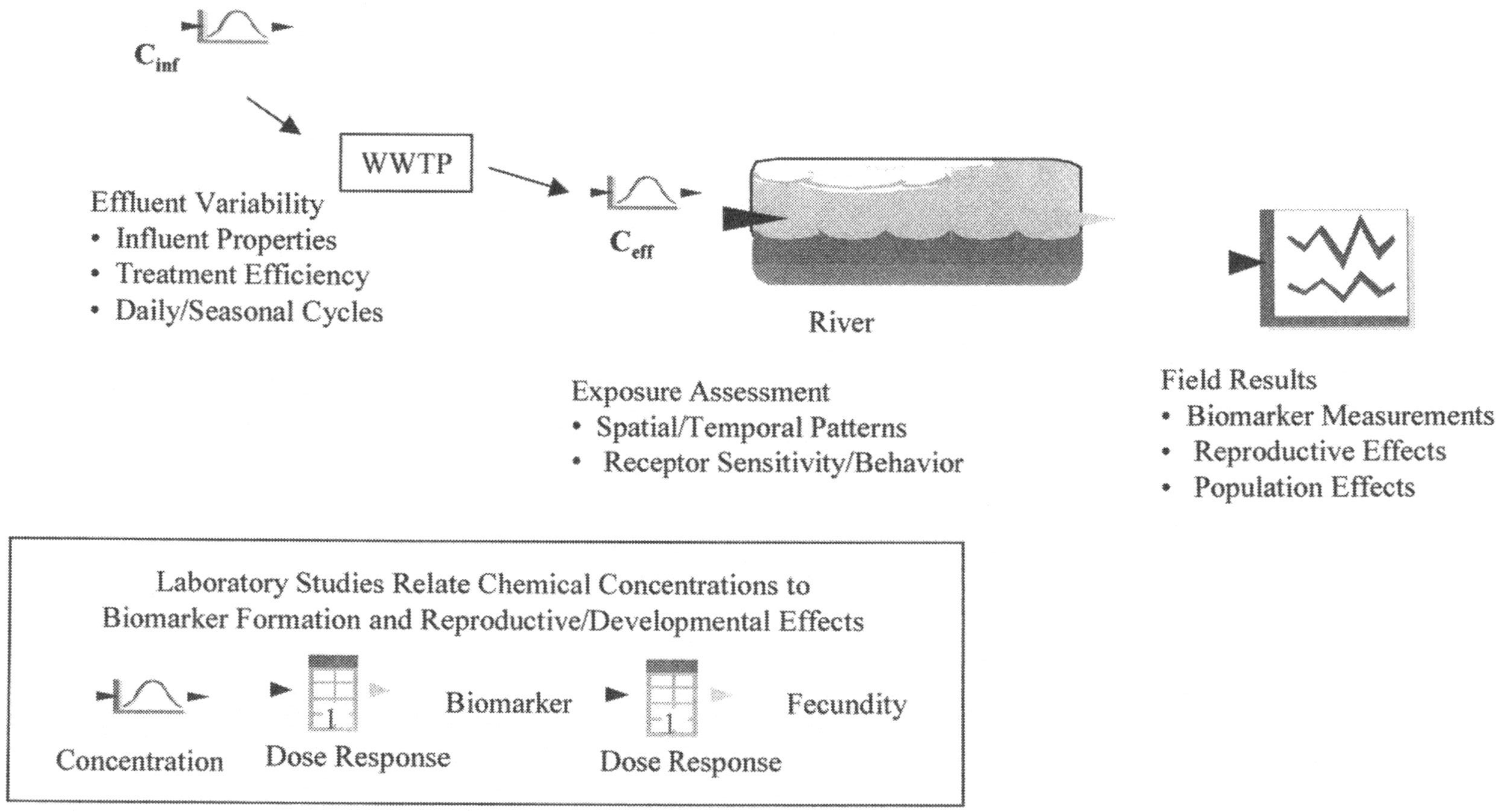

Figure 2-1. Conceptual Model - Estrogenic Compounds in Wastewater Treatment Plant Effluent: Evaluating Potential for Biomarker Formation and/or Adverse Reproductive Effects in Receiving Streams

2.2 Defining the Source Term: Effluent Characterization

Effluent characteristics can vary based on influent properties and treatment removal efficiencies. These properties commonly show daily and seasonal variability, and these characteristics are relevant to sampling strategies, exposure assumptions, and potential stream impacts.

Currently, there are three primary ways to characterize effluents for potential endocrine disrupting potential: chemical analyses, in vitro assays, and in vivo assays. All of these assays are in research phases. This can limit the laboratories' availability to perform the analyses and contribute to uncertainties, since the methods are not necessarily standardized and comparable. However, recognizing this, each method has advantages and disadvantages.

2.2.1 Chemical Analyses

The reporting of feminization of fish exposed to wastewater effluents in the early 1990s increased interest in improving analytical methods to verify the presence of estrogenic components in wastewater. Traditional analytical methods did not provide the tools to quantify trace concentrations in the presence of other organic matter in wastewater. This has improved with the use of gas chromatography/tandem mass spectrometry and new developments in immunochemistry (Sedlak et al., 2000).

One advantage of chemical analyses is that most people are familiar with this general approach that quantifies a specific chemical, leading eventually to an agreement as to the "protective concentration."

The potential analytical challenges may leave uncertainties. For example, the number of laboratories with experience in this area is limited; securing this expertise comes at high costs; and validation/replication may be undocumented. Given the temporal variability likely in effluents, a number of samples may be needed to evaluate patterns. In addition, the chemical concentration may not link directly to effects, since modifying factors of bioavailability may not be captured.

Domestic WWTP influents contain natural and synthetic hormones, which are removed or transformed to some extent during treatment. In addition, the treatment process itself may produce constituents with potential hormonal activity. The more commonly documented steroid compounds are natural estrogens — 17 β-estradiol (E2) and estrone (E1) — and synthetic estrogen, 17 β-ethynylestradiol (EE2). Other chemicals, notably alkylphenols and alkylphenolethoxylates, also have been reported in effluents and can behave as estrogen mimics. Several reported EDCs, like persistent organochlorines/pesticides (e.g. PCBs and DDT) might be of lesser importance for domestic wastewater discharges. Of greater interest are pharmaceuticals that may be relatively persistent (e.g. EE2 and mestranol as active agents in oral contraceptives).

Answering the critical questions about the significance and consequences of exposure to hormonally active agents requires understanding the levels of chemicals that are present in U.S. effluents and their potential impacts on receiving streams. WWTPs differ in their influent characteristics, treatment processes, and proportional flow in the receiving stream. Regional temperature differences also can influence treatment effectiveness and persistence in surface water and receiving streams. In addition, species of wildlife and their sensitivities to constituents will vary from receiving stream to receiving stream. Therefore, without additional baseline data, providing a balanced perspective on the overall effects of U.S. discharges is difficult.

The U.S. Geological Survey (USGS; Kolpin et al., 2002) recently published results of its reconnaissance of U.S. streams for pharmaceuticals, hormones, and other organic wastewater contaminants. Researchers detected EE2 in approximately 16%, E2 in 10%, and estriol in ~21% of the samples. These data suggest that estrogenic compounds are detectable in some locations, and may lead to further clarification on potential adverse effects in receptors.

2.2.2 In Vitro Assays

A number of in vitro assays being developed would provide an indication of the potential estrogenicity of an effluent, without necessarily requiring chemical identification. Zacharewski (1997) provided a critical review of in vitro assays that were used to identify and assess chemicals with estrogenic activity. These included competitive binding assays that measure the interaction of the estrogen receptor (ER) with the chemical of interest, cell proliferation, and estrogen receptor transcription assays. Of these, the most commonly applied to WWTPs is the yeast estrogen screen (YES) assay, which is a receptor transcription assay.

The YES assay is designed to show a positive response when an estrogenic chemical binds to an estrogen receptor molecule formed as a result of a gene inserted into the yeast. This produces expression of a "reporter gene" and ultimately a red color that may be measured by absorbance. The presence of several compounds could produce a cumulative positive response.

This assay also is in the research stage. The overall costs of materials are relatively low, and the assay could be conducted on a number of samples simultaneously, taking about 3 to 4 days for the color to develop. The assay also could capture anti-estrogenic activity, which could make it cost-effective once demonstrated for a particular application.

There are several limitations associated with the YES assay. The interpretation is not as straightforward as chemical analyses; however, it could capture additional information and provide a relative guide to the removal efficiency of the estrogenic components as well as spatial and temporal patterns of effluent in receiving streams. Several technical issues, including interferences and bacterial contamination, associated with applying this assay to environmental samples require experience and careful interpretation. In addition, the assay is specific to only one type of response. Others have recommended a potential suite of assays to address a range of responses that may be observed in in vivo assays that would not be identified in this screening (Garcia–Reyero et al., 2001).

With the advent of increasingly sophisticated molecular biology tools, new techniques will emerge that will provide alternatives for screening assays and these will become less expensive and applied more routinely. With the pressure to develop a screening program for thousands of chemicals, new techniques that may be extrapolated for use on effluents may emerge.

2.2.3 In Vivo Assays

Over the past few decades, the use of bioassays has increased as understanding of their applications, advantages, limitations, and ability to complement chemical measurements has improved. In the case of EDCs, the bioassay involves quantization of biological responses that have not necessarily been included routinely in traditional tests. Bioassays can measure specific responses and can be sensitive enough to distinguish small differences in analyte activities. These assays can be simple and rapid, providing an integrated measure of environmental effects.

Bioassays provide concentration–response relationships more directly than may be inferred from chemical tests or in vitro assays. For example, EE2 is one to two times more estrogenic than E2 as measured in the YES assay, but 25 times more estrogenic as measured by vitellogenin induction (Johnson and Sumpter, 2001). Therefore, the in vitro test does not address some issues, like metabolism or excretion rates, that can be identified in the bioassay.

Standard acute and chronic fish bioassays have evolved with consensus on protocols, endpoints, and data interpretation (pass/fail). The more complex lifecycle tests can capture the relevant reproductive endpoints potentially associated with EDCs, but these are expensive.

The Environmental Protection Agency has proposed the use of a short-term reproductive assay with fathead minnows to include as one component of the EDC chemical screening program (Ankley et al., 2001). This is a 21-day flow through test with endpoints that can include measurements of hormones and vitellogenin.

Other investigators are evaluating a range of approaches using different model organisms, exposure routes, durations, etc. Table 2-1 highlights some advantages and disadvantages of model organisms and selected studies using these organisms to evaluate EDCs.

Because a standard test is not yet available, a tiered approach to effluent screening could guide evaluation of the potential for a discharge to produce biomarkers in aquatic receptors in the receiving stream. A relatively short-term exposure (7 days) can produce elevated vitellogenin in receptors if the concentrations are above a threshold level. This is discussed in greater detail in Chapter 3.

As with the in vitro assays, general methods to characterize effects of EDCs potentially present in effluents are likely to evolve. For example, a bioassay has been reported for estrogens using transgenic zebrafish. The estrogenic substance would bind to DNA and produce a luciferase protein that can be quantified by measuring light activity, allowing live detection of organ-specific reactions.

2.3 Dose–Response

A significant body of knowledge is being developed on the response of various indicator organisms to estrogenic compounds. Several commonly measured endpoints, like vitellogenin, may not be the most sensitive endpoints. However, the range of the responses is significant and makes these endpoints useful gauges of potential for impacts.

As research proceeds, data will be generated that relate concentrations/biomarkers to critical issues like the magnitude and reversibility of the effect, impacts on overall reproductive success, and measurements of population effects. Biomarkers and receptor responses to estrogenic exposures are discussed in subsequent chapters.

Table 2-1. Overview of Common Test Species for Laboratory Studies

Test Species, Family name and Factors for Use as a Laboratory Test Species	Related Studies
Fathead Minnow (Cyprinidae) Advantages ◆ Rapid life cycle and ability to time the reproductive cycle. ◆ Extensively tested with large toxicological database. ◆ Historical use in regulatory programs and proposed as one of the target species to identify potential EDCs ◆ Sufficiently large to measure plasma vitellogenin, and steroids. ◆ Sexually dimorphic ◆ Widely distributed ◆ Disadvantage* is limitation in basic reproductive physiology and endocrinology.	Ankley, G.T. et al, 2001; Danylchuk, A.J. and Tonn, W.M., 2001; Harries, J.E. et al, 2000; Hemmer, J.M. et al, 2001; Jensen, K.M. et al, 2001; Kahl, M.D. et al, 2001; Kramer, J.J. et al, 1998; Lange, R. et al, 2001; Miles-Richardson, S.R. et al, 1999; Nichols, K.M. et al, 1999; Panter, G.H. et al, 1998; Panter, G.H et al, 2000;
Medaka (Adrianichthyidae) ◆ Recommended freshwater species for EDCs ◆ Rapid life cycle and ability to time the reproductive cycle. ◆ Small, plasma cannot be sampled for hormones or vitellogenin ◆ Male gonad differentiates during critical window, ~ 13 days post hatch, female gonads before hatching.	Foran, C.M. et al., 2000; Gray, M.A.,and Metcalf, C.D., 1997; Greenberg, M. E. and Bender, T. P., 1997; Gronen, S. et al, 1999; Metcalfe, C.D. et al, 2001
Zebrafish (Danio) ◆ Widely used in genetic and toxicity tests ◆ Well defined life history ◆ Reproduce rapidly ◆ Small, plasma cannot be sampled for hormones or vitellogenin ◆ * Passes a stage of juvenile hemaphrodism.	Fenske, M. et al, 2001; Holbech H. et al, 2001; VandenBelt, K., Verheyen, R., & Witters, H., 2001; Wang, H. et al, 2000
Sheepshead Minnow (Cyprinodontidae) ◆ Estuarine species ◆ Large toxicity database ◆ Easy to culture	Bowman, C.J. et al, 2000; Denslow, N.D. et al, 2001; Folmar, L. et al, 1997; Folmar, L.C. et al, 1999; Zillioux et al , 2001

2.4 Stream Characterization

One of the initial objectives of this study was to assess how widespread an EDC biomarker (like the commonly reported vitellogenin biomarker) may be, and factors that may incorrectly assign the presence of a specific biomarker to an adverse effect when it may occur with some frequency in natural populations. As a number of surveys are being conducted to measure these biomarkers, an awareness of the potential patterns and causes are needed.

Standardized tests of single aquatic organisms provide an efficient and cost-effective way to assess potential adverse effects under controlled conditions. Simple laboratory tests generally overestimate the magnitude and duration of exposure relative to that observed in more natural field systems, generally because of reduced persistence and bioavailability.

Many WWTP discharges to surface water have permits developed with a baseline of information on discharge patterns and stream flow. Watershed studies in the area may provide information on receptors, sediment, and surface water characteristics.

As additional data clarify the behavior of EDCs and pharmaceuticals in wastewater, approaches to evaluating effects will be refined. The fundamental models to define spatial and temporal patterns are applicable to these constituents. However, the concentrations are low and unique mechanisms of sorption and transformation may occur.

When the effluent activity is screened, an internal assessment can be made regarding the potential dilution and degradation for a receiving stream, and these results may be compared to studies showing how these levels relate to biomarker formation and adverse effects.

Other factors to be considered may include temperature ranges, residual sediment contamination, and/or other potential anthropogenic sources (e.g. from some agricultural practices).

The conceptual model for evaluating stream impacts may focus the approach to an assessment. This may range from concern about continuous "end-of-pipe" exposures to biomarker changes in natural populations outside of a mixing zone.

CHAPTER 3.0

BIOMARKERS/ASSAYS: USES AND LIMITATIONS

3.1 Purpose

The purpose of this section is to highlight the most common biomarkers used for studies of effluent discharges in aquatic environments, how these biomarkers relate to specific endocrine system functions, and uncertainties and analytical considerations.

3.2 Overview of Hormonal Control in Fish Reproduction

Reproduction in teleost fishes (fishes with a bony rather than a cartilaginous skeleton) is regulated by complex interactions of biotic and abiotic factors. Successful reproduction requires adequate nutrients for gamete production, integration of environmental cues with the hypothalamus–pituitary gland–gonad neuroendocrine cascade that brings about growth and maturation of gametes, integration of cues for expression of appropriate courting, mating, gamete release, and subsequent parental behaviors (see Figure 3-1).

Hormones are chemicals released into the blood in small amounts (concentrations 10^{-10} to 10^{-7} M) that elicit responses in cells at locations other than the endocrine gland where they are produced. These chemicals regulate a range of functions; for example, they serve as the mechanism to coordinate reproductive function with factors in the internal and external environment most conducive to offspring survival. This requires specialized cells to coordinate via chemical signals. For this reason, low concentrations of exogenous chemicals that interfere with any aspect of this process could result in various unanticipated physiological responses.

Because hormones carry critical information, their secretion must be turned on and off at precisely the right times. Components of the endocrine regulatory system include detectors to recognize a homeostatic imbalance, mechanisms to activate hormone secretions, an organ that recognizes and responds to the hormone, a detector to recognize that the effect has occurred and the signal can be shut off, a mechanism to remove the hormone from target cells and blood, and an apparatus to synthesize and replenish the hormone in the secretory cell.

Figure 3-2 illustrates aspects of hormonal control and physiological responses associated with reproduction in fish. Chemicals have the potential to disrupt endocrine systems by mimicking such endogenous hormones as estrogens and androgens by binding to their receptors and over-expressing genes at inappropriate times. Alternatively they can also inhibit hormonal activity by blocking receptors directly; interfering with the synthesis and function of receptors; or altering the production, metabolism, and clearance of natural hormones (Laws et al., 2000; Legler et al., 1999; Sohoni and Sumpter, 1998).

Currently, the most commonly reported measure to indicate exposure to estrogenic compounds is elevated concentrations of vitellogenin in the plasma of male fish. Vitellogenin is a precursor protein for egg yolk. It is synthesized in the liver of oviparous animals in response to estradiol, and is secreted into the blood where it travels to the developing oocyte. It is picked up by the developing oocyte via endocytosis, and then is cleaved into egg yolk proteins. The main use of vitellogenin is as a nutrient for the developing embryo, and, after hatching, for the fry. Both males and females have the gene for vitellogenin, but it normally is produced in highest

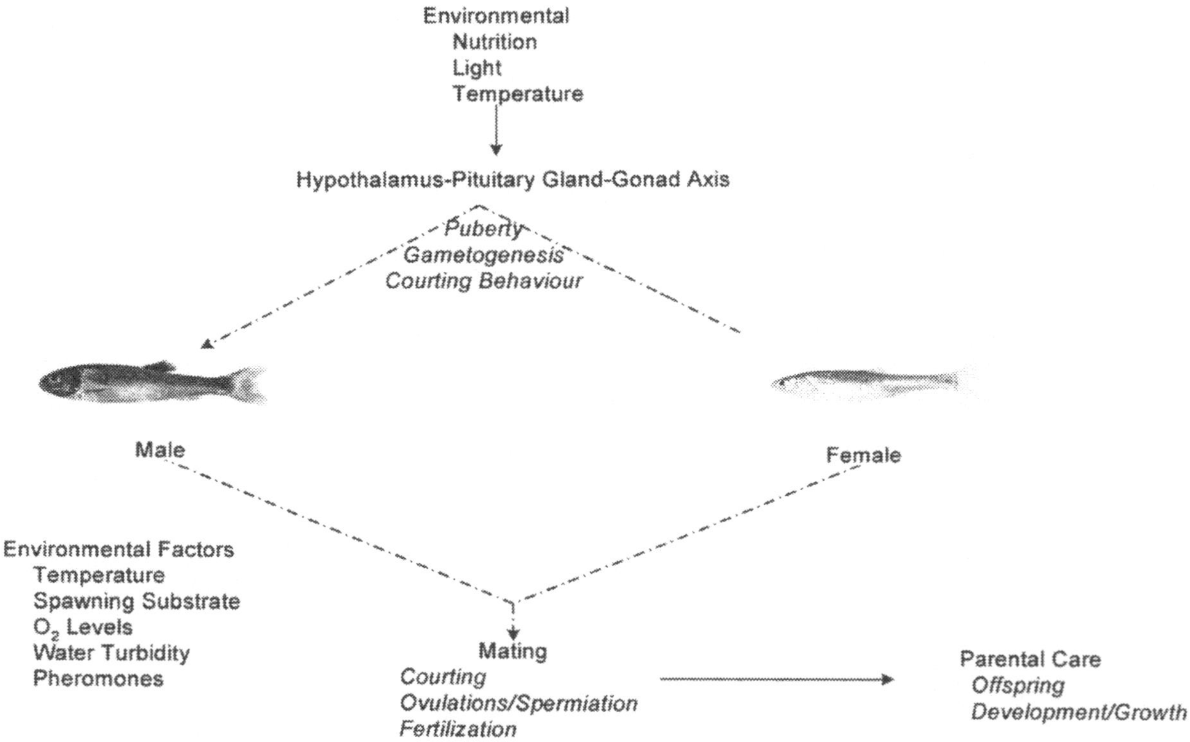

Figure 3-1.
Factors in Reproduction of Fish

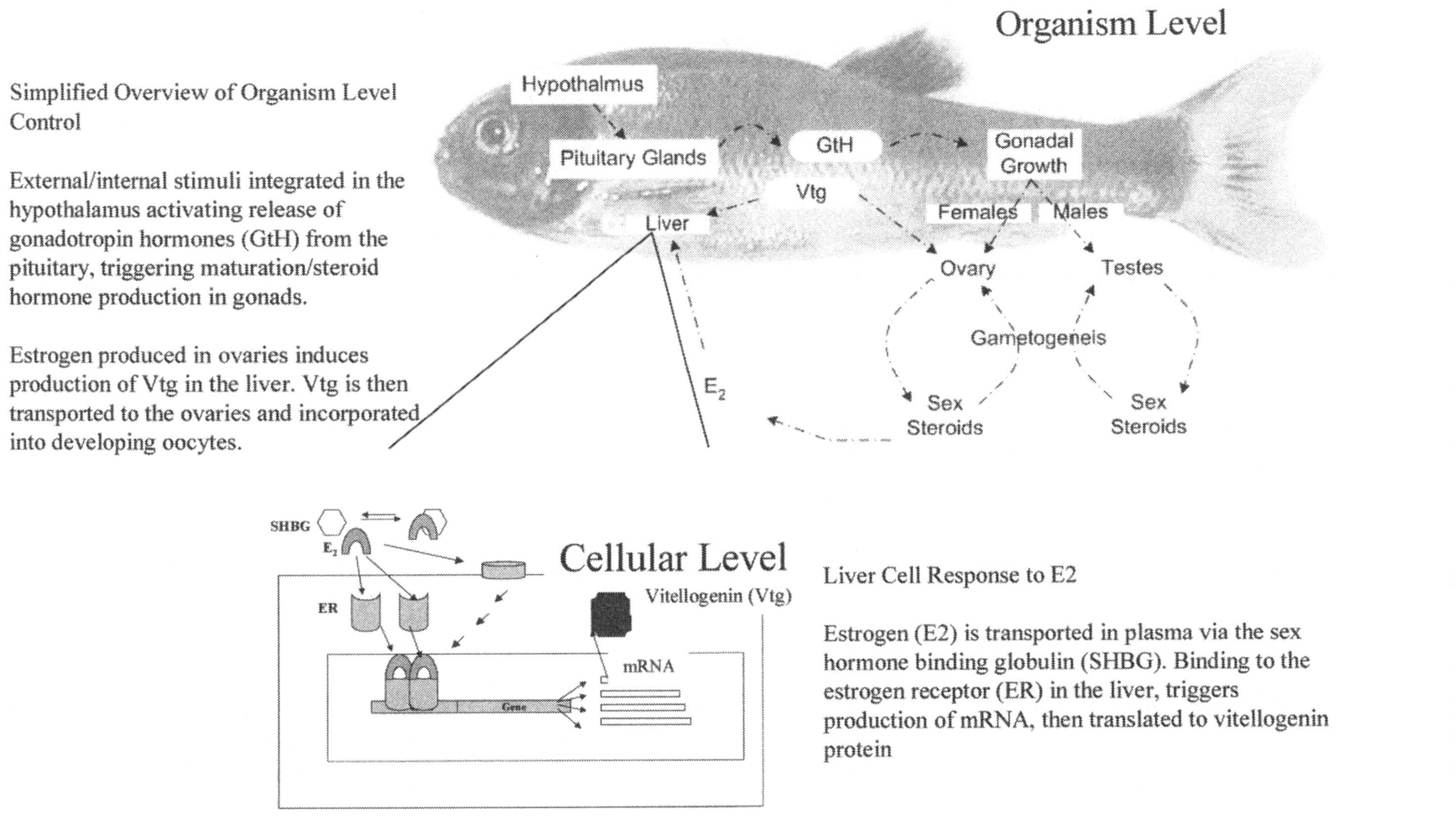

Figure 3-2. Fish Hormonal Control and Physiological Responses to Estrogenic Chemical Exposures

concentrations in mature females undergoing oogenesis. If males are exposed to estrogen or estrogen mimics, they will respond by producing vitellogenin. In this case, the vitellogenin simply accumulates in the blood until it is broken down by plasma proteases and excreted through the kidneys. Because the amount of vitellogenin in male plasma is proportional to the exposure (Folmar et al., 1999; Folmar et al., 1997; and Folmar et al., 1995), it has become a good quantitative measure and biomarker of exposure of fish to estrogenic substances in the water.

Estrogens act through the ER to induce the synthesis of vitellogenin. The first step in the process is the induction of vitellogenin messenger RNA (mRNA), which then is translated into the vitellogenin protein. Vitellogenin mRNA and protein serve as distinct biomarkers of estrogen exposure over time in male oviparous vertebrates. Vitellogenin mRNA is induced within 4 to 6 hours and peaks within 48 hours of a single exposure (Bowman et al., 2000; Bowman and Denslow, 1999). It is degraded quickly after 3 days (Bowman et al., 2000; Hemmer et al., 2001).

Vitellogenin protein in males, on the other hand, is long lived — lasting for 4 to 6 weeks after the termination of exposure, depending on the initial level of induction. Thus, the two assays are complementary, one can be used to monitor immediate induction and the other can trace an induction over time. Therefore, the presence of vitellogenin mRNA above baseline levels is indicative of a recent estrogenic exposure. Both the ER gene and the vitellogenin gene are up-regulated in a time- and dose-dependent manner by estrogen (Bowman et al., 2000), but at much different levels of induction. Comparing the two levels can help researchers interpret the impact of exposure on fish in the wild. In this study, the project team developed assays to measure both ER and vitellogenin mRNA gene expression in fathead minnows using a real-time polymerase chain reaction (PCR) technique. This enables researchers to determine whether or not caged fathead minnows have been exposed to estrogenic substances, and aids in the interpretation of field results.

Just as hormones vary during the reproductive cycle, vitellogenin levels in females also vary. In female largemouth bass, vitellogenin levels generally are very low in the summer and early fall. Soon after estradiol starts to increase, vitellogenin starts to be synthesized. Plasma levels normally reach their maximal levels by late February or early March, just prior to spawning. After spawning, the levels of both estradiol and plasma vitellogenin plummet back to minimum levels. Vitellogenin is cleared quickly from the plasma since it is actively sequestered into developing oocytes.

Males normally have low or undetectable amounts of plasma vitellogenin. Some potential causes for elevated levels of vitellogenin include: close contact with females that are exuding high levels of estradiol into their environment at the height of the season, eating a phytoestrogen, or coming into contact with another source of estrogen outside of the region where they are caught.

In field studies in the United States and elsewhere, when male fish have vitellogenin in concentrations that are higher than 10 μg/ml or when more than 10% of the male fish at a location have measurable levels of vitellogenin, generally the population has been exposed to an estrogenic contaminant (Folmar et al., 2001). Laboratory studies have shown that the concentration of plasma vitellogenin in males is proportional to the dose applied either by injection or in the water (Bowman et al., 2000; Folmar et al., 1999; Folmar et al., 1995).

Lower than expected levels of plasma vitellogenin in reproductive females also can be used as a biomarker, but in this case, as a marker of interference with the normal reproductive process. This marker is extremely useful and possibly more relevant than finding vitellogenin in males as a determinant of adverse effect. If females have low levels of plasma vitellogenin, their eggs likely will be of lower quality. While they may still spawn the same number of eggs and the eggs may hatch, the fry may not be viable. Female plasma vitellogenin levels are harder to interpret, however, since their levels normally change across the reproductive season. It is essential to determine normal seasonal patterns of fluctuation for plasma vitellogenin in females before using this biomarker effectively. Also, it is essential that control animals be collected in the same time period as impacted animals if their plasma levels are to be compared.

3.3 Traditional Assays of Reproduction and Developmental Effects in Fish

Many studies to date have focused on estrogenic effects attributed to natural and synthetic hormones and/or estrogen mimics. Disruption of the endocrine system by steroids/steroid mimics are targeted at adverse impacts on reproductive and developmental endpoints. Because of this, many of the assays and measurements included in traditional tests are relevant for evaluating potential adverse effects of constituents in effluents. Some of the standard tests and endpoints and more recently identified biomarkers are shown in Table 3-1. The biomarkers evaluated in this project are highlighted in this section.

Table 3-1. Common Endpoints Reported for Aquatic Effects of EDCs

Reproduction		Histology/Morphology	
	Spawning		Kidney condition
	Egg production		Liver condition
	Fecundity		Testis–ova development
	Hatching success		Pathological effects
			Secondary sex characteristics
Hormones			Presence of spawning tubercles
	E2		Gonadosomatic index
	11-KT		Hepatosomatic index
	Testosterone		Gross morphology
Molecular Markers — Proteins		General	
	Vitellogenin		Plasma
	mRNA		Hematocrit levels
	Zrp		Sex ratio
			Survival
			Mean length and weight
			Growth

3.4 Gonadosomatic Index (GSI) and Histopathology

One endpoint that commonly is included for screening studies to detect estrogenic/androgenic effects is the ratio of gonad weight to body weight (GSI = gonad weight/body weight X 100). This easily measured endpoint is applicable to both sexes of oviparous fish species.

GSI changes can be indicative of reproductive success. Significant variation in gonadal size can be measured throughout a reproductive cycle for many species and can be an indicator of reproductive maturity. This parameter is not specific to a particular mechanism of action, and may reflect a variety of factors like seasonal cycles as well as potential influence from exposures to EDCs or other contaminants.

Ankley et al. (1998) report that typical GSI values for reproductively active fathead minnows range from 1% to 2% for males and 8% to 13% for females. Fathead minnow GSI values for the 21-day flow-through study conducted in the project team's laboratory (Appendix B4) had an average GSI of approximately 1% for males and 7% to 10% for females. In these studies, short-term exposures at lower concentrations typically did not show GSI differences at the concentrations considered relevant for WWTP effluents.

According to Panter et al. (2000) this measurement is best taken in a high-growth period, and comparisons between experiments separated in time or using different batches of fish may be of limited value. Smith (1978) showed that male GSI increased rapidly, by a factor of three, during the 3 to 4 weeks preceding spawning. Therefore, exposures during periods of rapid gonadal development may be more effective in detecting hormonal interference by EDCs.

The GSI parameter measured in field studies also provides a guide to comparative reproductive status. Bass collected in the field effort for this study (Appendix E) showed variable measurements of GSI related to species, overall body weight, and season as shown in Table 3-2.

Table 3-2. Average Body Weight and GSI Collected in the North Georgia Study

	Body Weight (g)		GSI (%)	
Species	Spring	Fall	Spring	Fall
Females				
LMB	606	466	2.27	0.52
REB	97	97	6.82	0.82
SPB	275	254	1.74	0.44
Males				
LMB	260	340	0.18	0.12
REB	125	87	0.92	0.32
SPB	148	178	0.38	0.16

LMB = largemouth bass; REB = redeye bass; SPB = spotted bass.

Histology data on gonads are important for confirmation of gonadal phenotype, sexual development, and abnormalities like testis–ova.

A digital atlas of fathead minnow histology can be found at the following website: http://aquaticpath.umd.edu/fhm/. This site includes basic information on the fathead minnow, and illustrations of fathead minnow ovaries and testis at varying stages of maturity.

3.5 General Blood Chemistry

In addition to plasma hormones and vitellogenin, numerous other parameters have been measured to help evaluate the clinical condition. Blood characteristics differ among species, and can vary within a species with seasonal changes. Normal ranges for the species of interest would enhance the use of these data. Table 3-3 highlights selected constituents summarized by Folmar (1992).

Table 3-3. Summary of Blood Chemistry Measurements from Selected Studies

Species		Size or Age	Date (month)	Glucose (mg/dL)	Total protein (g/dL)	Hemato-crit (%)	Estradiol (ng/mL)	11-Keto-testosterone (ng/mL)	Testos-terone (ng/mL)
Common Name	Scientific Name								
Small-mouth Bass	*Micropterus dolomieui*	?	?		2.2-4.5				
Striped Bass	*Morone saxatilis*	2-20 kg	March - Aug.	44-181					
		22-79 kg			3.7±0.2	47.0±1.3			
Striped Bass FW	*Morone saxatilis*	1.8-10.9 kg	June	115±5					
		2-10 kg	March - Aug.		3.1-6.0	25-41			
Striped Bass SW	*Morone saxatilis*	47±15 g	July		4.1±0.2				
Common Carp	*Cyprinus carpio*	40-80 g	?	60					
Flounder	*Pleuronectes flesus*	94-179 g	?	35 ± 2		14.9±0.6			
		18-29 cm	June - Aug.			21.8±0.9			
Flounder, winter	*Pseudopleuronectes flesus*	118-376 g	Nov. - March	4.5±0.25		37±1.6			
		30-40 cm	March	5.34±0.4		19.1±1.54			
		?	?					5.5±0.9	
Goldfish	*Carassius auratus*	?	March - Apr.				0.3-0.6	5.5-8.7	6.5-9.7
		?	Oct.				0.3	0.4-0.6	0.6-2.2
Bluegill	*Lepomis macrochirus*	166±6 g	May					5.0±1.3	4.0±0.7

Data as summarized in Folmar (1992). SW, saltwater; FW, freshwater.

A few routine parameters include:

- Hematocrit (the percent red blood cell content),
- Proteins (which have a role in homeostasis and blood coagulation, vitamin and hormone transport, and specific immunity), and
- Glucose/lactate. Glucose levels, which can vary with species, size, age, sex, and nutritional and reproductive status, typically would not indicate metabolic dysfunction, but can be altered by hypoxia, handling, exercise, and disease. Lactate levels are indicative of oxygen limitation, and can be influenced by sampling procedures and physical exertion at the time of sampling.

Changes in blood composition have been used to assess the clinical condition of animals. Blood characteristics can differ among species and change significantly with seasons. The normal ranges for a given species would be needed, requiring the use of controls or reference sites (Folmer et al., 2001).

The percent red blood cell content (hematocrit) is easily estimated following centrifugation. Measurements of hematocrit for fathead minnows in a reproductive assay (Appendix B-4) ranged from 25 to 56, but differences were not significant among treatment groups.

3.6 Molecular Measures of Endocrine Balance: Proteins/Hormones/mRNA

Many of the traditional measurements included in fish lifecycle tests (including histology, secondary sex characteristics, and fecundity) are further supported by measurements of plasma hormones, proteins, and/or mRNA.

Englebienne (2000) presents an overview of immune and receptor assays, including method validation and data analysis. At present, no standardized methods are available for vitellogenin analysis. That is, no single approach has been subjected to in-depth investigation and evaluation for sources of inaccuracy. It is expected that once this occurs, it will become a reference method showing reproducible and accurate results across laboratories, and will become a method that is used routinely by a number of laboratories where a bias, when compared to a reference method, will be established. In addition to analytical validation, the baseline data on normal ranges of these biochemical markers need further characterization.

3.6.1 Plasma Vitellogenin Measurements in Fish

Plasma vitellogenin levels can be measured by enzyme-linked immunosorbent assay (ELISA), radioimmunoassay (RIA), and immunoblotting (Western). ELISA is the assay of choice because of the large number of samples that can be quantified at one time with a high level of sensitivity. Vitellogenin is a large (300 to 600 kDa) protein with two subunits (160 to 200 kDa each). Because of its size and other attributes, vitellogenin can be purified from estrogenized fish plasma.

The widespread use of this assay as a biomarker for exposure to estrogenic compounds has resulted in the development of antibodies. For example, University of Florida researchers have developed a panel of monoclonal antibodies that are specific for the vitellogenin of several fish species:

American Eel	Largemouth Bass	Sheepshead Minnow
Brown Bullhead	Sturgeon	Smallmouth Bass
Carp	Bluegill Sunfish	Striped Bass
Fathead Minnow	Medaka	Walleye Pike
Goldfish	Northern Pike	White Mullet
Killifish	Pinfish	Yellow Perch
Cunner	Rainbow Trout	Sucker
Swordfish		

While generally it is shown that relative vitellogenin concentrations may be internally consistent for a study, comparisons across studies show reported vitellogenin values that differ significantly. When obtaining data for vitellogenin quantitation, the report should include information on the antibody source/type, reference material, matrix effects/interferences, detection limit, sensitivity and linearity, and repeatability/reproducibility. When comparing results from various studies, the relative response of the method should be considered, recognizing differences in the techniques used.

A monoclonal antibody recognizes one site on the vitellogenin protein; a polyclonal antibody recognizes many. If a vitellogenin molecule has been degraded or fractured from freeze thawing or other sources, a different result can be obtained in the assay especially if the assay uses a polyclonal antibody. In its work, the project team measured vitellogenin in fathead minnows and bass.

3.6.2 Zona Radiata Proteins

Several papers have indicated that zona radiata proteins may be a more sensitive biomarker than vitellogenin (Arukwe et al., 1997; Gillesby and Zacharewski, 1997). There are at least three different zona radiata proteins that come together to make up the chorion layer around the developing egg: Zrp-α, β, and γ. Like vitellogenin, the zona radiata proteins are synthesized in the liver under the influence of estrogen. The proteins are secreted into the blood and, like vitellogenin, travel to the female gonad where they are used to make the vitelline envelope that surrounds the egg. The synthesis of Zrp-β appears to be more sensitive to low estradiol levels than vitellogenin, and it has been developed into an alternative biomarker for estrogenic response (Arukwe et al., 1997). As with vitellogenin, specific antibodies need to be developed for each species. So far, this assay has been used only by a limited group of laboratories. It was not evaluated in this study.

3.6.3 Sex Steroid Levels

Plasma hormones have been reported as an endpoint in field and laboratory studies of endocrine disruption. Typically, these are the reproductive steroid hormones in part because there are commercial kits available for these assays. Several factors can cause changes in hormone levels, including alterations in synthesis, secretion, metabolism, excretion, or binding to plasma proteins. This contributes to the complexity of interpreting the root cause.

In female fish, the important steroid hormones are E2 and testosterone (T). In males, the important hormones are these as well as 11-keto-testosterone. The concentration of these steroid hormones in plasma varies according to the natural reproductive cycle of the fish. Actual levels vary considerably from one animal to another at the same reproductive stage, but, in general, females have high estradiol and low androgen levels while males have the opposite — high androgens and low estradiol. A useful measure is to take the ratio of the hormones within each fish. Typically, females are expected to have an E2/T ratio greater than 1, and males are expected to have a ratio less than 1.

For smaller fish species used in laboratory testing, it is difficult to obtain sufficient plasma for this analysis and more sensitive methods may be developed. Therefore, these data typically are reported more for larger species.

Seasonal or circadian rhythms will result in significant variability of hormone levels. A baseline of data on hormone cycles and the numerous factors that may influence these hormone levels is needed to interpret the data appropriately.

3.6.4 mRNA

At the cellular level, mRNA levels reflect gene expression and subsequent protein production. Various methods currently used to measure mRNA levels include Northern blotting, reverse transcription-polymerase chain reaction, and copy DNA micro arrays. The ability to measure mRNA provides insight into the underlying genetic regulatory processes.

Measurements can be made for specific mRNA and protein pairs. For example, as indicated in Figure 3-2, vitellogenin mRNA is translated to the vitellogenin protein; therefore, these levels are related. Emerging techniques for large-scale mRNA expression measurements, such as DNA chips, may simultaneously reflect the expression of a broad array of genes.

A primary action of estrogen or estrogen mimics is to induce the expression of genes by binding to soluble ERs and inducing the synthesis of mRNAs for specific genes. Two types of genes that are responsive in this manner are those that encode vitellogenin and the ER itself. In the liver, the mRNAs for vitellogenin and ER are used as templates for the synthesis of vitellogenin protein (which then is secreted from the liver into the blood stream) and additional ERs (which stay within the hepatic tissue), respectively. The response at the mRNA level is quite fast, within 12 to 24 hours, and reaches maximum response 48 hours after a single exposure. With no further stimulation, the induced mRNAs are degraded and are not used for further protein synthesis. Thus, the response at the mRNA level is immediate and short term. This is in contrast to the response at the protein level. Proteins, which are made from the template mRNAs, typically do not reach their peak concentration until 3 days after induction by a single exposure. Vitellogenin is particularly stable, so these two measurements would reflect potentially different exposure time periods.

Several studies have shown the expression of mRNA and subsequent protein levels for a specific interaction. The process of interpreting mRNA levels, alone, as a biomarker may require additional evaluation because protein interactions and secretion can produce differences between protein and mRNA levels. For example, it is possible to have genes expressed and the regulation of the response controlled after transcription (mRNA formation) or after translation (protein production). Therefore, to place this biomarker in the context of protein production and subsequent adverse impacts requires the collection of supporting data.

3.7 In Vitro Assays for Effluent Screening

A number of assays have been evaluated as potential chemical screening tests to identify the potential to mimic the effect of naturally occurring estrogens. Zacharewski (1997) provided a critical review of in vitro assays that were used to identify and assess chemicals with estrogenic activity. These included competitive binding assays that measure the interaction of the ER with the chemical of interest, cell proliferation, and ER transcription assays. In vitro assays can be cost effective, rapid, and targeted for specific mechanisms.

3.7.1 Yeast Estrogen Screen (YES) Assay

Of the array of in vitro assays available, the one most commonly used for effluent screening is the YES assay. WWTPs treat a number of synthetic organic compounds; in the process, some of these chemicals may become completely or partially degraded and/or sorbed to biosolids. Treated effluents may have low levels of several organic compounds that can show estrogenic activity in receiving streams. The YES assay is a tool that can be used to evaluate the cumulative, bioavailable, and estrogenic activity in effluents. Desbrow et al. (1998) used this assay to identify the active fraction in seven effluents, showing that the activity correlates with the fraction containing steroidal hormones.

The yeast strain *Sacharomyces cerevisiae* has been used to investigate receptor–ligand interactions. Advantages of using yeast include the lack of ERs, media with no steroids, and a system that allows insertion of proteins and reporter genes. The yeast strain developed at GlaxoSmithKline integrated the DNA sequence of the human ER into the main chromosome of the yeast. The yeast cells also carry a "reporter" gene. Upon binding an active ligand with the receptor, the reporter gene produces a substance that generates a red product that can be quantified by absorbance at 540 nm (see Figure 3-3). The methods used by the project team's laboratories were based on the work of Routledge and Sumpter (1996). (The yeast was obtained from J.P. Sumpter, Department of Biology and Biochemistry, Brunnel University, Uxbridge, Middlesex UB8 3PH, United Kingdom. Glaxo developed the strain and signed an agreement with Sumpter not to commercialize the strain.)

When evaluating the use of the YES assay for chemical screening, methods may vary by incubation time, mode of chemical addition, solvent type, initial concentration of yeast cells, and "creeping" (response observed in adjacent wells of the plate affecting blanks). While this assay can measure an estrogenic response, it also is possible to use both an estrogen and an anti-estrogen in the assay and detect a potential inhibitory effect.

3.7.2 Applications for Water/Wastewater Samples

The YES assay was used in various laboratory studies and field samples during the project team's investigation. The development of the standard curve using estradiol was straightforward and consistent with published results. Over the course of 2 years, more than 96 YES assay plates were developed and the approach to sample concentration and data interpretation evolved.

For application of this assay to water samples with low concentrations of estrogenic compounds, the samples require concentration on C18, with subsequent solvent extraction. Samples typically are prefiltered with a 0.45 μm glass-fiber filter. For solids, a solvent is used to extract hydrophobic estrogenic compounds for subsequent testing.

For initial solutions with concentrations of E2 at 1 nM (270 ng/L), the aqueous solution could be used directly with the yeast media, and could develop a strong response without requiring concentration by solvent extraction or C18. However, this is an upper bound on concentrations and not likely to be found commonly in effluents or river samples.

Because of varying approaches to sample handling and reporting, the project team translated the dose–response to measurements of mass of E2 per well of the plate, and prepared standards by directly applying the methanol standard to the plate as well as spiking the E2 standard into water, and taking the sample through the concentration process to assess detection limits and recoveries. In this process, the project team did not observe loss of signal, but, in fact, found response enhanced. This was repeated independently in both laboratories.

To evaluate active fractions in WWTP effluents, Desbrow et al. (1998) concentrated 20-L water samples by solid phase extraction. Fractions were eluted from the column with solutions of increasing percentage of methanol. Activity, as measured by the YES assay, was present in extracts from 50% to 85% methanol.

Researchers observed that the development of the standard curve with E2 was straightforward with use of this assay. The two bioassay laboratories quickly developed approaches that worked with laboratory water and/or E2 methanol standard application to the plate. The assay was used in several evaluations of chemical stability and activity with storage. Interferences were noted in several wastewater and river samples, requiring additional method development to concentrate and/or process samples.

Although this assay can provide rapid feedback on a number of samples and comparative estrogenicity, some method development time should be included before application. Furthermore, it is recommended that data be presented with the standard curves, scan of the plate, and include spike recovery information.

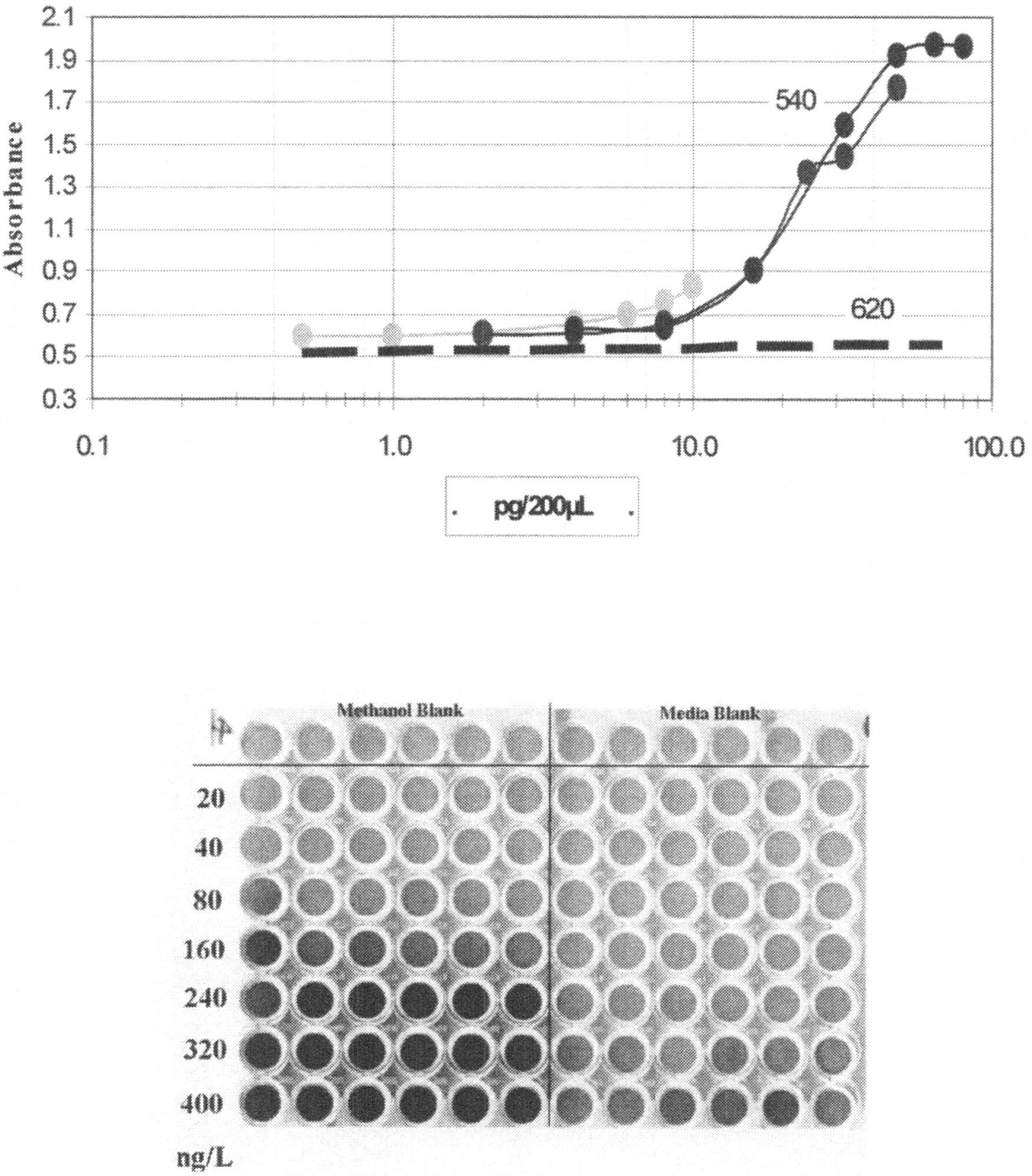

Figure 3-3. Yeast Estrogen Screen (YES): Illustration of Standard Curve and Associated Plate

CHAPTER 4.0

LABORATORY STUDIES OF BIOMARKER DEVELOPMENT AND REPRODUCTIVE EFFECTS

4.1 Purpose

Effects of EDCs in effluent have been reported in various studies using a number of different fish species. Laboratory studies to evaluate the effects of specific estrogens, most commonly estradiol or ethynylestradiol, have led to the development of information on dose–response under controlled conditions for fathead minnows, sheepshead minnows, and medaka. EDCs also have been detected in effluents, and several studies have reported induction of vitellogenin (Harries et al., 1996; Harries et al., 1997; Jobling et al., 1998; Tyler and Routledge, 1998; Rodgers–Gray et al., 2001).

Laboratory studies performed over the past 2 years provide a context for framing the question of potential formation of biomarkers in fish exposed to effluents, and the significance of this for adverse impacts in receiving streams. The project team performed a series of experiments to evaluate laboratory practices and procedures for screening effluent. Additional studies were performed to evaluate the potential reproductive effects associated with these biomarkers.

The results of the project team's studies are compared with published data on potential adverse effects, and biomarker formation resulting from exposures to estrogenic compounds.

4.2 Preliminary Protocol Development Studies

Because the physiological measurements targeted in this study develop in response to extremely low concentrations (ng/L) of chemicals, the project team evaluated the potential for chemical losses from sorption or degradation and contributions from laboratory containers and cleaners to affect results.

During an initial evaluation of potential contributions from plastics and cleaners, tanks containing these materials were set up to run in parallel with 17β-estradiol (E2) control and negative control solutions.

Vitellogenin was present in plasma samples at levels significantly above the negative controls for several of these materials (see Figure 4-1). The concentrations of cleaners that were tested were approximately 10 mg/L — much higher than trace levels that may result from residuals in tanks after cleaning. Because the plastics were placed in the tanks in pieces, they also reflect an upper bound on the potential contribution in a routine process. However, these tests did increase awareness of the importance of considering potential contributions from routine laboratory materials.

The potential losses from sorption to plastic and glass, and potential estrogenic contribution from plastic were evaluated using the YES assay (Figure 4-2). The decrease in estrogenicity was greater in plastics than glass, and may be of greatest concern at the lower concentration ranges.

The YES assay produced a significant response at solution concentrations above 0.1 nM without concentration of the water samples on C18 columns. The stability of effluents spiked with 1 nM of E2 was tested using the YES assay. Samples stored at 4°C showed a reduction in response of approximately 50% after storage for 18 days, while all estrogenic response for spiked effluent stored at 20°C was lost in this time period. This suggests that losses would occur in the tanks over the several-day static-exposure period.

The project team used nominal concentrations for the exposures, and monitored feed solutions using the YES assay. Studies by others on the differences in nominal and measured concentrations suggest reductions will occur, but these may be less than 50% (depending on the chemical and conditions). When fish were added to tanks with 1 nM of E2, the estrogenic response in the YES assay became negligible. However, if the water in the tank was renewed with control water, vitellogenin levels would be much lower than they would be in a similar tank without any water change. Studies by others suggest the chemical would be measurable; the increased plasma vitellogenin levels that would be present if water was not replaced suggest the presence of estrogenic activity. This suggests that interference may inhibit the YES assay response.

4.3 Fathead Minnow Exposures to Estradiol

A series of experiments was conducted over the 2-year study to evaluate the levels of vitellogenin in fathead minnows in response to E2 exposure. These tests included various static, static-renewal, and flow-through exposures, with concentrations ranging from 0.1 to 1 nM (27 to 272 ng/L). The vitellogenin data from all of these various studies are illustrated in Figure 4-3.

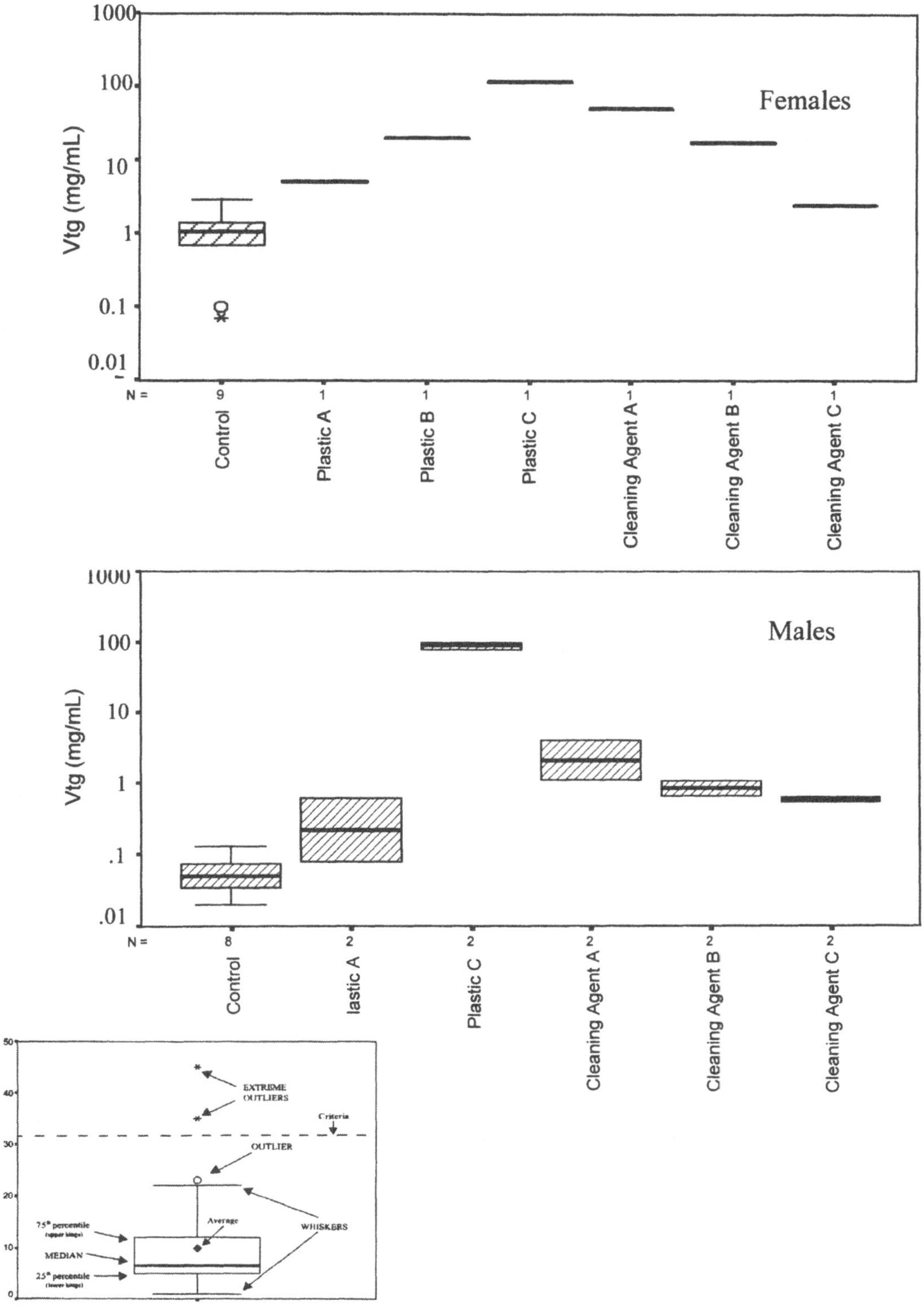

Figure 4-1. Plastics and Cleaners: 10-day Static Non-Renewal Exposure of Fathead Minnows

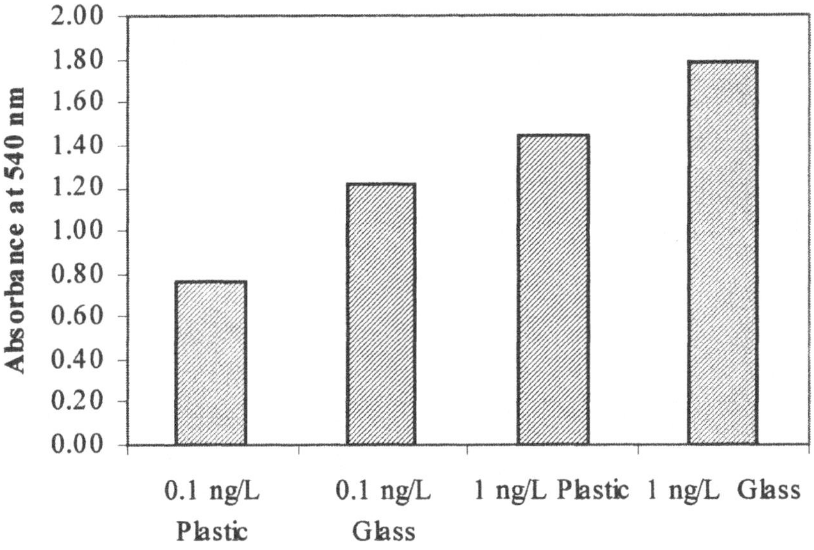

Potential Loss from Sorption of E2 Solutions in Glass and Plastic Containers

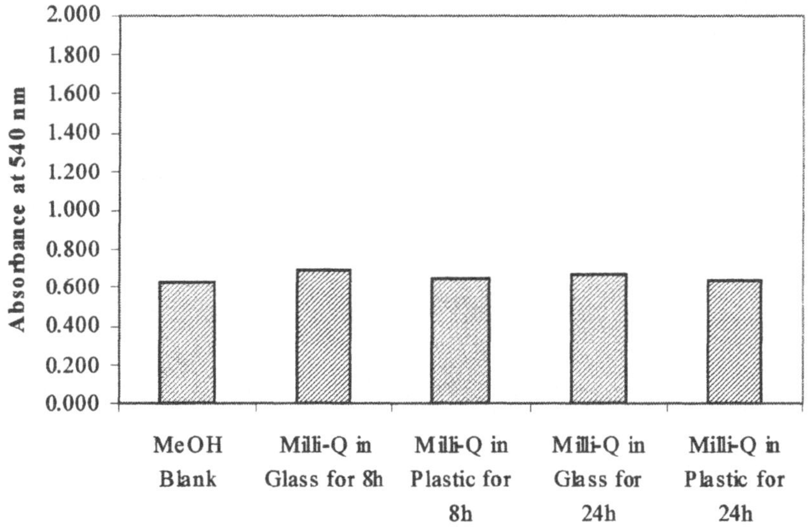

Potential Positive Interference From Storage of Water in Glass and Plastic Containers

Figure 4-2. Solution Storage = Potential Losses From Sorption and Contributions From Storage Containers Using the YES Assay.

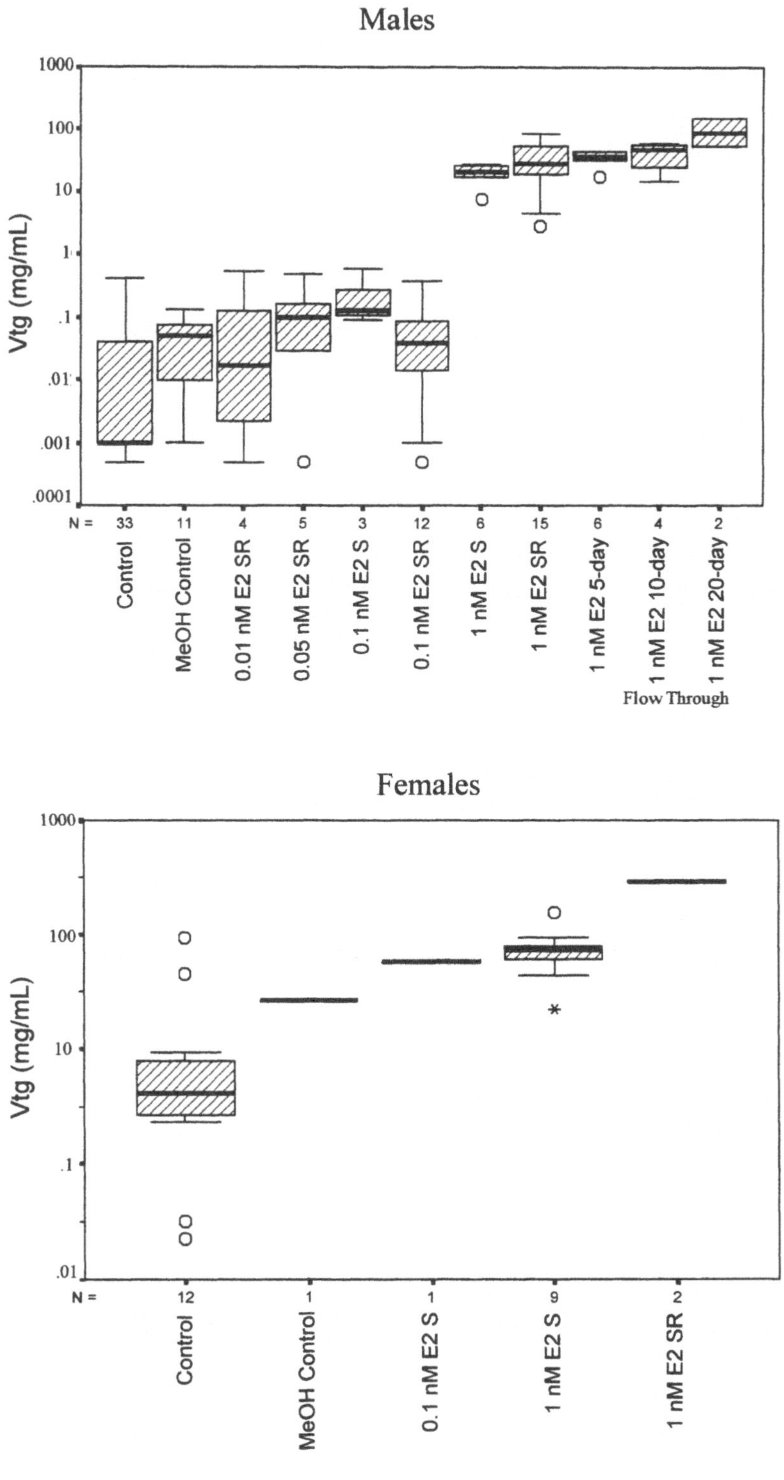

Note: S = Static: SR = Static Renewal

Figure 4-3. Increases in Plasma Vitellogenin Levels in Fathead Minnows Exposed to E2.

If the concentration of estrogenicity of the effluent is near or above 0.1 nM (27 ng/L), a positive response would be detectable in a 7-day static renewal. Concentrations may increase over a 21-day continuous exposure, so the shorter-term test may not represent a maximum. The challenge of effluent screening is to collect grab samples that would be more representative of peak concentrations that may be present intermittently; a flow-through system would better capture these peaks. The project team reached these conclusions based on a series of exposures conducted over several months.

As an initial stage in the evaluation, the project team performed a flow-through test with 1 nM of E2 (270 ng/L). Samples were collected at 5-, 10-, and 20-day intervals. The results of this test were compared to vitellogenin levels after 10-day static and static renewals exposures at 1 nM, and also static renewal for a 0.1-nM concentration of E2.

A significant response is observed with a static test with no renewals at high E2 concentrations (see Figure 3-3). Vitellogenin concentrations increased with continuous exposure; however, a significant response was observed in as short a time as 1 week. The project team's observations during these initial exposures included transitions in males' secondary sex characteristics after 21-day exposures at the high E2 concentrations.

A primary advantage of a flow-through assay is that it better integrates temporal patterns in the effluent.

4.4 Evaluation of mRNA as a Biomarker of Exposure

Using mRNA to measure changes in gene expression following exposure to xenoestrogens is, potentially, a more sensitive measurement of change than the measurement of the protein. In addition, because of more rapid formation and turnover of mRNA compared to protein, elevated mRNA levels would reflect more recent exposure.

Because fathead minnows are asynchronous spawners, estrogen and vitellogenin levels would fluctuate during the reproductive season. Generally, vitellogenin levels in females during this period may fluctuate from less than 1 to more than 20 mg/mL. Females were used in this experiment to evaluate impacts at the estrogen receptors by increasing (E2) or decreasing [anti-estrogen (ICI)] production of the mRNA.

To compare vitellogenin protein and mRNA, female adult fathead minnows were exposed to four compounds at 1-nM static concentrations — E2, ICI 182-780 (anti-estrogen), 5-androstan-17-methyl-17-ol-3-one (androgen), and flutamide (anti-androgen). Responses to these compounds were compared to responses in the tri-ethylene glycol (TEG) solvent control. Fish were exposed to each compound in separate tanks (20 fish per tank) and four fish were collected at the following time intervals — 2, 4, 6, and 24 hours and 7 days after exposure (see Appendix B5).

Only E2 was expected to show a significant increase in vitellogenin message and protein; this is consistent with the observed responses (Figure 4-4). These two biomarkers show similar patterns over the course of 7 days. Additional evaluation of these patterns over a range of exposure concentrations may elicit a better understanding of the relative responses, particularly at low doses near the threshold levels.

The androgen and anti-androgen were not expected to directly affect vitellogenin or estrogen receptor alpha mRNA, because activity would occur via binding to different receptors. However, these were included in the assay to determine if changes were noted in the estrogen receptor response. Other receptors may be cloned and optimized for this assay, and be used to detect patterns of expression associated with these other compounds.

4.5 Biomarkers and Reproductive Success

A 21-day flow-through bioassay test using adult fathead minnows focused on evaluating the ecological significance of estrogenic compounds so that the biomarker responses could be placed in the context of a more definitive test with reproductive endpoints directly measured.

The flow-through system (see Figure 4-5) was designed to expose adult fathead minnows to two endocrine-disrupting compounds. The acclimation time to establish consistent reproduction in the various exposure tanks was several months before addition of the test compounds. The system was adapted from methodology outlined in the USEPA manual, "Methods for Measuring the Acute Toxicity of Effluents and Receiving Waters to Freshwater and Marine Organisms" EPA/600/4-90/027F (August 1993). Responses to exposure solutions of E2, the anti-estrogen ICI (182,780), and an E2–ICI mixture were compared to a solvent control.

Throughout the study, the project team monitored egg production; at the end of the study, it collected blood samples and analyzed them for vitellogenin, and measured the gonadosomatic index (GSI), spawning, egg production, and fertilization.

Male and female fathead minnows exposed to E2 and the E2–ICI combination produced significant amounts of vitellogenin above controls (see Figure 4-6). The GSI values were not significantly different from controls, except for decreased GSI for females exposed to the E2–ICI mixture.

Although vitellogenin increases were significant in solutions with E2, egg production was reduced significantly only in solutions containing ICI (Figure 4-7). Egg production was reduced in the E2 test group, but not significantly. Fertilization was not adversely effected in any of the exposure groups.

The 0.5-nM exposure concentrations in this study were much higher than levels anticipated in effluents. This suggests that the increase in vitellogenin as shown in the E2 exposures did not adversely affect the reproductive endpoint to the extent observed with the anti-estrogen. Although the eggs were fertilized, this study did not extend to evaluate potential effects in offspring.

Plasma Vitellogenin Protein

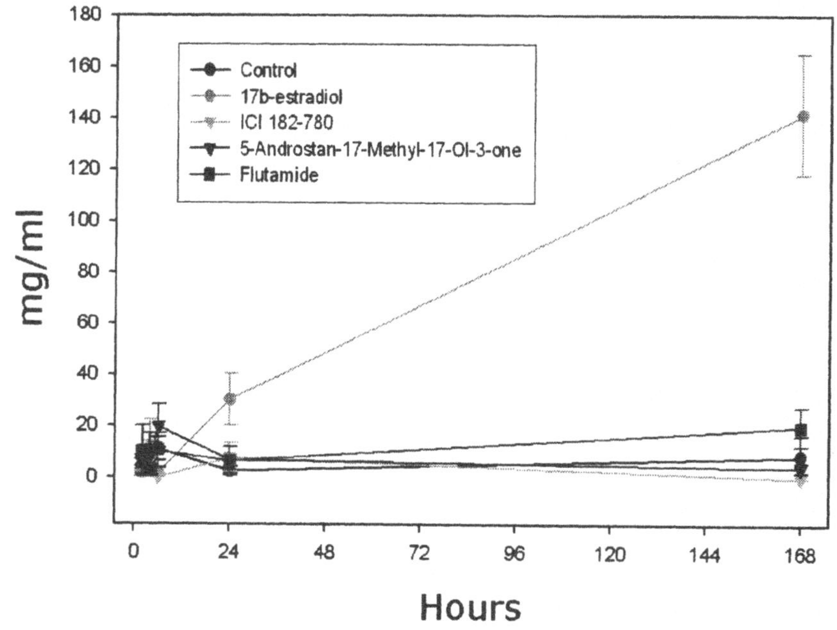

Vitellogenin mRNA Expression

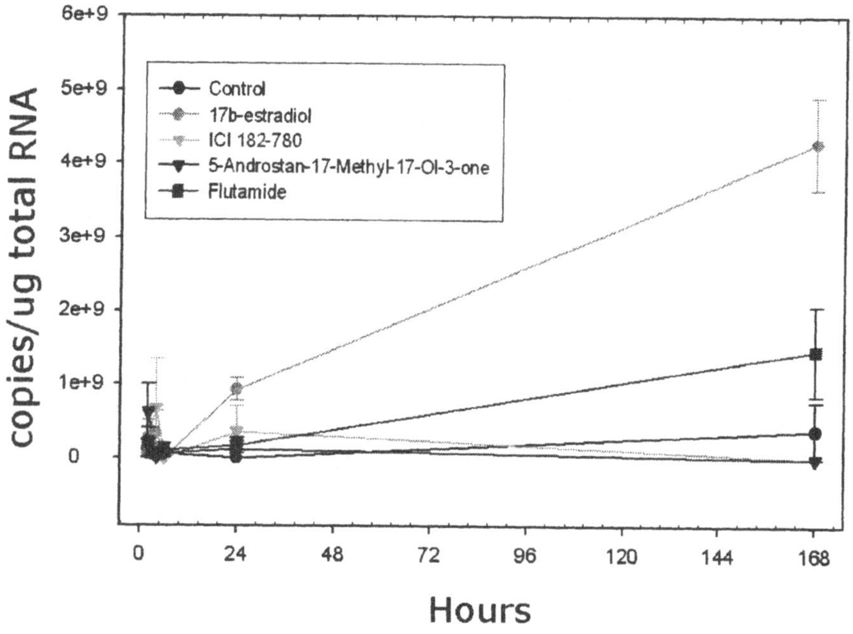

Note: Test Compound Concentrations

Figure 4-4. Comparison of Vitellogenin Protein and mRNA Following Static Exposures to (Anti-) Estrogen/Androgens

Figure 4-5. Apparatus for 21-day Flow-through System for Fathead Minnow Exposures to Measure Biomarker and Reproductive Endpoints

This type of detailed analysis is more costly than simple biomarker measurements. However, it supports observations that elevated vitellogenin levels alone do not necessarily adversely affect reproduction in adult fish. The potential adverse impacts at other windows of exposure are not included in this assay.

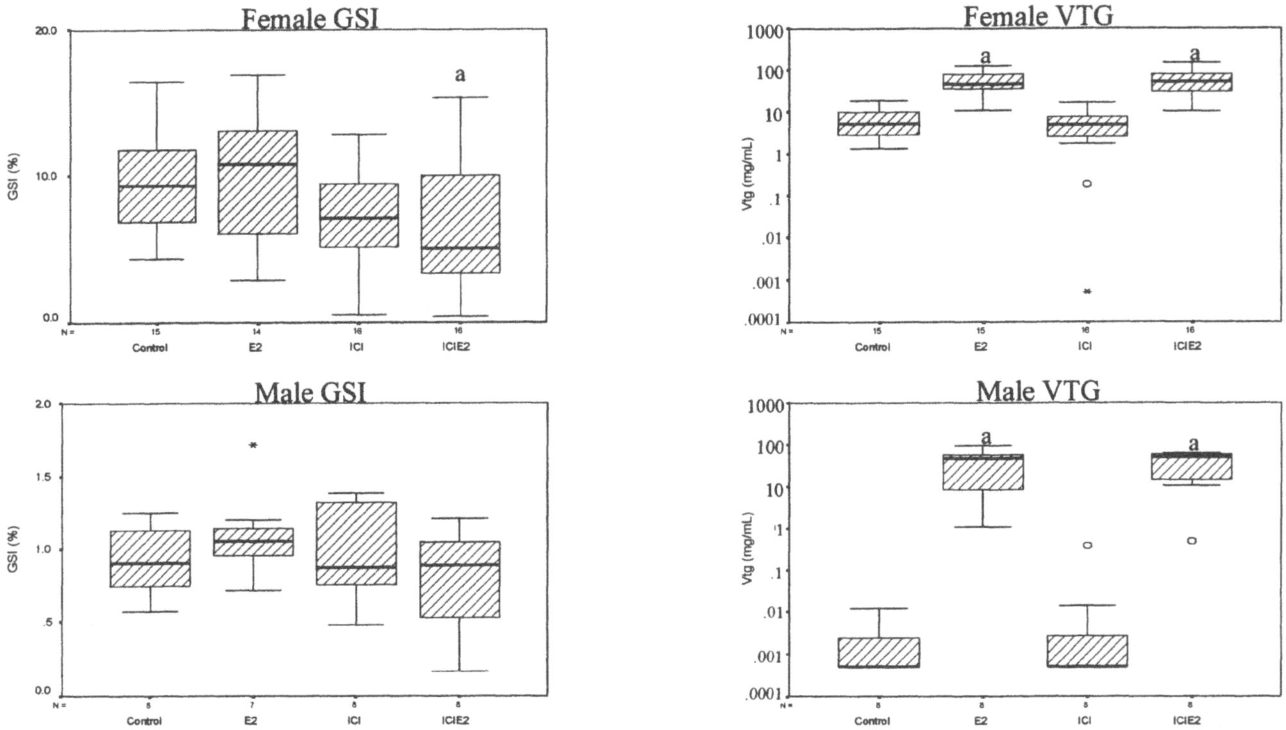

Figure 4-6. Short-term Fathead Minnow Reproductive Study - 21-Day Flow-through Study Comparing Plasma Vitellogenin (Vtg) Levels and GSI for Control, E_2, Anti-estrogen (ICI) and Mixture. E_2 control, triethylene glycol solvent control; E_2, 17β-estradiol (0.5 nM); ICI, ICI#182,780 (antiestrogen) (0.5 nM); ICIE2, Mixture 0.5 nM ICI +0.5nM.

[a] Statistically different from control.

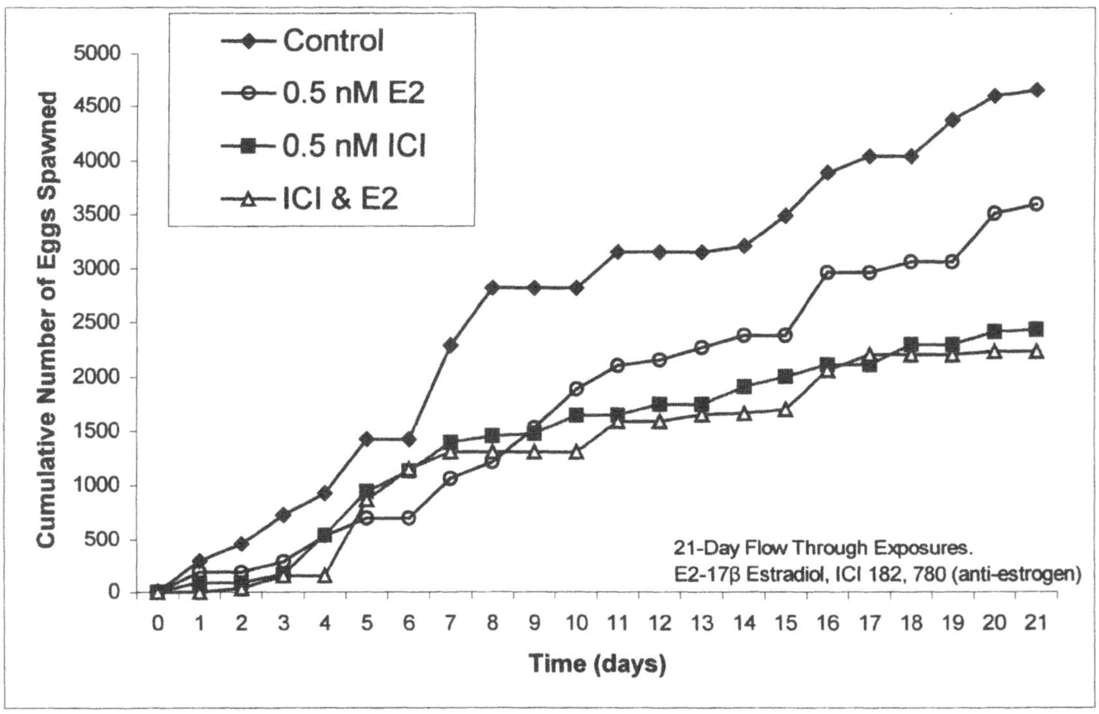

Figure 4-7. Effects of E2 and ISI on the Cumulative Fecundity of Fathead Minnows

4.6 Effects of E2 and EE2 on Selected Fish Species

When selecting a specific target species and biomarker, the results can be placed in the context of other published studies. Therefore, the project team reviewed studies that characterize effects from exposure to estrogens associated with effluents.

Toxicity data on the effects of E2 and EE2 on fish were compiled from recent literature (Lange et al., 2001; Metcalfe et al., 2001; Miles–Richardson et al., 1999; Panter et al., 2000; Scholz and Gutzeit, 2000; Thompson et al., 2000; Thorpe et al., 2000; Tyler et al., 1999; and Zillioux et al., 2001). Tables 4-1 and 4-2 show the lowest observed effect level (LOEL) and no observed effect level (NOEL) values of E2 and EE2 for each species-endpoint, respectively. Geometric mean was used to calculate the average of the species-endpoint with more than one value. Maximum acceptable toxicant concentration (MATC) then was calculated using the geometric mean of NOEL and LOEL. The species-endpoints were ranked (based on the MATC values) on a scale of 0 to 100 to determine their relative sensitivity to E2 and EE, with 0 being the most sensitive and 100 the least sensitive to the compounds. The MATC values then were plotted against their sensitivity ranks as shown in Figures 4-8 and 4-9.

Table 4-1. Sensitivity Ranking of the Effects of 17β-Estradiol on Selected Fish Species Based on the Maximum Acceptable Toxicant Concentration

Species	Exposure Time, days	Endpoint	Sensitivity Ranking	LOEL (ng/L)	NOEL (ng/L)	MATC (ng/L)
Japanese medaka[a]	100 d	Intersex (testis–ova) induction	7	10	1	3
Rainbow trout[f]	14 d	Vtg. induction	14	9	3	5
Fathead minnow[b]	14-19 d	Histopathologic effects (females)	21	27	17	21
Fathead minnow[b]	14-19 d	Vtg. induction	29	28	17	22
Japanese medaka[a]	100 d	Eosinophilic fluid in major organs, body cavity	36	100	10	32
Japanese medaka[a]	100 d	Sex ratio (more females)	36	100	10	32
Carp[e]	30 d	Vtg. induction	50	50	25	35
Fathead minnow[c]	21 d	Decreased gonadosomatic index in males	57	120	60	85
Fathead minnow[b]	14-19 d	Histopathologic effects (males)	64	135	68	96
Rainbow trout[f]	14 d	Increased hepatosomatic index	71	244	70	131
Channel catfish[d]	21 d	Vtg. induction	79	1000	100	316
Japanese medaka[d]	21 d	Vtg. induction	79	1000	100	316
Fathead minnow[b]	14-19 d	Reduced male secondary sex characteristics	93	2700	540	1207
Bass[d]	21 d	Vtg. induction	100	10000	1000	3162

References:
[a] Metcalfe et al., 2001; [b]Miles–Richardson et al., 1999; [c]Panter et al., 2000; [d]Thompson et al., 2000 ; [e] Tyler et al., 1999; [f]Thorpe et al., 2000;.
Vtg. = vitellogenin; LOEL = lowest observed effect level; NOEL = no observed effect level, and MATC = maximum acceptable toxicant concentration.

Table 4-2. Sensitivity Ranking of the Effects of 17α-Ethynylestradiol on Selected Fish Species Based on the Maximum Acceptable Toxicant Concentration

Species	Exposure Time, days	Endpoint	Sensitivity Ranking	LOEL (ng/L)	NOEL (ng/L)	MATC (ng/L)
Japanese medaka[b]	100 d	Intersex (testis–ova) induction	5	0.1	<0.1	0.05
Sheepshead minnow[d]	59 d	Atresia of postvitellogenic ova	9	2	0.2	0.6
Sheepshead minnow[d]	59 d	Fibrosis of testis (males)	9	2	0.2	0.6
Fathead minnow[a]	28 d	Gonad histology (F1)	18	4	1	2
Fathead minnow[a]	56 and 172 d	Intersex (testis–ova) induction (F0)	18	4	1	2
Fathead minnow[a]	56 d	Juvenile standard length (F0)	18	4	1	2
Fathead minnow[a]	172 d	Lack of sexual differentiation in males	18	4	1	2
Fathead minnow[a]	56 d	Sex ratio (more females, F0)	18	4	1	2
Japanese medaka[c]	2 m	Aromatase detected in testis of XY males	41	10	1	3
Japanese medaka[c]	2 m	Reduced egg production	41	10	1	3
Sheepshead minnow[d]	59 d	Intersex (testis–ova) induction	50	20	2	6
Sheepshead minnow[d]	59 d	Reduced egg production	50	20	2	6
Fathead minnow[a]	28 d	Larval standard length (F0)	59	16	4	8
Fathead minnow[a]	172 d	Vtg. induction (F0)	59	16	4	8
Fathead minnow[a]	56 d	Juvenile survival (F0)	68	64	16	32
Japanese medaka[b]	100 d	Eosinophilic fluid in major organs, body cavity	68	100	10	32
Japanese medaka[b]	100 d	Sex ratio (more females)	68	100	10	32
Japanese medaka[c]	2 m	Sex reversal (male to female with ovary)	68	100	10	32
Sheepshead minnow[d]	59 d	Hatching success	86	200	20	63
Sheepshead minnow[d]	59 d	Reproductive success	86	200	20	63
Sheepshead minnow[d]	59 d	Sex reversal	86	200	20	63
Sheepshead minnow[d]	59 d	Survival	100	400	200	283

References:

[a] Lange et al., 2001; [b] Metcalfe et al., 2001; [c] Scholz and Gutzeit, 2000; [d]Zillioux et al., 2001;

d = days, m = months;m Vtg. = vitellogenin; LOEL = lowest observed effect level, NOEL = no observed effect level; and MATC = maximum acceptable toxicant concentration.

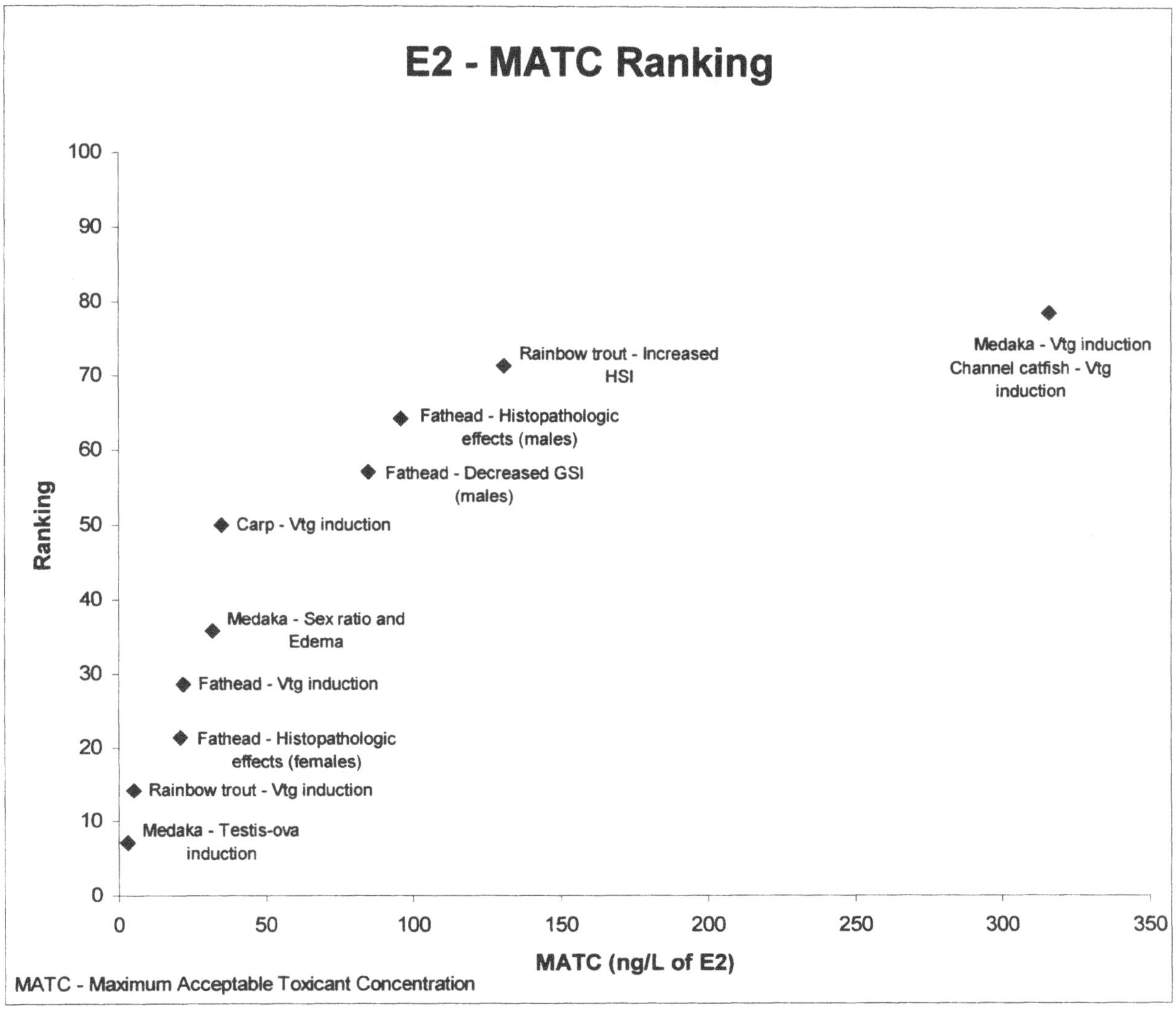

Figure 4-8. Sensitivity Ranking of Different Endpoints for Various Species of Fish in Response to Exposure to 17β-estradiol. Lower Ranks Indicate More Sensitivity to the Test Compound.

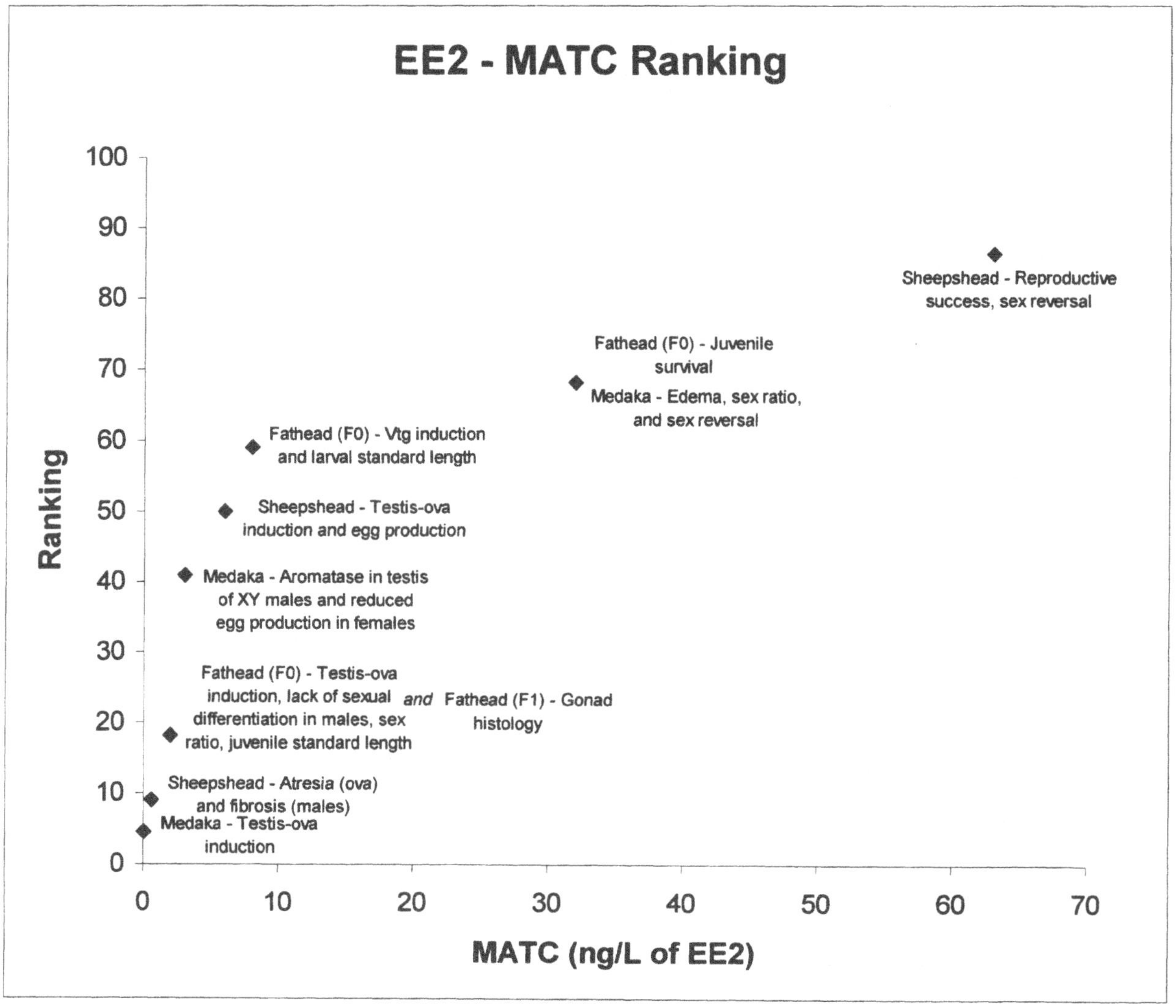

Figure 4-9. Sensitivity Ranking of Different Endpoints for Various Species of Fish in Response to Exposure to 17α-Ethynylestradiol. Lower Ranks Indicate More Sensitivity to the Test Compound.

Although this ranking is preliminary and would be revised as additional data are obtained, it provides an idea of the relative response and concentrations at which effects were observed. The following limitations should be noted in using these tables.

- Experiments were conducted by various laboratories using different materials and methods.
- The range of concentrations used in the tests may present a bias. The difference between test levels in some studies was as high as a factor of 10 (e.g., 0.1, 10, 100, 1000, etc.). The LOEL for a test with a wide interval in concentration may not be the approximation of the actual LOEL. For example, if the LOEL from a test was determined to be 100 ng/L, and the NOEL was the next lowest level (10 ng/L), the actual LOEL might have been 20 or 80 ng/L, but could not be determined from this test since the range between test levels was too wide.

- Tests were performed with different exposure times. It may not be appropriate to compare results among the tests with different exposure times since toxicity is a factor of both concentration and exposure time.
- Most publications presented results based on nominal concentrations, but some presented the data based on measured concentrations.

The MATC for induction of vitellogenin by E2 is approximately 21 to 22 ng/L, which is higher than the MATC for rainbow trout. Testis–ova in medaka is the more sensitive endpoint, based on exposures from 0 to 100 days post-hatch as compared to shorter adult exposures for trout and fathead minnows.

CHAPTER 5.0

EFFLUENT CHARACTERIZATION

5.1 Purpose

Several studies have indicated that WWTP effluents can contain natural and/or synthetic chemicals that can produce estrogenic responses in fish. Many of these studies report elevated levels of vitellogenin in male fish as a primary biomarker of exposure to xenoestrogenic compounds observed in wild or caged fish placed in the receiving streams.

Chapter 2 highlighted potential approaches to characterize effluents for their potential to produce a detectable biomarker and/or endocrine-disrupting impact in receiving streams. Effluent characterization can include the use of chemical analyses, in vitro assays, and in vivo assays; however, the emphasis in this project was potential for biomarker development with in vivo assays.

This section provides an overview of issues to consider when implementing an effluent screening study to evaluate potential biomarker formation, and presents the results of effluent screening performed in this study. One goal of this study was to evaluate approaches for effluent screening so that a facility may better understand the potential for discharges to impact receiving streams. Emphasis was placed on a simple in vivo screening test that would parallel potential observations made during field studies in receiving streams. The YES assay was evaluated with selected effluent samples, but was not the focus of the screening program.

5.2 Objectives of Effluent Screening

Before designing and implementing an effluent-screening program, program objectives should be clarified and focused. Effluent characterization for EDCs can be very costly and may produce ambiguous results. Published studies address a range of biomarkers/endpoints, target species/ages, chemicals of potential interest, and in vitro assays.

Currently, the predominant biomarker associated with WWTP effluents is the development of elevated vitellogenin in male fish associated with estrogenic compounds in effluents. This sensitive biomarker, which can develop relatively quickly (within days) following exposure to estrogens, has been measured in numerous large-scale surveys and, if reported in the receiving stream, may infer adverse impacts. Therefore, understanding the potential for vitellogenin formation may be a factor for plant managers. While chemical and/or in vitro assays may be used to infer potential biomarker formation, in vivo testing is most direct.

The issue of the significance of the formation of a biomarker like vitellogenin should be coupled with emerging data on the potential for significant adverse effects to individuals or populations. A screening assay provides preliminary insights on the magnitude of the response that may be used with emerging studies of relevance for critical endpoints like reproductive success.

5.3 In Vivo Effluent Characterization

There are several benefits to understanding effluent properties under controlled conditions before biomarkers are detected in a facility's receiving stream. Many of the same issues associated with acute and chronic toxicity testing for effluents are considerations for effluent screening to evaluate potential biomarker response. This section highlights considerations for effluent screening, which include exposure conditions and duration, sample collection and handling, target species, and endpoints.

5.4 Exposure Conditions and Duration

Static, static-renewal, and flow-through procedures are three modes of exposing organisms to effluents in laboratory tests. The preferred approach is to use flow-through conditions. Establishing the exposure apparatus at the WWTP, so effluent can be replaced continually at a predetermined rate throughout the duration of the test, can best integrate variability in effluent composition and minimize losses of active compounds in static and/or static-renewal tests.

Static and/or static-renewal tests are a common compromise in toxicity tests to balance realism with cost-effectiveness. For its study, an initial objective of the project team was to evaluate the potential to develop a simple, inexpensive screening approach for effluents using fathead minnows as a model organism. The goal was to optimize exposure duration and limit effluent volume. Effluent screening for estrogenic responses then could be conducted in the laboratory in a fashion similar to standard acute or chronic exposures, but with additional endpoints.

Exposure duration is a function of the target species and endpoints. Increases in male fathead minnow plasma vitellogenin can be detected within several days of exposure to estrogenic compounds present above threshold levels (the level below which no response is observed). If concentrations are near the threshold, one or two renewals can enhance the response. While 7 days would be adequate for vitellogenin induction, it may not capture other endpoints of potential interest, like differences in GSI and/or secondary sex characteristics, unless estrogenic compound concentrations were very high. However, vitellogenin could be used as the more sensitive endpoint.

Preliminary tests with juvenile fathead minnows suggest that a longer exposure duration may be needed with additional renewals. This was attributed to delayed response for individuals without prior exposures to estrogens.

The USEPA proposed protocol for screening EDCs is a 21-day reproductive study with adult fathead minnows that would be performed as a flow-through experiment, and would include such endpoints as reproductive success. This test would be much more costly than a 7-day screening, but would allow biomarker measurements to be correlated directly with potential population effects.

5.5 Sample Collection and Handling

Assuming flow-through exposures are cost prohibitive, effluent samples would be collected for static-renewal screening. As with whole effluent toxicity tests, the type of sample collected to represent an effluent can influence the test's outcome. For effluents with temporal variability, a single grab sample may over or underestimate the response; therefore, composite samples may be preferred. However, for this screening, it would be more preferable to obtain samples at peak estrogenic activity than to obtain a composite sample. An evaluation of the intermittent exposure of fathead minnows to estradiol published by Panter et al. (2000) concluded that response to intermittent exposures to higher concentrations can be greater than estimated from integration of concentration and time (continuous exposure at lower concentration).

The project team followed EPA protocols for storing and handling samples for effluents screened in its study. The goal is to keep storage and handling to a minimum to limit decreases or increases in activity. Storage of positive control solutions of E2, and E2-spiked effluents were monitored using the YES assay, and activity was maintained for samples stored at 4°C.

Because of unique considerations for evaluating estrogenic responses, the potential for sorption to vessels and/or contributions of estrogenic compounds from plastics should be evaluated before testing; positive and negative controls should be similarly handled.

5.6 Target Species

Several animal models currently are being evaluated for screening chemicals for potential endocrine disruption. Species of greatest convenience for laboratory exposures may be correlated to species relevant for the receiving streams as comparative data continue to accumulate. For example, fathead minnows may be handled more easily in laboratory settings, and bass or trout may be of greater interest in the stream. A database that helps to extend source characterization exposures to the field is being developed to improve researchers' ability to extrapolate from species to species.

As part of its studies, the project team compared induction of vitellogenin in adult fathead minnow exposures to induction in juvenile trout and juvenile fathead minnows. Juvenile fathead minnows were less sensitive, and juvenile trout were more sensitive in the exposures conducted.

In this case, sensitive is defined as the lowest concentration at which an increase in vitellogenin would be measurable. While the levels of plasma vitellogenin are a function of exposure to E2, there appears to be a threshold concentration below which no response occurs.

An important consideration in test result variability may be the organism's pretest history, which may affect its sensitivity to an effluent sample. It seems to be a trivial matter to obtain fathead minnows, because of their routine use in bioassay laboratories. While this may be true of juvenile fatheads, many suppliers do not have large quantities of adult minnows for sale (these are used as brood stock). Adult minnows obtained from ponds are more likely to be diseased, and, in some cases, males initially received in the laboratory have elevated levels of vitellogenin. This could have resulted from the close proximity of males and females, shipping in plastics, or other unknown causes. Shipping adult fathead minnows also is not as simple as shipping juveniles for traditional assays.

In the future, suppliers may opt to supply adults as well as juveniles. However, in the interim, laboratories should maintain specimens in controlled conditions and test males for vitellogenin before effluent exposure. It is not essential that males be free of vitellogenin, since the positive response would be determined by increase in vitellogenin in effluent versus negative controls. However, it is possible that the sensitivity of the organism increases with prior activation of the mechanism to produce this protein.

Juvenile trout obtained for the effluent exposure experiment discussed below also had some base level of vitellogenin prior to effluent exposure; the history of these specimens was unclear. It is possible that they originated from a pond where hormone treatments were used. Again, this could affect the sensitivity of the organism in assessing estrogenicity of an effluent, but may not truly reflect effects on trout in natural conditions for organisms not reared in trout farms.

Therefore, the history of the specimens used in an assay may influence the results and have differing responses in the assay, and may account for some of the variability that may be reported in studies.

5.7 Biomarkers

Vitellogenin induction likely is to be a primary biomarker of interest because of the frequency of its use in association with studies of effluent discharges. Therefore, before selecting a species for the exposures, it should be verified that the vitellogenin assay is available and routine. Additional costs would be incurred if the assay needed to be developed before the effluent exposures were performed. In addition, numerous methods are being used such that the relative responses may be comparable, but the reported levels from one laboratory may not be equivalent to those of another.

The emphasis on vitellogenin is due to the fact that numerous effluent studies have shown elevated levels during exposures of males to estrogenic compounds. It is not necessarily the most sensitive endpoint. For example, Metcalfe et al. (2001) report that development of testis–ova in medaka (from hatch to 100 days posthatch) will occur at lower concentrations than elevation of vitellogenin level in adults. It also is likely that new assays will be developed and new endpoints will be identified. However, at present, vitellogenin is a critical parameter to measure for these studies and is likely to be more sensitive than GSI. If a positive response is noted, additional characterization may be needed (or considered in the estimates of field exposures).

5.8 Laboratory Effluent Screening Results

A number of WWTPs agreed to provide a grab sample for use in an effluent screening program. After a series of control exposures to E2 as described in Chapter 4, several sets of effluents were collected and screened using the static-renewal exposure. During the course of this study, 15 effluents were screened in the laboratory, and support was provided for a 21-day flow-through effluent exposure study for another WWTP (see Appendix B2).

Figure 5-1 shows effluent screening results for seven wastewater samples exposed in August 2000. These samples included six domestic WWTP effluents and one industrial effluent. Responses are compared with responses to negative (laboratory water) and positive (0.1-nM E2) controls. One of the six WWTP samples showed a significant response above the positive control, and two others were marginally above the negative control but below the response in the E2 control. This was considered a positive sign that facilities may show varying levels of responses — recognizing that a single grab sample may not represent peak activity.

A second set of exposures was performed in late September with five WWTP and three industrial effluents. All samples, including the 0.1-nM E2, showed a negative response in this assay. Because this was the first time that a positive response was not observed in the 0.1-nM E2 controls, various aspects of the test were reviewed. A review of the comprehensive data set seemed to indicate that significant differences exist in fathead minnows obtained from suppliers, and this may represent seasonal differences in sensitivity for adults obtained from ponds. Therefore, the lack of response in this set of effluents may reflect this decreased sensitivity.

Generally, it is preferable to conduct a flow-through exposure; however, this would need to be done at a facility because of the large volume of effluent required. The project team had an opportunity to support an effluent flow-through study, providing materials and methods, and vitellogenin analyses. The results from this exposure are illustrated in Figure 3-1. For this effluent, vitellogenin levels in fathead minnows did not increase above controls; however, significant increases were noted for exposures to juvenile trout.

The juvenile trout were obtained from a trout farm and controls had detectable levels of vitellogenin. The history of these specimens was not clarified, but may represent all female stocks produced via hormone exposures. In preliminary experiments with juvenile fathead minnows in its laboratory, the project team found that no vitellogenin was present initially, and that a positive response required additional renewals and exposure durations as compared to adults. This has been reported as “estrogen memory” — once exposure to estrogens has occurred, subsequent exposures may show higher and more rapid responses.

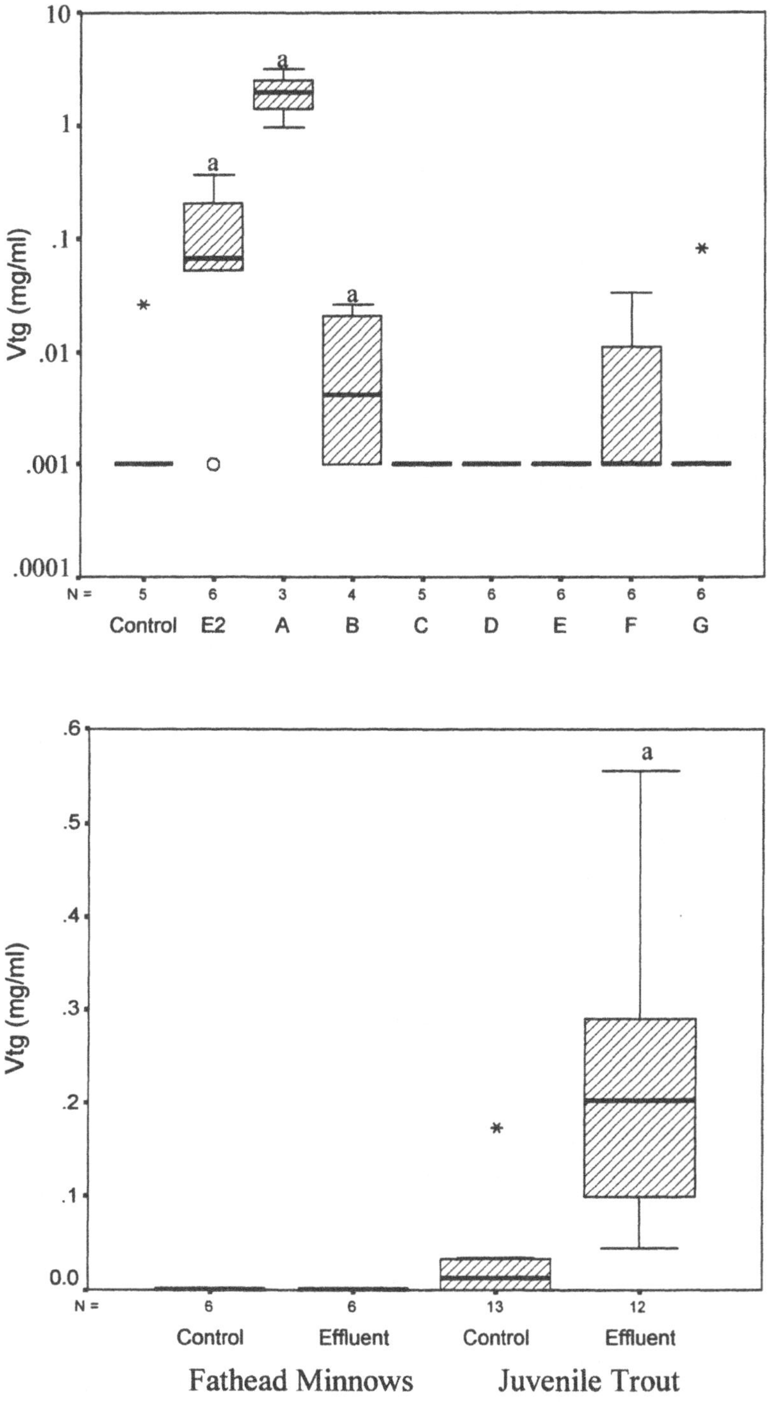

[a] Different from control at a = 0.05

Figure 5-1. Effluent Screening Evaluation: Plasma Vitellogenin in 7-day Static-renewal Fathead Minnow Exposure and Flow-through Effluent Exposure.

Juvenile trout have been reported to respond to E2 exposures with induction of vitellogenin at lower concentrations than reported for fathead minnows. Because of the baseline of vitellogenin in juvenile trout, it is not clear if this enhanced response is representative of responses in natural populations.

5.9 Effluent Screening Conclusions

Levels of steroid hormones reported in wastewater effluents are likely to show significant temporal variability. Ranges of steroid hormones can vary, but E2 concentrations reported in the literature frequently are below the threshold level for induction of vitellogenin, currently estimated near 0.1 nM (27 ng/L).

Several studies have shown induction of vitellogenin in male fish, including adult fathead minnows. The measurement of this in vivo response in fish following effluent exposure integrates the effects of numerous chemicals that may be present. The key findings from the current study include the following.

- Male adult fathead minnows exposed to effluents can have increased vitellogenin levels when they are exposed to estrogenic effluents above threshold levels, as identified for E2.
- Because of effluent variability, additional samples should be included with a goal to obtain samples at times when concentrations are likely to peak. The peak concentrations, perhaps more than the total exposure integrated exposure concentrations from a composite, would be most likely to represent a response of caged fish exposed to effluents.
- There may be seasonal factors beyond effluent characteristics that should be documented before effluent exposures. In addition, it is critical to use both positive and negative controls to validate the assays and organism sensitivity.
- Juvenile trout can show detectable increases of vitellogenin in plasma at concentrations of estrogenic substances that are not observed with fathead minnows. This suggests that trout is a more sensitive species for this endpoint.
- Additional evaluation of seasonality and baseline conditions of fish to be exposed should be documented in further studies

CHAPTER 6.0

DESIGN AND INTERPRETATION OF FIELD STUDIES

6.1 Purpose

Laboratory studies provide information on effluent characteristics and can identify the potential for biomarker formation (including temporal patterns). To evaluate impacts in receiving streams, the magnitude and time course of the discharge traditionally is merged with environmental properties (including receiving water volume, biota type, physical and chemical properties, and water-renewal rate) and the properties of the chemicals that may interfere with natural hormonal functions.

This section addresses issues in designing and implementing field studies to evaluate potential endocrine disruption in receiving waters. As indicated in Chapter 2, the objectives of a field study should be clearly defined in advance, and should include approaches to evaluate the causes and significance of physiological indicators. However, because an increasing number of studies report hormonal activity in effluents, and surveys that may be conducted may show a distribution of vitellogenin induction in watersheds, a facility also may need to develop a fundamental understanding of the implications and uncertainties of these studies, and develop approaches to communicate these complex issues with the public.

The observations, conclusions, and recommendations in this section are based on the North Georgia field study performed as Phase 2 of this project (see Appendix C), with supporting/supplemental information taken from published literature. The study area (see Figures 6-1 through 6-3) included three stations in the Conasauga River and four stations in the Chattahoochee River sampled in the spring and fall of 2001. Table 6-1 provides an overview of the field effort. This study provided a mechanism to evaluate and compare methods while gathering data for several distinct land uses.

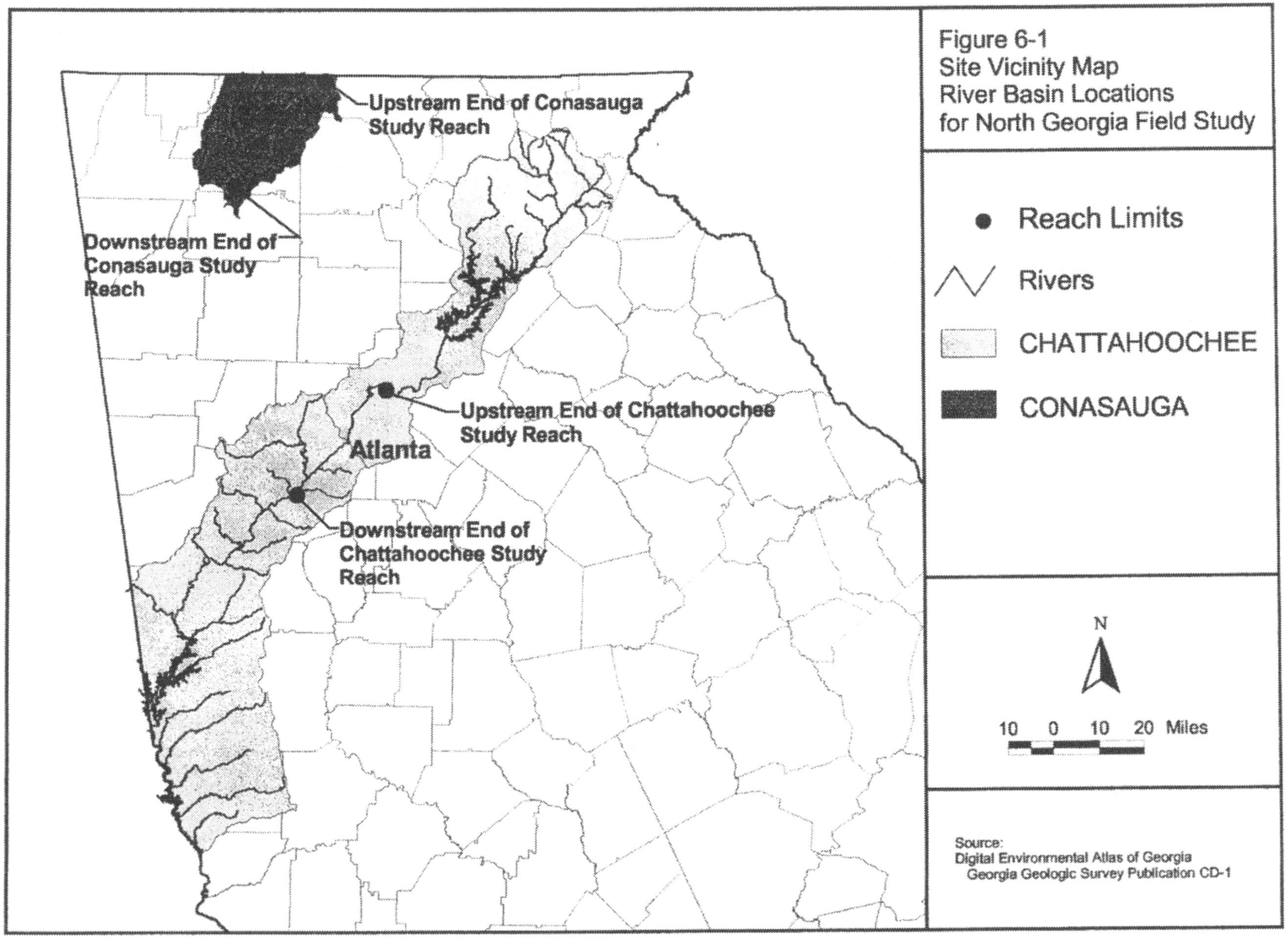

Figure 6-1
Site Vicinity Map
River Basin Locations
for North Georgia Field Study

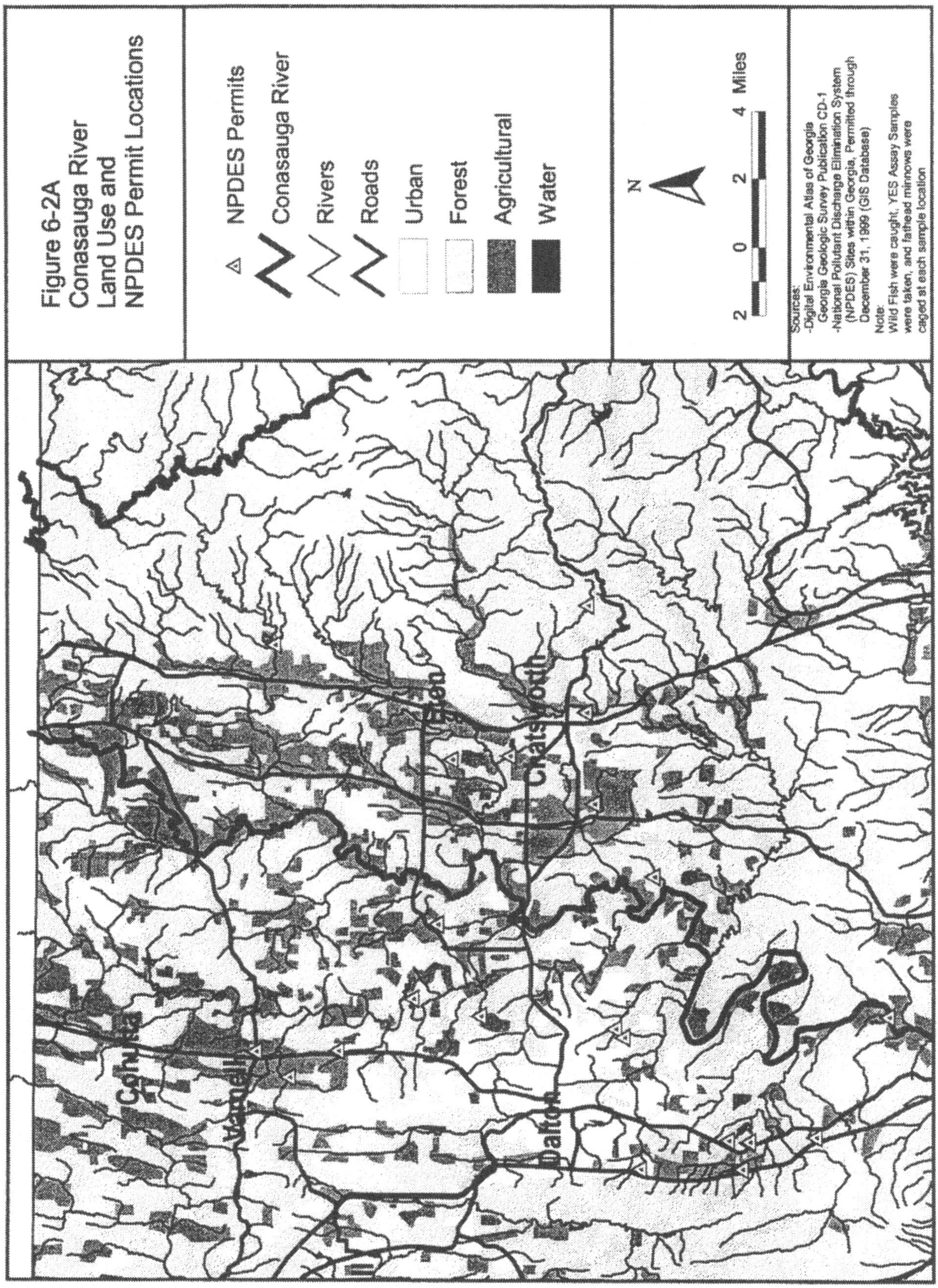

Figure 6-2A
Conasauga River
Land Use and
NPDES Permit Locations

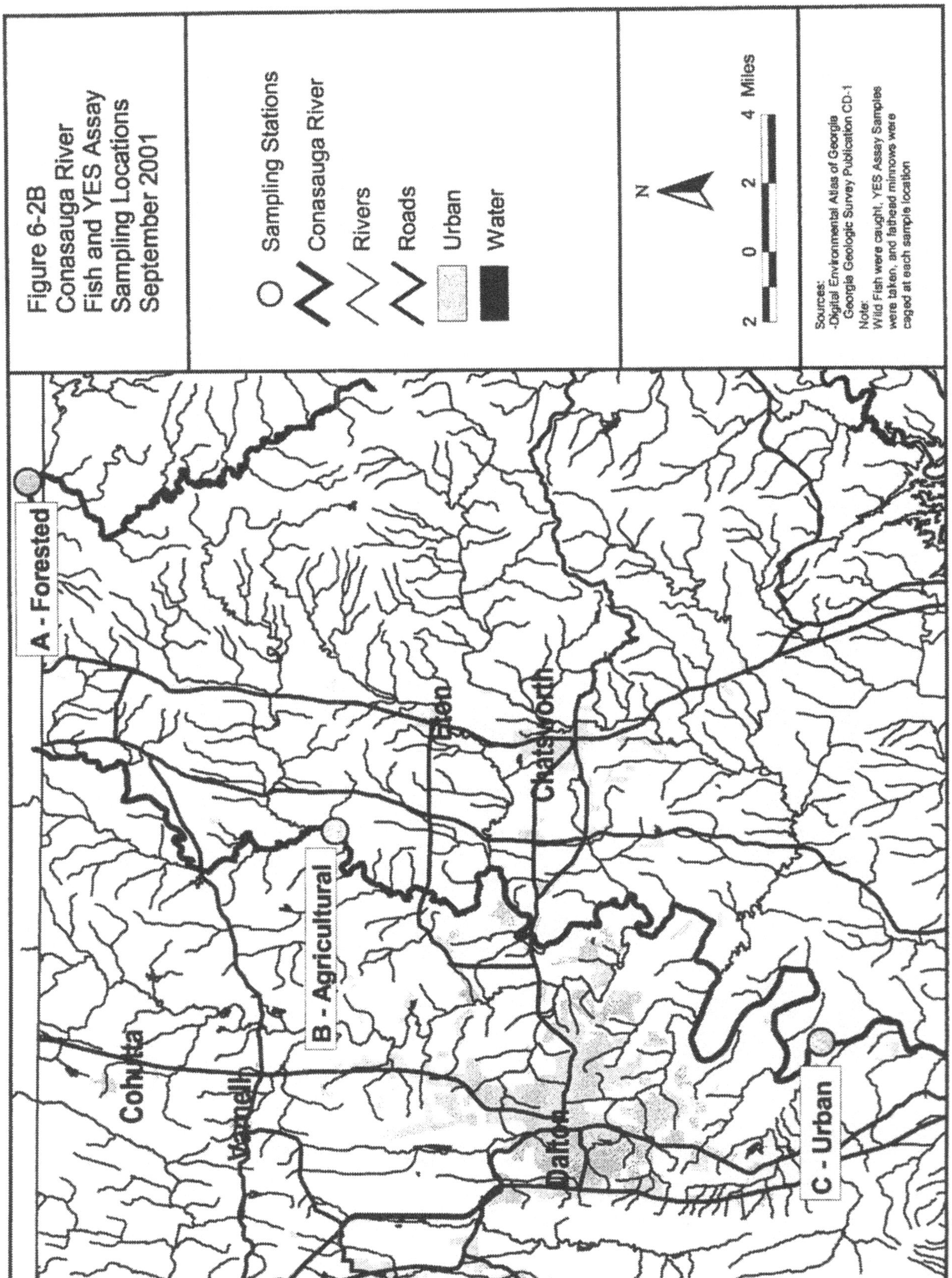

Figure 6-2B Conasauga River Fish and YES Assay Sampling Locations September 2001

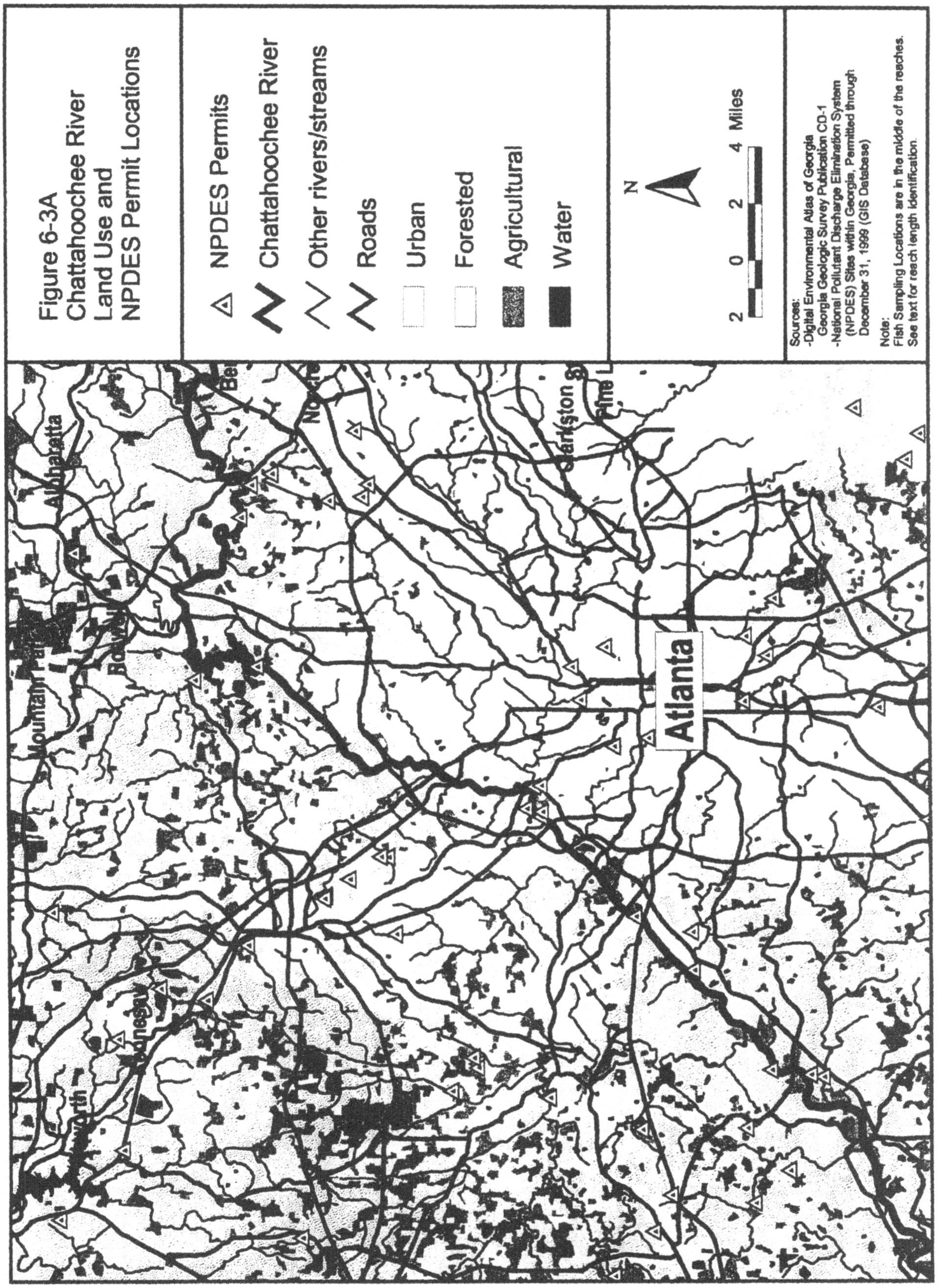

Figure 6-3A
Chattahoochee River
Land Use and
NPDES Permit Locations

Figure 6-3A
Chattahoochee River
Land Use and
NPDES Permit Locations

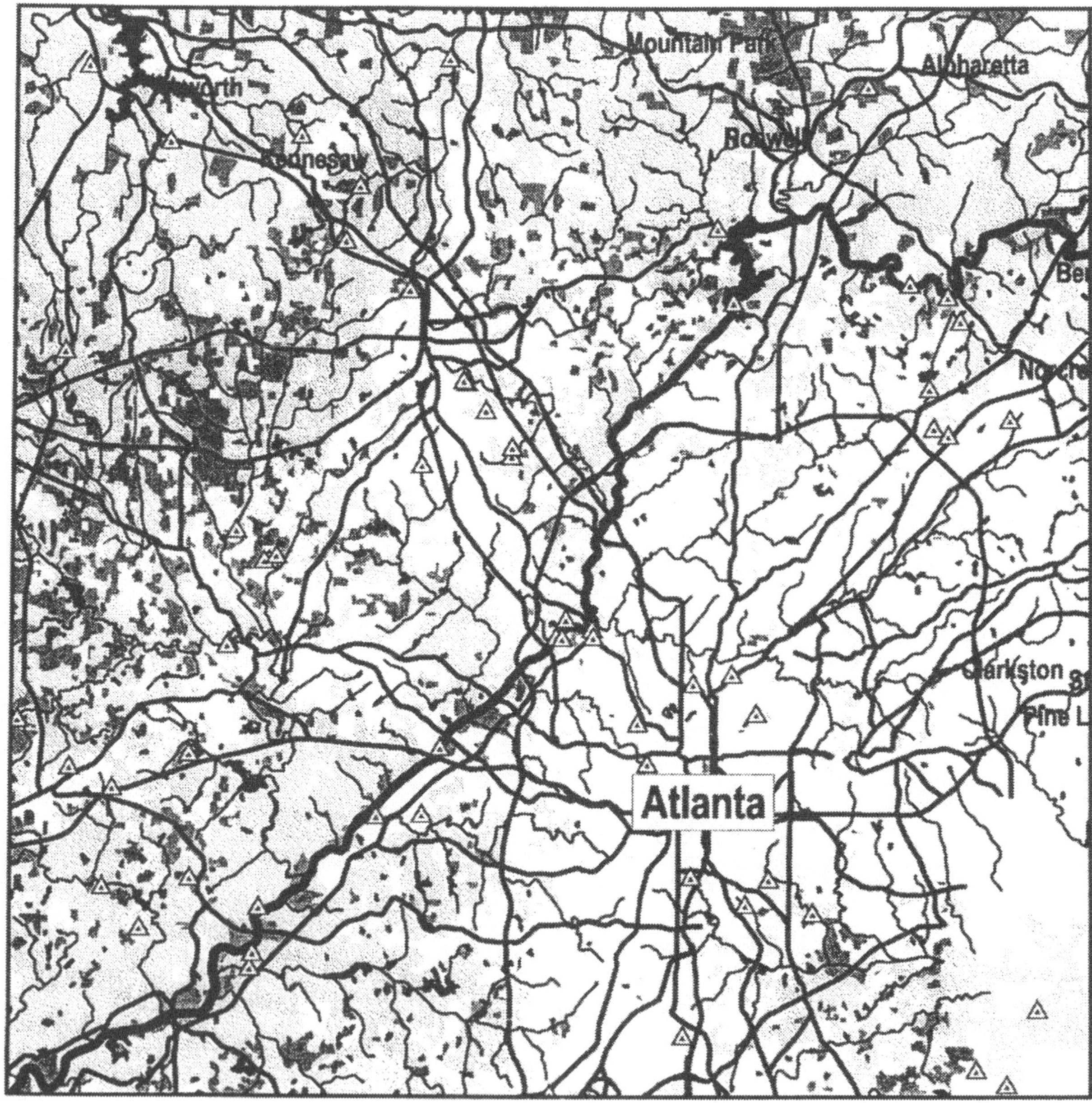

NPDES Permits
Chattahoochee River
Other rivers/streams
Roads
Urban
Forested
Agricultural
Water

N

2 0 2 4 Miles

Sources:
-Digital Environmental Atlas of Georgia
Georgia Geologic Survey Publication CD-1
-National Pollutant Discharge Elimination System
(NPDES) Sites within Georgia, Permitted through
December 31, 1999 (GIS Database)

Note:
Fish Sampling Locations are in the middle of the reaches.
See text for reach length identification.

Table 6-1. Summary of Field Sampling During Spring and Fall 2001 Field-sampling Events

Station	Season	Caged Fathead Minnows (# Survived/ # Processed/ Total Caged)	Wild Fish Collection Method	Species Collected (number: spring/fall)	Surface Water Tests (a)		
					YES Assay		Hormones (b)
Conasauga River Stations (Figure 4-2)					Spring	Fall	
A. Forested Station	Spring	39/20/40	Angling	Redeye Bass (19/19)	2	1	1
	Fall	30/26/30		Spotted Bass (1/2)			
B. Agricultural Station	Spring	30/20/40	Angling	Spotted Bass (19/10)	2	1	
	Fall	15/15/30		Largemouth Bass (1/2)			
C. Urban Station	Spring	19/19/40	Electrofishing	Spotted Bass (19/21)	2	1	1
	Fall	39/24/40		Largemouth Bass (1/0)			
Chattahoochee River Stations (Figure 4-3)					Spring	Fall	
D1. Station #1	No fathead minnows caged at the Chattahoochee Stations		Electrofishing	Largemouth Bass (15/8)	1	1	1
				Spotted Bass (5/12)			
D2. Station #2			Electrofishing	Spotted Bass (12/14)	4	5	4
				Largemouth Bass (7/4)			
D3. Station #3			Electrofishing	Spotted Bass (14/7)	3	5	4
				Largemouth Bass (6/10)			
D4. Station #4			Electrofishing	Spotted Bass (17/0)	3	4	3
				Largemouth Bass (0/21)			

Notes:
(a) number of samples near that station.
(b) Fall only, analysis by GC/MS/MS (see Section A6.1).
In the fall sampling, the species of two bass collected at Station #1.were not identified, not included in table.

Each facility will have specific issues to address. Target species of concern will vary, available background data on the watershed and/or species may vary, and resources and technologies may place constraints on designing and implementing a field-sampling effort. This section highlights issues associated with target species, biomarkers and assays, background, seasonality, and environmental fate.

6.2 Selecting Target Species

Several of the more important considerations in selecting target species include availability of a baseline of data, established methods for biomarker analysis, sensitivity, and relevance/importance for the receiving stream in question.

Published studies have demonstrated estrogenic effects in surface waters using caged and wild fish. Caged fish can be useful to represent exposures at a specific location, integrating the exposure over a defined period (e.g., 1 to 3 weeks). Wild fish can integrate exposures over a wider range and exposure period. Table 6-2 summarizes field studies that have been conducted to evaluate potential effects of endocrine disrupting compounds (EDCs).

Table 6-2. Summary of Field Studies for EDCs

Ref.	Location (Contaminant)	Species Used or Sampled	Sampling Approach	Biomarkers/ Measurements	Comments/ Conclusions
Allen et al. (1999)	Four contaminated estuaries in the U.K.: Tyne, Crouch, Thames, and Mersey	Flounder, Platichthys flesus	Field sampling by trawling	Vtg. Concentration in blood, inter-sexuality	Fish from one or more sites in each estuary had significantly elevated concentrations of Vtg. in their blood compared to those from control sites. Approximately 20% of male fish from the Mersey estuary contained oocytes in their testes, but this intersex condition was not seen elsewhere.
Batty (1999)	The Hawsbury–Nepean River system, Australia (sewage effluent)	Mosquitofish, Gambusia affinis (male)	Field sampling (method not specified)	Male secondary sexual characteristic — growth and development of modified anal fin (GP)	GP length was reduced in size in fish sampled downstream from a WWTP discharge point, when compared to fish sampled upstream of the discharge point.
Burse et al. (2000)	Elevenmile Creek, FL (paper-mill effluents)	Bluegill sunfish, Lepomis macrochirus	Field sampling (method not specified)	Steroid hormone levels, cyp450 induction, and Vtg. Level	Vtg. was not induced in males, suggesting that the effects were not estrogenic. Males from this creek had higher T levels, when compared to those from a reference bluegill sunfish site.
Giesy et al. (1997)	Saginaw River and Saginaw Bay, MI (contaminated by various industries — PCDFs, PCDDs, and PCBs)	Walleyes, alewives, yellow perch, rainbow smelt, gizzard shad, emerald shiners, and common carp	Fish collected by electro-shocking, trawling, or plankton nets	Tissue concentrations of PCBs, PCDDs, PCDFs, PCNs, PCDTs, and 2,3,7,8-TCDD equivalents	TCDD-EQ and TEq conc. in forage fishes and small and large walleyes did not, within the range of accuracy and precision of the bioassay and the instrumental analyses, differ. In midsized walleyes, common carp, and alewives TCDD-EQ conc. were significantly greater than conc. of TEq calculated from the conc. of PCDD, PCDF, and no- and mo-PCB congeners.
Goodbred et al. (1996)	25 sites in U.S. streams within 11 NAWQA study units, representing a wide range of environmental settings typical of major U.S. regions	Common carp, Cyprinus carprio (adults of both sexes)	Sampled w/ pulsed DC electro-shocking (from electro-shocking boat)	E2, 11-KT, Vtg., and gonadal histopathology	Significant site-to-site differences in biomarkers and the presence of significant correlations between biomarkers and contaminants indicate fish in some streams may be experiencing endocrine disruption.
Harries et al. (1996)	River Lea, U.K. (treated sewage effluent); reservoirs in southeast U.K. (raw water storage reservoirs)	Rainbow trout, Oncorhynchus mykiss (adult male)	Cages: left onsite, 3 weeks for River Lea; 6 weeks for the reservoirs	Increase in plasma Vtg. in male fish	An increase in plasma Vtg. was observed in fish placed at the River Lea for up to 15 km (with statistically significant up to 4.5 km downstream). The significant response was decreased to 3 m and 1.6 km later in the year when the dilution factor increased (more rainfall). There was no

Ref.	Location (Contaminant)	Species Used or Sampled	Sampling Approach	Biomarkers/ Measurements	Comments/ Conclusions
					significant increase in plasma Vtg. in fish placed in the reservoirs.
Harries et al. (1997)	U.K. rivers in Kent, Sussex, Essex, Essex/Suffolk, and West Yorkshire (WWTPs)	Rainbow trout, Oncorhynchus mykiss (adult male)	Galvanized steel cages (left onsite for 3 weeks)	Induction of Vtg. synthesis in male fish, inhibition of testicular growth	Fish placed in the neat effluent or close to the point of effluent discharge had marked and rapid increases in plasma Vtg., particularly in the river receiving effluent from wool-scouring mills. Effects were observed up to several kilometers downstream.
Hashimoto et al. (2000)	Tokyo Bay, Japan (industrial and domestic sewage effluent)	Male flounder, Pleuronectes yokohamae	Fish captured by dragnet	Serum Vtg. levels in male fish, serum concentration of E_2 and T, and histological appearance of gonads	Statistically higher concentration of serum Vtg. was observed in wild male flounder collected from Tokyo Bay compared to fish from control site. Testes of three of 20 males collected in inner Tokyo Bay contained small number of oocytes.
Jobling et al. (1998)	Eight U.K. rivers (effluent from WWTPs)	Roach, Rutilus rutilus	Electro-fishing or netting (60 to 100 adult per location)	Intersexuality, Vtg. concentration in plasma, GSI	Gonad histology revealed that a large proportion of males were intersex (having both male and female gonadal characteristics). There was a weak, but highly significant, positive correlation between the intersex index and plasma Vtg. Concentration. GSI in fish sampled at sites receiving the effluent was less than those in fish at control sites.
Larsson et al. (1999)	Swedish WWTP receiving mainly domestic wastewater	Rainbow trout, Oncorhynchus mykiss (juvenile of both sexes)	Cages (left onsite for 2 weeks)	Vtg. in plasma; 17α - ethinyloestradiol, estrone, and E2 in the bile	A widely used synthetic estrogen affects the endocrine systems of fish exposed to sewage effluent water.
Lye et al. (1999)	Tyne and Tees estuaries in northeast U.K. (Tyne is industrialized/urbanized; Tees is highly industrialized)	Flounder, Platychthys flesus (sexually mature male)	Field sampling collected by beam trawl (over 1-week at Tees)	Vtg. induction in male	High levels of Vtg. found in fish from Tyne (n=6, length >25 cm). The catch was very poor at Tees despite an extended trawl period (n=11) and was restricted to small fish (>10 cm), too small to bleed for Vtg. analysis.
McNatt et al. (2000)	Spring Creek (Fenholloway River tributary), Taylor County, FL (paper-mill effluent)	Eastern mosquitofish, Gambusia holbrooki	Field sampling (method not specified)	The presence of GP (male elongated anal fin) in female fish	Female fish living below the effluent exhibited elongated anal fins (resembling male GP, but the tips lack termination complexes). Female fish also had a smaller maturation size, fewer and smaller embryos, and mature ova per female compared to those from a reference site.

Ref.	Location (Contaminant)	Species Used or Sampled	Sampling Approach	Biomarkers/ Measurements	Comments/ Conclusions
Munkittrick et al. (1994)	Twelve Canadian pulp mills (including BKME from mills using chlorine and sulfite)	White sucker, Catostomus commersoni	Fish captured using gill-net sets (10.2 cm; Espanola collections used 9.5- 10.2-cm mesh)	Organ weight, hepatic mixed-function oxygnenase activity, plasma steroid levels, and liver dioxin levels	This survey has shown that induction of hepatic EROD enzymes and depressions of plasma sex steroid levels during gonadal growth are found downstream of several pulp mills and that these changes are seen at some mills without chlorine bleaching and at mills that have secondary treatment.
Munkittrick et al. (1997)	North American pulp mills (effluents at various BK pulp mills)	Wild fish populations	Field sampling (not specified)	Delayed sexual maturity, smaller gonads, changes in fecundity, and a depression in secondary sexual characteristics	In earlier studies, these biomarkers were observed in wild fish populations at large BK pulp mills with primitive pulping or little effluent treatment. Improved wild fish performance has been recorded at five sites after the mills modernized their waste treatment and pulping processes. However, the improvements were not complete and were not universal. Reduced gonad size and depressed sex hormones are still seen at one mill with 100% chlorine dioxide substitution and secondary treatment.
Nagler et al. (2001)	Hanford Reach of Columbia River, central Washington state (suspected EDCs).	Spawning Chinook salmon, Oncorhynchus tshawytscha	Field sampling (not specified)	Female with male genetic marker (XY chromosomes)	84% of phenotypic females had XY chromosomes. The authors speculated that the sex reversal was due to either ED compounds in water or temp. changes in the river caused by water-management protocols at dams upstream from the spawning ground.
Nichols et al. (1999)	Central Michigan WWTP effluents	Fathead minnow, Pimephales promelas (adult male and female)	In situ cages for 3 weeks	Vtg., E2, and T in plasma	No trends in hormone concentrations, male secondary sex characteristics, or gonad histology could be attributed to putative endocrine disrupter exposure in WWTP effluent. The results indicate that the risk for estrogen antagonist exposure below these central Michigan WWTPs is small.
Orlando et al. (1999)	Escambia River (coal-fired electric power plant and chemical company effluents) and Blackwater River (reference site)	Largemouth bass, Micropterus salmoides	Sampled by electro-shocking	Vtg., E2, and T in plasma	Lower plasma T and proportionately higher liver mass in males from contaminated site suggest that reproductive steroid levels may have been altered by increased hepatic enzyme activity. The presence of vitellogenic males indicates that an exogenous estrogen source was present in Escambia River.

Ref.	Location (Contaminant)	Species Used or Sampled	Sampling Approach	Biomarkers/ Measurements	Comments/ Conclusions
Tyler and Routledge (1998)	Rivers in England and Wales (effluent discharges from WWTP)	Rainbow trout, Oncorhynchus mykiss (adult male) and wild fish population	Cages for rainbow trout and field sampling for wild fish	Induction of Vtg. in male trout and the occurrence of intersex in wild fish population	Vtg. induction in male trout occurred up to 5 km downstream from the WWTP discharge point. Wild fish populations have been shown to be intersex (manifestation of endocrine disruption induced by estrogen exposure).
Williams et al. (1998)	Miramichi River (industrial site) and Kouchibouguac River (reference site), New Brunswick, Canada (industrial effluents, e.g., pulp and paper mills)	Atlantic tomcod, Microgadus tomcod	Fish captured using open-water nets, and switching to fishing nets through ice during winter	Sex steroid levels (T & 11-KT in males & T & estradiol in females), gonad weights, vitellogenic oocyte diameter, and hepatic levels of cytochrome P4501A mRNA (CYP1A mRNA)	Female fish in Miramichi River showed significantly smaller vitellogenic oocytes, lower plasma T, estradiol, and relative gonad weight than fish in Kouchibouguac River. Miramichi CYP1A mRNA levels were higher than Kouchibouguac in the fall and lower in the spring sample.
Zhou et al. (1999)	Piles Creek and Tuckerton (control site), N.J. (contaminated with heavy metals and organic materials)	Mummichog, Fundulus heteroclitus	Fish captured by Gees minnow traps for 2 consecuti ve days	Thyroid histology and thyroid hormone (thyroxine and triiodothyroni ne) levels.	When compared to Tuckerton fish, Piles Creek fish had larger thyroid follicles and greater follicle epithelial cell heights, and contained higher plasma thyroxine levels, but no difference in triiodothyronine.

GP = gonadopodium, 11-KT = 11-ketotestosterone, WWTP = wastewater treatment plant, T = testosterone, BK = bleached kraft, NAWQA = National Water Quality Assessment Program, conc. = concentration, EROD = ethoxyresorufin O-deethylase, BKME = bleached kraft mill effluent, ED = endocrine disrupting, CYP450 = Cytochrome P450A, PCDF = pentachlorodibenzo-p-furan, PCDD = pentachlorodibenzo-p-dioxin, PCB = polychlorinated biphenyls, PCN = polychloronaphthalenes, PCDT = polychlorinated dibenzothiophenes, 2,3,7, 8-TCDD = tetrachlorodibenzo-p-dioxin, TEQ = toxicity equivalents, P450 1A mRNA = combination of P450 1A and messenger RNA, and CYPIA mRNA = cytochrome P450 1A mRNA.

6.2.1 Caged Fish

The more common target species caged in streams and/or reservoirs to evaluate impacts of EDCs in surface water include rainbow trout and fathead minnows. In the United Kingdom, Harries et al. (1996 and 1997) and Tyler and Routledge (1998) showed vitellogenin increases for rainbow trout caged for 3 weeks in streams with wastewater discharges. Larsson et al. (1999) caged juvenile trout for 2 weeks to evaluate wastewater effluent effects and report estrogenic effects in Sweden. Adult fathead minnows were caged for 3 weeks in Michigan WWTP receiving streams, but no trends were noted. Several of the published studies have used caged trout. It has been suggested that trout may respond by vitellogenin induction at lower concentrations of estrogenic compounds than fathead minnows. Nichols et al. (1999) and

Hemming et al. (2001) reported vitellogenin induction in the first station of a wastewater treatment wetland where male fathead minnows were caged for a 3-week period. In their study, Nichols et al. (1999) reported reduced survival. Survivals were reported from 3% to 83%; however, contributing factors included escape, predation, and some vandalism.

In the North Georgia field study, fathead minnows were caged for 2 weeks at each of three locations in the Conasauga River. Because fathead minnows were the primary model organism for the laboratory studies, laboratory data could be used to compare and interpret data from the field exposures. A total of 124 fathead minnows were processed — 59 in the spring and 65 in the fall. As shown in Table 6-1, transient toxicity (50%) was observed at one of the three stations in the spring (urban) and one of three stations in the fall (agricultural). The potential for vandalism was considered when selecting locations for caging fish in the North Georgia study.

Conditions in the North Georgia field study area (including temperature and oxygen) were not considered optimal for supporting trout.

6.2.2 Wild Fish

Streams, estuaries, lakes, bays, and reservoirs receiving a variety of domestic and industrial effluents have been sampled in the United States, Canada, United Kingdom, Sweden, Japan, and Australia. Common test species include flounder (*Platichthys flesus*), mosquitofish (*Gambusia affinis*), rainbow trout (*Oncorhynchus mykiss*), fathead minnow (*Pimephales promelas*), largemouth bass (*Micropterus salmoides*), bluegill (*Lepomis macrochirus*), and white sucker (*Catostomus commersoni*).

A primary factor in determining which species of wild fish to use is an analysis of resident fish populations to ensure that the chosen species is represented consistently throughout the regions being evaluated. Flounder was selected as a representative marine species distributed in the estuaries where effluents were discharged. The flounder vitellogenin response was about one order of magnitude less sensitive than rainbow trout in laboratory exposures to estrogen. Flounder responses in estuaries may show moderate responses in some estuaries, and may be highly induced on another estuary receiving effluents from similarly large populations. This suggests that causative agents may differ. Flounder spend their lives in association with sediments, so they can show more extreme responses compared to trout (Allen et al., 1999).

Recognizing that target species may vary over an area, researchers should document and take into consideration differences in physiology, habitat, feeding behaviors, and other factors that may influence responses.

Bass was a target species in the North Georgia study. A commercially important species, largemouth bass have been the focus of several studies that are improving the understanding of biomarker patterns in their seasonal reproductive cycles (Sepulveda et al., 2001 and Gross et al., 2001). Analytical techniques have been developed and lifecycle information is evolving. A total of 266 bass were processed (131 in the spring, 135 in the fall) from samples collected at seven stations. A single species of bass could not be targeted, due to variations in physical and chemical characteristics between the headwaters and lower reaches of the Conasauga River.

The three species of bass collected in the North Georgia study (redeye, spotted, and smallmouth bass) have different habitat preferences based on water temperature, flow velocity, and substrate. Generally, the redeye bass occurs in cooler, swifter headwaters; largemouth bass prefer warmer, more sluggish waters; and spotted bass prefer conditions in between the other two. Some overlapping does occur, and areas containing either spotted and redeye bass or spotted and largemouth bass were common. Spotted, redeye, and largemouth bass are closely related species whose food resources overlap; all three species feed primarily on insects, crayfishes, and fishes.

6.3 Sample Collection Methods

Various sampling approaches have been used in published studies, including electroshocking, trawling, gill nets, and minnow traps. Because of the presence of endangered species, electrofishing was not allowed in the two northernmost stations in the North Georgia study, so fish were collected at these stations by angling. Sample collection methods have been reported to potentially influence plasma hormone levels, as a result of stress induced during collection and handling. Capture and confinement stress has been reported to potentially affect reproductive physiology and plasma hormones; however, differences among control sites could be identified (Jardine et al., 1996). In the North Georgia study, seven of the ten highest testosterone measurements were reported in samples caught by angling as opposed to electrofishing. Collection methods are not expected to influence levels of vitellogenin or other endpoints.

6.4 In Vivo Biomarkers and Assays

To the extent possible, the most robust data set provides the foundation for a weight-of-evidence approach to evaluating potential adverse effects in the wild. Because of the complexity of factors in the field, a range of observations and measurements can improve interpretation of the results and place the observed responses in the context of other influences in the receiving streams.

Currently, an increase in plasma vitellogenin in male fish is one of the more frequently reported biomarkers of exposure to estrogenic compounds in effluents. Other physiological indicators include intersexuality, as documented by gonad histology and plasma hormone levels. In addition, an assay has been developed to measure changes in the expression of two particular genes involved in reproduction to examine the effects of exposure at the molecular level. Chapter 3 presented an overview of biomarker assays; this section highlights observations from field efforts.

6.4.1 Vitellogenin

Vitellogenin is an egg yolk precursor protein produced as part of the normal reproductive cycle in female oviparicus species. This protein is found in males at low levels, but can be induced with exposure to estrogenic compounds. Vitellogenin is one of the more commonly reported biomarkers in field studies to evaluate potential exposures to EDCs, particularly estrogenic compounds. Generally, it is considered most sensitive in male fish where natural levels remain low or below detection limits. Reductions of vitellogenin levels in female fish may also indicate potential hormonally mediated responses. When comparing results across studies, analytical methods may be an important factor in the concentrations reported until the methods are standardized. However, the relative responses may be a more important consideration at this time.

For the North Georgia study, the available vitellogenin assay had been validated for largemouth bass. Therefore, the cross-reactivity and relative response of the assay to redeye and spotted bass needed confirmation prior to completion of the assays. It is possible that at this time, several assay-validation procedures should be clearly developed before implementing the field effort.

Figure 6-4 summarizes vitellogenin levels in bass included in the North Georgia study. These data show that vitellogenin levels are much higher in the spring. Differences across stations were not significant.

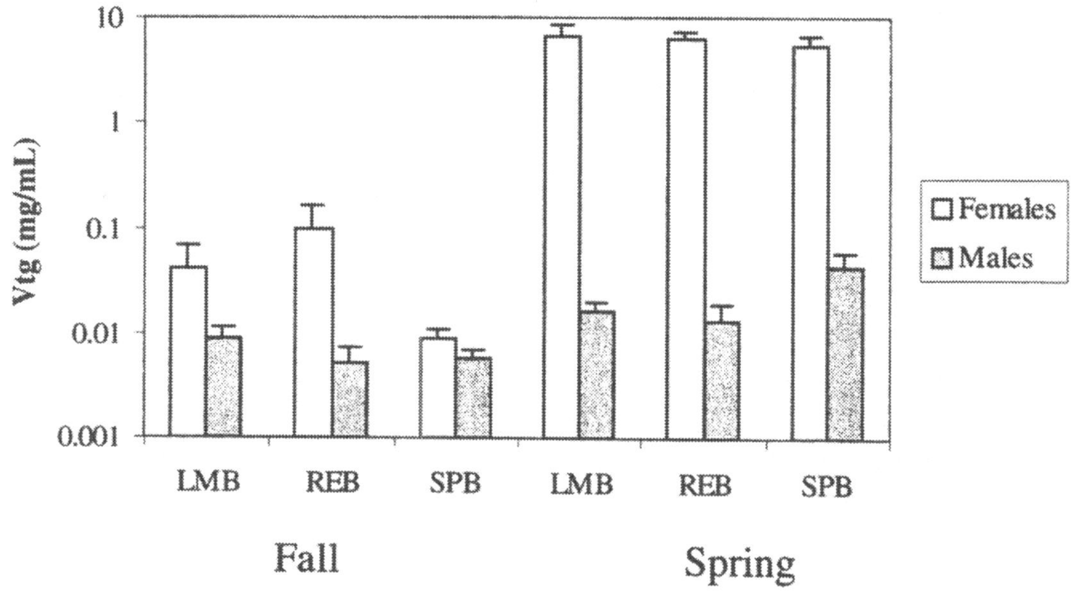

LMB = Largemouth Bass; REB = Redeye Bass, SPB = Spotted Bass

Figure 6-4. Comparison of Vitellogenin Levels in Bass Species for the Fall and Spring.

6.4.2 Sex Ratios/Sexual Differentiation and Phenotype

Total and organ (liver and gonads) weights, fish length, and observations on secondary sex characteristics can be important supporting information. Endpoints like size of breeding tubercles, color changes, or gonadopodium alteration are easy to identify, but may be somewhat subjective.

Sexual differentiation can be an indicator of endocrine disruption and has been reported as a sensitive endpoint for early-life exposures in medaka. Environmental factors like water temperature can affect sex determination and ratios in fish, and the model species should be well characterized with respect to the regulation of sexual differentiation (Ankley et al., 1998).

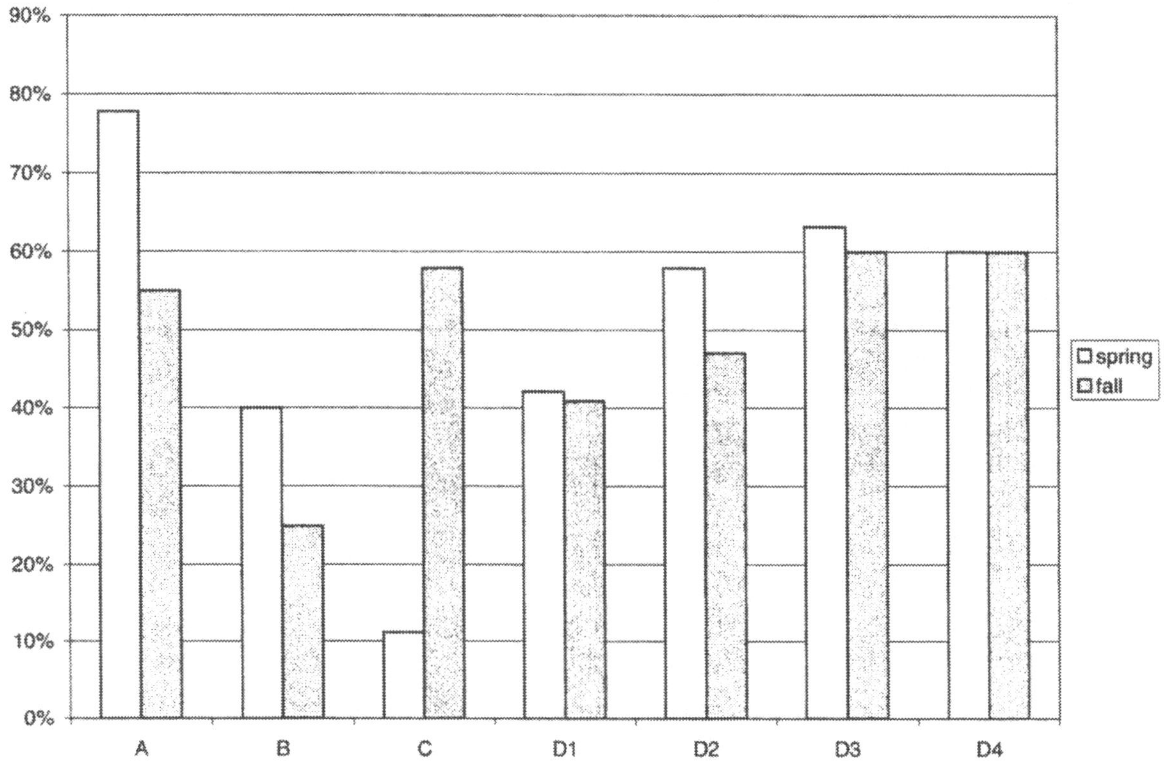

Figure 6-5. Percent Female Bass Collected at Each Sampling Station. Note: A=Forested; B = Agricultural; C = Urban. (Conasauga Stations shown in Figure C-3. D1 = Station #1; D2 = station #2; D3 = Station #3; D4 = Station #4 (Chattahoochee Stations shown on Figure C-2.)

Altered sex ratios have been reported for fish populations impacted by effluents. The ratios for bass for each of the seven stations in the North Georgia study are shown in Figure 6-5.

This parameter is considered to be highly variable. While most bass caught at Station C (Conasauga urban/mixed-use station) in the spring were males, this illustrates that this type of skewing may occur randomly.

6.4.3 Plasma Steroid Concentrations

As indicated in Section 2, plasma sex steroid concentrations can vary as a result of several mechanisms of action, and can fluctuate through the course of the reproductive cycle and time of day.

11-Ketotestosterone (11-KT) has been identified as a more androgenic male hormone than testosterone in teleosts, and more important in the reproductive physiology of male fish. Testosterone, however, has been reported in several studies, in part because the kits are more readily available. Testosterone also has a role in controlling spermatogenesis, and similar quantities of 11-KT and testosterone may be present (Folmar et al., 1996).

Plasma steroid concentrations — 17β-estradiol (E2) and testosterone — were measured in bass in the North Georgia study using ELISA kits that use monoclonal antibodies to mammalian E2 and testosterone. Figure 6-6 summarizes the average levels for spring and fall.

6.4.4 Messenger RNA (mRNA)

This biomarker can be used as a probe for vitellogenin induction; however, this assay is used less routinely because of the specialized training and equipment necessary. Several recent studies have included this assay. Fathead minnow mRNA was measured in the North Georgia field study.

Figure 6-7 shows the levels of fathead minnow vitellogenin and vitellogenin mRNA in the spring sampling event. These measurements illustrate similar trends, with much higher levels for females than males and similar patterns for the three different sites where these minnows were caged. The vitellogenin mRNA represents more recent exposures (previous few days), whereas the vitellogenin protein levels may integrate response from pre-caged conditions and the 2-week exposure period.

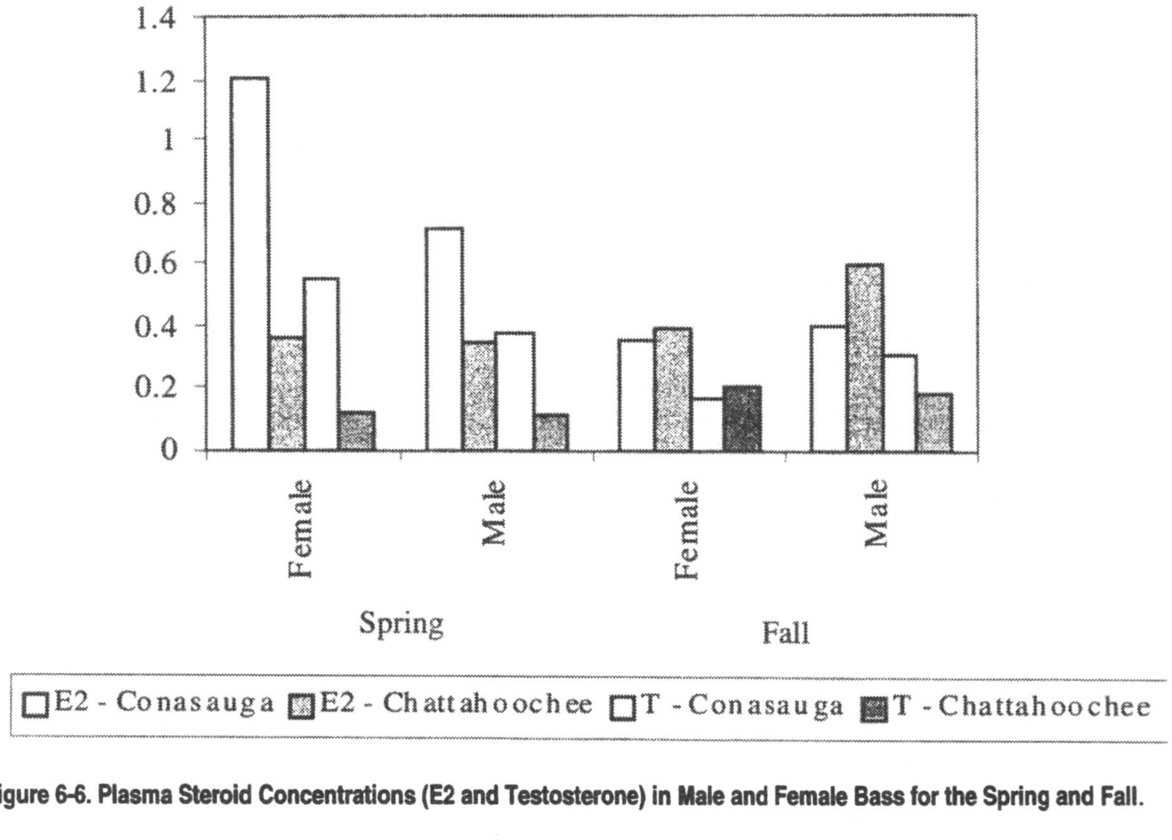

Figure 6-6. Plasma Steroid Concentrations (E2 and Testosterone) in Male and Female Bass for the Spring and Fall.

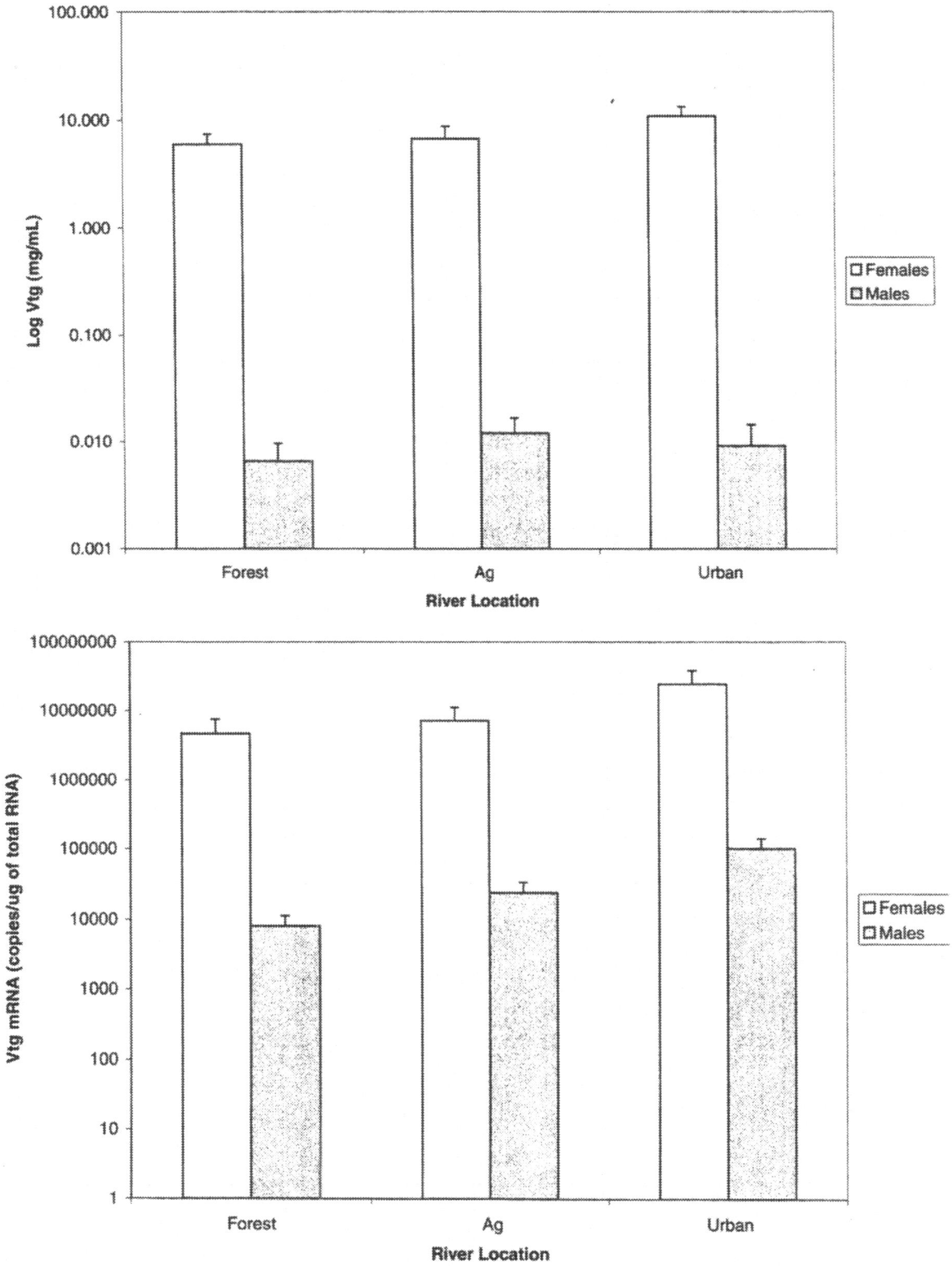

Figure 6-7. Comparison of Vitellogenin Protein and Vitellogenin mRNA in Fathead Minnows Caged 2 Weeks in the Conasauga River (Spring 2001).

6.5 Surface Water and Sediment Testing

Environmental media samples may be collected and evaluated. These samples may include physical/chemical characterization of sediment and water samples, hormone analyses, and in vitro assays.

In addition, studies suggest that fathead minnow exposures to sediment samples in the laboratory may provide insight into cumulative impacts from potentially variable aqueous releases.

Characterization data include DO, pH, temperature, and other standard parameters that influence contaminant fate. Sediment data of interest may include particle size distribution and organic carbon content. These data, collected with information on sources and flows, may influence the interpretation of attenuation mechanisms and factors other than estrogenic compounds that may be adversely impacting receptors. This becomes increasingly important, as studies in different areas of the country are compared.

Surface water samples may be concentrated in the field and analyzed in the laboratory for the parameters of interest. Water samples collected in the fall for the North Georgia study included 14 samples concentrated by solid-phase extraction and analyzed by gas chromatography tandem mass spectrometry; traces of estradiol and estrone were detected in selected samples. Because these analyses can be costly and are not standardized, and since they reflect potentially transient occurrences, they may be included in a field assessment to the extent practical.

The in vitro assays are promising because they can be sensitive and integrate response from several estrogenic compounds. Additional refinement of the analyses and evaluation of interferences will continue to increase their use as a cost-effective screening tool.

6.6 Background

Incidences of intersex conditions or elevated plasma vitellogenin in background sites and/or natural populations not influenced by anthropogenic estrogen sources are important considerations in interpreting results. In studies conducted over the past 2 years, vitellogenin has been detected with some frequency in male plasma. As a composite of all control fathead minnow samples from laboratory studies, vitellogenin was detected in more than 40% of samples. Concentrations generally were low, with an average of 0.015 mg/mL (April 2000 through August 2001) as compared with average levels of approximately 8 mg/mL in the female controls. Supplies of juvenile trout used for effluent flow-through evaluations also showed a baseline level of vitellogenin that may be associated with pond rearing of these fish. Other studies that have used fathead minnows from ponds also have reported vitellogenin detection frequency ranging from 14% to 62% for males (Nichols et al., 1999).

The selection of reference sites can be important in evaluating background conditions. In the North Georgia study, the reference site was located in a wilderness area where anthropogenic sources of estrogenic compounds were thought to be minimal. Plasma vitellogenin was detected in male bass in the reference site 60% of the time in the spring and 20% of the time in the fall.

The vitellogenin mRNA analyses indicate that estrogenic substances may be present in this area; however, it is uncertain if this represents contributions from natural sources (e.g. wildlife impacts) or an unidentified anthropogenic source.

Low levels of intersexuality (1.11%) observed in roach fishes (Rodgers–Gray et al., 2001) were not related to dose of effluent exposure or the time period of exposure (even after 4 months), even though statistically significant vitellogenin increases were detected.

6.7 Seasonality

The design of field studies to evaluate the potential adverse effects of EDCs to receiving streams should consider the typical patterns of biomarkers during reproductive cycles and the various environmental factors that may influence these natural levels.

In evaluating the laboratory and field studies for this project, it appears that seasonal factors influence biomarker levels and, possibly, their responses to external estrogen sources.

Fathead minnow gametogenic and spawning activity naturally occurs from about May through August. Under laboratory conditions, temperature and light cycles can be maintained to produce constant reproductive activity. However, fathead minnow suppliers may use natural ponds as sources for adults; therefore, their responses to exposures may be influenced by their natural cycles.

Figure 6-8 shows the patterns of vitellogenin in male control fathead minnows over the 2-year study. Vitellogenin concentrations appear to be at their lowest in the fall, when the highest occurrence of values was below detection limits.

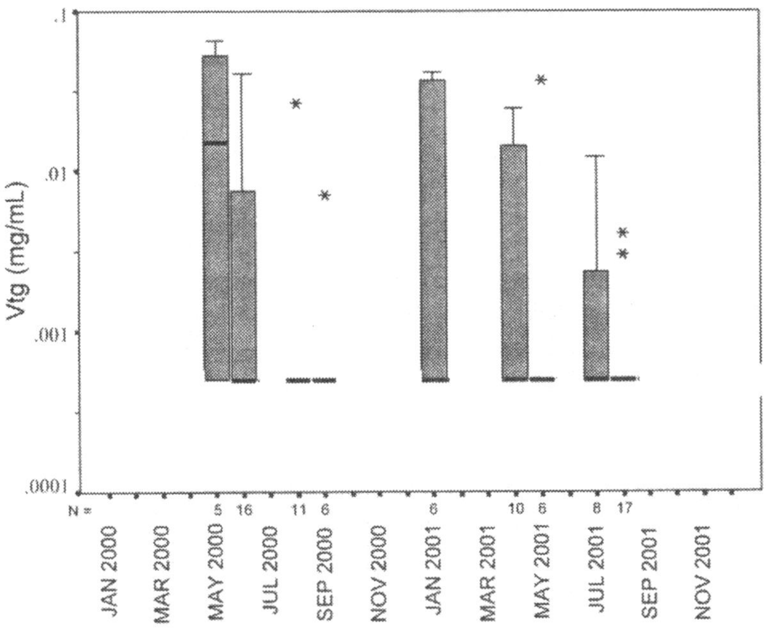

Figure 6-8. Overview of Plasma Vitellogenin Levels in Control Male Fathead Minnow Over the 2-year Study.

Table 6-3 summarizes the seasonal patterns of vitellogenin and GSI for bass collected in the North Georgia study. Female vitellogenin levels were more than two orders of magnitude higher in the spring than in the fall, while male vitellogenin and GSI measurements were two to six times higher in the spring.

Table 6-3. Seasonal Patterns of Average Vitellogenin and GSI in the North Georgia Study

Season	Average Vtg. (mg/ml)		Average GSI (%)	
	Female	Male	Female	Male
Spring	6.03	0.031	3.18	0.34
Fall	0.034	0.007	0.53	0.17

One preliminary assumption that the project team made was that it would be easier to distinguish a positive vitellogenin induction in the fall because of lower natural background levels. However, the 0.1-nM E2 positive control fathead minnow exposures in September 2000 did not show a positive response compared to comparable tests. This E2 value is near the reported threshold for vitellogenin induction. Although additional confirmatory data are needed for verification, the observed patterns suggest that there may be reduced sensitivity (vitellogenin induction) in the fall to equivalent exposures to estrogens.

These seasonal patterns represent a benchmark field study on the effects of wastewater treatment plant effluent to receiving waters. These patterns showed significant vitellogenin induction in caged adult male rainbow trout near outfalls and up to 4.5 km downstream in July (Harries et al., 1997). Overall background vitellogenin concentrations were lower in samples taken in November; only two locations showed significant differences from controls. This was attributed to changes in effluent characteristics and dilution; however, species sensitivity may be another contributing factor.

If the target endpoint of interest is intersex conditions and/or gonadal abnormalities that represent permanent physiological changes, there may be advantages to conduction evaluations in the fall when gonadal tissues are not in active reproduction and intersex conditions are more easily identified.

6.8 Environmental Fate

A number of conditions in receiving streams may influence the responses to EDCs that may be present in a discharge. Assuming the same potential for estrogenic response of the effluent, these effects will be attenuated in the receiving streams by several mechanisms, including dilution, degradation, and/or sorption.

Many EDCs are hydrophobic and will adsorb to sediments. Male fathead minnow vitellogenin levels in a 7-day static exposure to 1 nM of E2 were reduced by a factor of three in the presence of control sediment. For steady-state conditions, effluent/sediments would reach equilibrium and additional removal by in-place sediments no longer would influence surface water column reductions. Suspended solids and other transient organic matter would be expected to reduce impacts.

Figure 6-9. Vitellogenin Results for Sediment Screening.

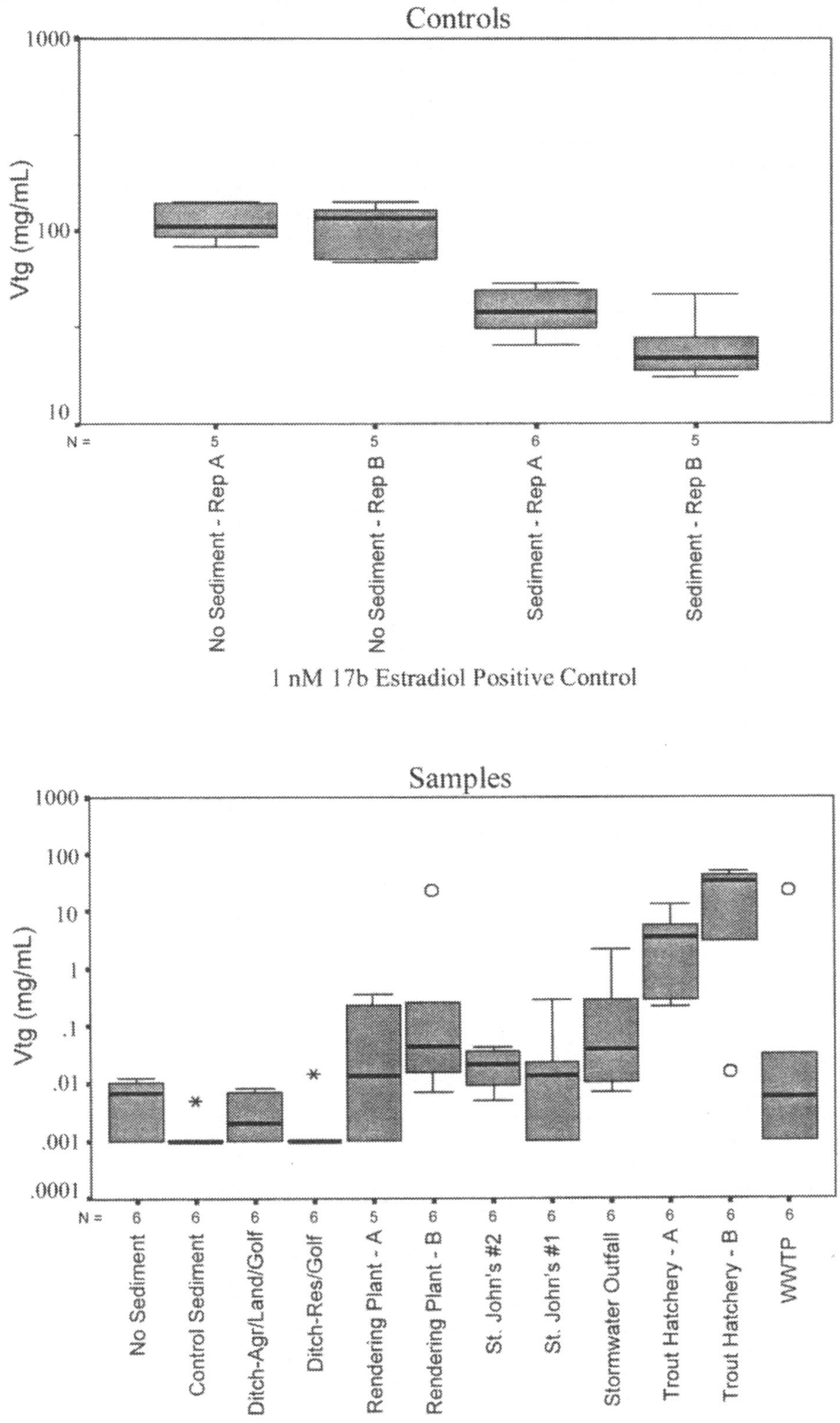

Conversely, sediments could be evaluated as a potential source to the water column. Of 10 sediments screened in a static laboratory fathead minnow study, three showed responses that were elevated significantly above the negative controls (see Figure 6-9). These data suggest that while water concentrations may be transient, it may be possible to evaluate residual activity accumulated in sediments as a result of ongoing discharges or regional influences.

CHAPTER 7.0

ECOLOGICAL SIGNIFICANCE

7.1 Introduction

One challenge of ecological risk assessment is clearly defining, in operational terms, the environmental value(s) to be protected and the methods to be used to measure significant changes. Suter (1993) states that "a physiological change is biologically significant if it affects a property of the whole organism such as survival or fecundity; a change in fecundity of individuals is biologically significant if it affects the number of species, the productivity, or some other property of the ecosystem."

Generally, the goal of ecological risk assessments is to maintain healthy and sustainable wildlife populations, with endpoints including survival, fecundity, egg viability, egg-hatching success, etc. Endocrine disruption occurs when a chemical alters the function of the endocrine system and, consequently, causes adverse health effects in an intact organism or its progeny. This definition emphasizes the mechanism by which the effect occurs, and that the level and type of change results in an adverse effect.

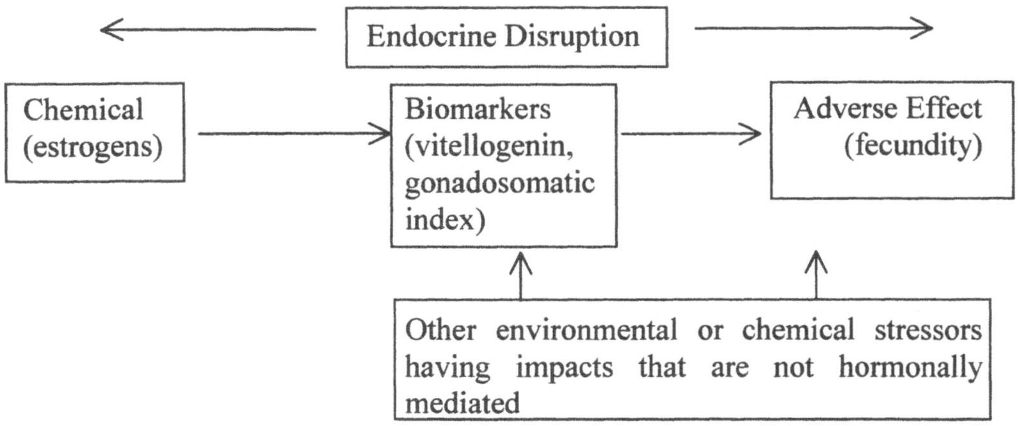

Because effluent properties and receiving-stream characteristics vary, it is possible that in some percentage of cases, discharges may have measurable adverse effects on the reproductive success of receptors in the receiving stream via endocrine disruption.

Biomarker response differences between target areas and reference locations may be measured, but these may not have significant effects on individuals or populations. In some cases, the biomarker differences may be associated with a range of environmental factors that are not specifically hormonally mediated. These indicators may include changes in biomarkers (e.g. steroid hormones, elevated levels of vitellogenin in male fish) and/or estrogenic responses using sensitive in vitro assays.

Various surveys currently underway are measuring chemical concentrations, performing in vitro screening assays on effluents, and measuring biomarkers as indicators of potential adverse effects. This increases the need to clarify these observations in the context of potential significant adverse effects. Recognizing the range of reproductive strategies of various fishes and the multitude of hormones that control reproduction, it is not possible, with available data, to comprehensively document all possible relationships between chemicals, biomarkers, and reproductive success. Therefore, this section discusses the specific relationships evaluated in this study.

This project explored aspects of the presence of estrogenic activity in effluents, biomarker formation, background patterns, and relationships of biomarkers and reproductive success. This chapter highlights some of the following observations from this study, and includes reviews of other published studies that may assist in interpreting and communicating this complex issue.

- Exposure patterns (concentration and time) determine potential biomarker formation and adverse effect. Many current laboratory studies to determine dose–response are performed with continuous exposures that may not represent spatial–temporal patterns in surface water and/or receptor movements.
- There may be multiple stressors in receiving streams that may influence biomarker responses and/or show toxic effects that are not necessarily hormonally mediated.

7.2 Estrogenic Chemicals in and/or Estrogenic Response (In Vitro Assays) of Surface Water/Effluent

One consideration in determining ecological significance relates to the concentrations and spatial distribution of the estrogenic compounds. New analytical methods are being developed to detect low levels of steroid hormones in effluents and/or surface water (see Chapter 2). For this discussion, the focus is on E2, its degradation product E1, and EE2. These hormones have been measured in several studies, typically characterizing effluents. Table 7-1 summarizes chemical measurements reported for effluents compared to the recent surface water data summarized by Kolpin et al. (2002).

These three parameters are highlighted because of their relevance to estrogenicity of effluents, and because they are examples of the potential range in behavior and potency; their chemical properties are shown in Table 7-2.

Table 7-1. Comparison of Published Concentrations of Estradiol, Estrone, and Ethynylestradiol in Effluents and Surface Water

		E2				E1				EE2		
Reference	Country	Freq %	Min	Max		Freq %	Min	Max		Freq \%	Min	Max
Summary Concentrations, ng/L												
Kolpin et al., (2002).	USA	10	1	93		7.1	8	112		5.7	23	831
Baronti et al., (2000).	Italy	100	0.35	3.5		100	2.5	82.1		73.3	0.30	1.7
Desbrow et al, (1998).	England	100	2.7	48		100	1.4	76		33.3	0.2	7
Median Concentrations, ng/L												
Belfroid et al, (1999).	Nether-lands	0.9				4.5				<0.3-1.8		
Ternes et al, (1999).	Germany	<1				9				1		
Ternes et al, (1999).	Canada	6				3				9		
Ternes et al, (1999).	Brazil	<0.2				7				1		

Studies comparing the frequency of detection and range of detected concentrations in surface water data Kolpin (2002) and effluents. E1, estrone; E2, estradiol; EE2, ethynylestradiol.

Recently published USGS data include the broadest survey of chemical concentrations of these constituents in the United States. This study shows much higher concentrations of EE2 than reported in previous studies. This synthetic hormone is slightly more estrogenic than E2 for the YES assay, but substantially more potent for in vivo systems. The reported concentrations exceed the MATC for one or more endpoints in more than 15% of samples (see Figures 4-8 and 4-9). Further validation of these results is necessary. Furthermore, it should be noted that these results represent observations at one location at one time and may not accurately reflect spatial and temporal patterns.

In the field study for this project, chemical analyses were performed only on 14 samples collected in the fall sampling event in the two North Georgia watersheds. Estrone was reported at a concentration of 8.5 ng/L at one location, with a trace level of E2 present in that sample below the quantization limit. Estrone was present, but below quantification limits, in two other samples. The quantified detect and trace levels were in samples from the Chattahoochee River.

Estrone is an oxidation product of E2, and the ratio of E1 to E2 has been reported to be high in sewage outfalls (Colucci et al., 2001). This is consistent with observations in the fall samples in the project team's study. However, the USGS data did not show these two constituents in the same samples, and levels of the two constituents showed similar concentration profiles rather than the expected higher concentrations of E1.

Table 7-2. Physical Chemical Properties of Estrogenic Hormones

Compound	Structure	Physical Chemical Properties	Estradiol Equivalents[c]
Estradiol – E2[a] CAS #: 50-28-2 MW: 272.39	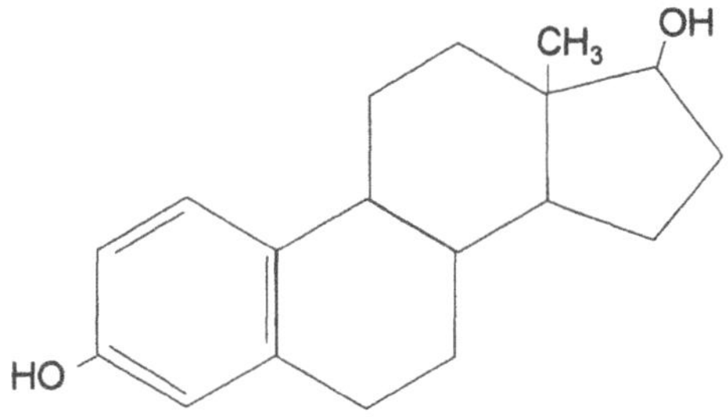	Log K_{OW}: 3.94 S_w: 13 Ref: [a]	in vivo: 1 in vitro: 1
Estrone – E1[a] CAS #: 53-16-7 MW: 270.37	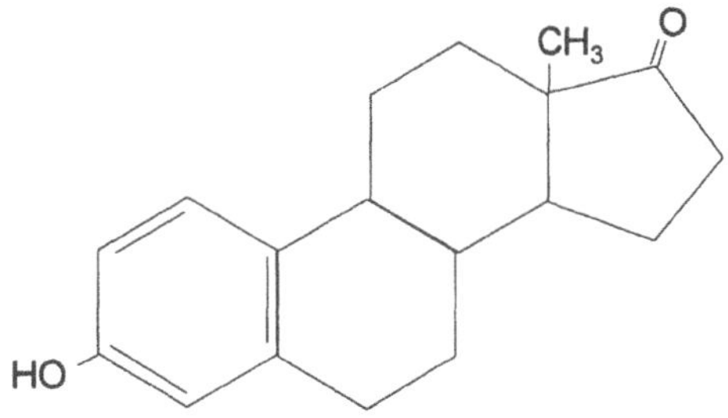	Log K_{OW}: 3.43 S_w: 13 Ref: [a]	in vivo: 0.5 in vitro: 0.5
Ethynylestradiol – EE2[b] CAS #: 57-63-6 MW: 296.4	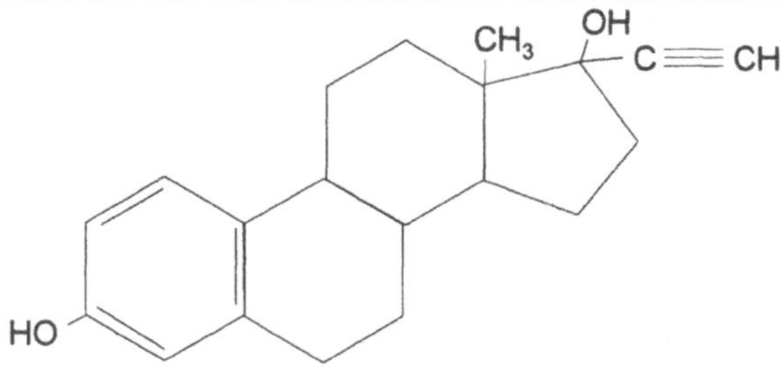	Log K_{OW}: 3.67 S_w: 4.8 Ref: [b]	in vivo: 25 in vitro: 1-2

[a]Colucci et al, 2001.
[b]Colucci and Topp, 2001.
[c]Johnson and Sumpter, 2001.
in vivo: Vtg response in trout; in vitro: YES assay
MW, molecular weight; Sw, Solubility in Water (mg/L); CAS, Chemical Abstract Services; Kow, octanol partition coefficient.

The YES assay is a tool that can integrate estrogenic response of multiple estrogenic compounds in water/effluent samples. In the spring, all water samples in the field study had responses higher than the negative control. Response was lower in the fall, and appeared to have a component that inhibited the estrogenic response in the spiked samples as compared to control water. The project team did not confirm specific spatial distribution patterns in its samples.

Additional chemical and/or YES assay measurements in the future will add to our understanding of the spatial and temporal patterns of estrogenic constituents. It is likely that these patterns, combined with the movement of receptors, produce intermittent and variable levels of exposure to estrogenic compounds that may be present.

7.3 Exposure Patterns and Biomarker Response

There are numerous possible modes of action of chemicals that may interfere with normal hormonal activity. Biomarkers that measure different biochemical and/or physiological responses are emerging, and several of these used together may provide the best understanding of responses to effluent exposures.

Biomarkers can vary significantly in their sensitivity, time to respond, reversibility, and specificity. Therefore, following an exposure, different biomarkers form at different times and their relationships to adverse reproductive effects also differ. This section illustrates the exposures that result in significant detectable changes in two biomarkers: vitellogenin induction and GSI.

7.3.1 Vitellogenin

Vitellogenin is a biological marker representing specific interactions between estrogen and estrogen receptors in the liver. Increased vitellogenin production can represent estrogen exposures.

During its study, the project team attempted to determine the lowest amount of exposure that would show a detectable vitellogenin increase. This would determine the smallest effluent volume needed for a laboratory screening protocol. With a 1-nM (272-ng/L) E2 concentration, which is much higher than what is anticipated in effluents, vitellogenin concentrations were elevated above controls after only 24 hours of exposure (see Figure 4-4) or 2 days in a static exposure that was followed by 5 days in clean water.

Thresholds for vitellogenin induction have been reported. For E2, the MATC was estimated at 5 ng/L for trout and 22 ng/L for fathead minnows. In this study, exposure concentrations near the threshold for vitellogenin induction in fathead minnows (~ 0.1 nM) showed a significant response for males in a static or static renewal during spawning periods.

These studies illustrate that the vitellogenin biomarker is extremely sensitive and can be induced readily, particularly during natural cycles of vitellogenesis. In addition, the protein has a long half-life, meaning that vitellogenin levels will not decrease significantly when transferred to clean water. This enhances the potential to detect elevated levels at locations where only intermittent exposures to estrogenic chemicals may occur.

7.3.2 GSI

This easy-to-measure biomarker is not specific for any particular mechanism of action, but laboratory studies have shown that exposures to estrogenic compounds can result in decreased GSI.

The effluent-screening protocol in this project indicated that vitellogenin showed a greater response in a shorter period of time. Therefore, in a 7-day exposure, no significant changes in GSI were observed, particularly at lower concentrations in a static or static-renewal assay. This suggests that vitellogenin may be elevated with no significant change observed in GSI. Further, in the short-term reproductive study (see Chapter 4), 21-day exposures at 0.5-nM

E2 concentrations did not result in significant changes to GSI in males or females; however, when an anti-estrogen also was present, female GSI decreased significantly.

These observations are consistent with observations by Panter et al. (2000) that intermittent exposures produced greater-than-expected responses from simple integration over time. In other words, continuous exposures at 120 ng/L produced vitellogenin levels similar to intermittent exposures at these concentrations. Their study also showed that changes in GSI are easier to detect during a normal phase when the gonadal growth rate is high. They also showed that continuous exposure at 120 ng/L, and 50% intermittent exposures showed total inhibition of gonad growth. Males exposed intermittently to E2 (25% E2, 75% control water) showed no evidence of inhibition.

These observations emphasize that different biomarkers can show different response times, and that biomarkers that may relate more directly to reproduction (GSI) respond only after additional exposure (longer duration and/or higher concentrations).

7.4 Background

Understanding the normal levels and patterns of biomarkers is important in distinguishing their level of impact. Vitellogenin was detected in 60% of samples (male bass collected at a wilderness area) in the spring and 20% in the fall. This is similar to what was detected in samples in downstream locations. Vitellogenin levels typically were low — much lower than in females. This detection frequency is similar to the 47% reported by Sepulveda et al. (2001). Since male fish have circulating levels of E2, some low background levels may be normal.

Vitellogenin was detected in a variable percentage of male fathead minnows in the negative controls. This may be attributed partially to conditions during shipping; however, detectable levels also were found in adult males used in the short-term reproductive study after they had been conditioned in the laboratory for several months. This detection frequency is lower near the end of the spawning season.

One initial hypothesis was that differences in vitellogenin levels would be more apparent for exposures when the background levels in males were lowest. This would reduce variability in the control samples in distinguishing changes compared to positive controls. However, the project team's data suggest that vitellogenin induction may be inhibited during these periods of low plasma vitellogenin levels.

Intesex conditions may occur with some frequency in normal populations. Although this was not a focus of the project team's study, researchers did note that ovo–testis were not observed during their review of slides of bass gonadal tissue.

7.5 Adverse Effects

The previous sections suggest that different biomarkers respond to different concentrations and exposure durations.

Several laboratory studies have demonstrated that exposures to estrogens or estrogen mimics result in elevated levels of plasma vitellogenin and changes in other biomarkers, including GSI, can be observed. However, it is critical to consider the exposure concentrations/durations where these were correlated, the relationship to fecundity, other effects noted during the assay, and the relevance of these concentrations to issues of effluent exposure.

The project team's reproductive study showed that continuous fathead minnow exposure to 0.5 nM (136 ng/L) of E2 for 21 days resulted in a significant increase in vitellogenin, and a slight (insignificant) reduction in cumulative egg production. This concentration is much higher than anticipated in effluents and surface waters. Vitellogenin induction would occur with E2 concentrations much lower than used in this exposure.

The EE2 levels reported in the USGS survey were much higher than reported in other effluents studies. This synthetic hormone is much more bioactive than E2, and at the reported levels, might suggest potential for significant adverse effects.

Several studies have reported reproductive impairment with exposure to estrogenic compounds. Some of these are at very high concentrations, where a number of toxic effects may occur.

While this study suggests that the vitellogenin biomarker in adults may not reflect significant adverse effects to the individual or to reproductive success, it may suggest that additional evaluation is needed for sites with more sensitive species and/or additional evaluation is needed at other critical stages in development.

Identifying significant adverse impacts in field studies is challenging. There are many sources of variability, and a large sample size is needed to detect significant differences. In the project team's study, transient acute toxicity was noted for the fathead minnows. The source of this and implications related specifically to endocrine disruption are unclear. Many receiving streams may be identified as impaired, resulting in multiple stressors that can influence levels of hormones and reproductive success. Deriving cause and effect under these conditions may not be possible. It is challenging to collect sufficient data to understand exposure patterns (spatial and temporal fluctuations), and it is not always clear that the measurable differences in biomarkers are significant factors in population effects in receiving streams.

CHAPTER 8.0

CONCLUSIONS AND RECOMMENDATIONS

8.1 Conclusions

Based on its study, and considered in the context of published data, the project team has developed the following responses to probable questions about the subject of potential endocrine disruption in wastewater treatment plant receiving waters.

What is the likelihood that my effluent would produce a detectable biomarker?

Biomarkers and effluents vary. Vitellogenin is the most commonly reported biomarker associated with estrogenic constituents in effluents. Many effluents typically may be below the concentrations that result in increased vitellogenin levels in fathead minnows. However, because only occasional elevations of estrogenic compounds may produce detectable levels of vitellogenin, it is possible that even occasional discharges of elevated hormone concentrations could result in biomarker formation. Juvenile trout show increases in this biomarker at lower concentrations than fathead minnows.

What other sources (natural or anthropogenic) may be present in this watershed that may also produce these effects?

The reference site in the project team's North Georgia study is located in a wilderness area. Vitellogenin was detected in 60% of the male bass in the spring and in 20% in the fall. Concentrations were much lower than observed in females. Males have circulating levels of estrogen as well as testosterone, and some base level of vitellogenin may be normal.

Acute toxicity to caged fathead minnows was observed in one location each season. The remaining fish did not show differences in vitellogenin. However, this suggests that several stressors may be present in watersheds that can potentially impact a number of biomarkers. Watersheds that show impairments should consider a range of factors that can influence overall health and reproductive success.

Other contributions relate to food source and a number of specific activities, including agriculture (particularly animal feed lots) and aquaculture. Sediments may contribute to the exposures, and may play a role in reducing the water concentration.

What is the potential area that is affected from my discharge?

Steroid hormones may have half-lives of hours to days, depending on receiving stream characteristics. For discharges with limited dilution, studies have shown elevated levels of vitellogenin in caged fish located kilometers downstream from the discharge point. Conversely, activity was removed a short distance from the discharge point in a treatment wetland with sheet flow.

How long would a fish be exposed before a response could be observed? Is it reversible?

Vitellogenin levels can be elevated following a brief exposure to estrogen compounds above threshold levels. This may be hours or days, depending on the concentration. These concentrations would decrease when exposure is eliminated; however, it may take several weeks to return to normal levels.

Reversibility is a function of the magnitude and type of impact. Formation of ova–testis in medaka, exposed from hatch to 100 days post hatch, would not be reversible, and this could occur at concentrations lower than required for induction of vitellogenin.

Were the fish collected at an appropriate time in their reproductive cycle? How does this influence the observed measurements?

Many studies include seasonal collection of fish, particularly in the spring and fall. Based on its observations, the project team concluded that wild fish may be more sensitive to estrogens (as measured by vitellogenin increases) in the spring. However, other biomarkers (e.g. histopathology to detect intersex conditions) may be better evaluated in the fall when the gonads are not in active reproduction.

What is the range of the fish? What species are most sensitive?

The behavioral patterns and ranges of fish may influence the overall exposure pattern. Many fish have a limited range; if they are present near an outfall, their exposures reflect the temporal variability of the discharge and stream conditions.

There are ranges in sensitivity, and again these differ by biomarker of interest. For vitellogenin, trout are more sensitive than fathead minnows, flounder, and carp.

What is the possibility that the observed histology is an artifact (i.e., occurs at some frequency in natural populations)?

This is related to understanding a baseline of normal conditions and variability in individuals in a population.

How accurate/precise are the assays? What common problems are encountered?

While the underlying technologies are evolving, many biomarkers and in vitro screening assays currently are in research stages, both to fully develop the methods and to interpret the results. The U.S. Environmental Protection Agency is in the process of developing and validating methods for chemically screening endocrine disrupting compounds. Consequently, different laboratories may vary the protocols and additional work remains to verify the accuracy and precision of these techniques. As kits are prepared and used, some of these assays will become more standard.

If a biomarker is detected, will reproduction be impaired? What is an appropriate endpoint?

The definition of endocrine disruption includes the concept that the effect is adverse at the organism or population level. Some biomarkers are sensitive indicators at the biochemical level that an exposure has occurred, but may not directly relate to a more relevant endpoint like reproductive success. That is, estrogenic compounds may induce vitellogenin at low concentrations, but adverse effects to reproduction may occur only after more extended exposures to higher concentrations. Furthermore, compounds like anti-estrogens may interfere with reproductive success and not show elevations in vitellogenin. Therefore, the hypothesis and a range of biomarkers are needed to understand the implications of a change in biomarker. Once these patterns are understood, biomarkers may aid in documenting spatial and temporal patterns.

It has been shown that the period from hatch through gonadal development may be a time of critical sensitivity. Adverse effects may occur at concentrations that do not adversely impact adults. The reproductive biology of the target species of interest must be understood in detail to evaluate the potential impacts of this in a receiving stream.

8.2 Recommendations for Further Research

Seasonal Responses

- Data from the project team's study suggest that vitellogenin induction is significantly decreased in the fall (post spawning). The endocrine system regulates the reproductive cycle with an integration of a number of hormones. In cycles where vitellogenin production is inappropriate, this pathway may be reduced by other feedback mechanisms. This could be verified in laboratory studies, and would influence the interpretation of field data on the potential impacts of effluents.
- Evaluate the potential role of homeostasis in controlling adverse effects on oocyte development and/or maturation during periods in which adults appear less responsive to external sources of estrogenic compounds.

Messenger RNA (mRNA)

- Increasingly, the use of mRNA is being studied as a biomarker that may be more sensitive than measurements of proteins. While these can be correlated in some experiments, the relationship of the message and protein may require additional evaluation. Specifically, it has been shown that mRNA and proteins do not always correlate because there are other mechanisms to reduce protein formation and/or accelerate its breakdown.

Effluent Patterns

- Increase the database on effluent seasonal and diurnal patterns as they may relate to biomarker formation and reproductive impairment.

YES Assay

- Further validate the use of the YES assay on a range of environmental media. This includes interpreting relative responses, quantifying responses, and validating interferences.

APPENDIX A

BIOMARKER ASSAY PROTOCOLS

APPENDIX A1

PLASMA VITELLOGENIN PROTEIN

A1.1 Fish Processing For Vitellogenin Protein

Fish are anesthetized, blotted dry, and placed on a cutting board. Weight and length are taken at this time. For smaller fish use a sharp razor blade to cut off the tail mid-way between the base of the tailfin and the anal fin. For larger fish a syringe and needle are used. Blood is collected in a heparinized capillary or a vacuutainer tube containing heparin. The tube is sealed and place immediately in a centrifuge. After centrifugation, the blood is clearly separated into red blood cell and plasma components. The plasma is removed and placed into an appropriate container, which is to be stored at, then stored at -80°C until analyzed.

A1.2 Vitellogenin Assay

A1.2.1 Background

Female fish produce vitellogenin, the egg yolk precursor protein, as part of the normal cycle of oogenesis. Plasma vitellogenin is not normally present in males but the gene for the protein is present and can be activated in males by exposure to estrogen or estrogen mimicking chemicals. Plasma vitellogenin levels can be measured by ELISA, radio-immunoassay (RIA), and immuno blotting (Western). ELISA is the assay of choice because of the large number of samples that can be quantified at one time with a high level of sensitivity. Vitellogenin is a large (300-600 kDa) protein with two subunits (160-200 kDa each). Because of its charge size and other attributes vitellogenin can be purified from estrogenized fish plasma using anion-exchange chromatography. Purified vitellogenin from each specific fish species of interest is use to construct standard curves for ELISA.

A1.2.2 Procedure

Known amounts of purified vitellogenin are placed in a 96 well microtiter plate to serve as a standard curve that will later be analyzed along with unknown samples. Since there is a significant "plasma protein effect" male plasma must be added to standards at the same concentration as the unknowns. A standard curve is performed ranging from 5 to 1000 mg/mL giving 13 points including a zero reading. Adding male plasma to the standards ensures the matrix of the known samples is consistent with that of the unknown. Unknown samples are run at the same dilution as the male plasma in the standard curves. Unknown samples and standards are loaded 50 μL to a well (run in triplicate for each unknown and standard). After the plates have incubated overnight at 4°C they need to be washed four times with PBSTZ (phosphate buffered saline containing Tween-20 and Azide) using a standard automated plate washer. Next the plates are blocked with 360 mL per well of BSA-PBSTZ (containing 1% bovine serum albumen (BSA) and Aprotinin, a protease inhibitor) for 2 hours at room temperature. This blocks all the binding sites on the plate that are not already occupied by protein. Next the primary monoclonal antibody (specific for the species of interest) is diluted into BSA-TBSTZ to the appropriate level depending on which plate dilution issued (for example, 1.0 μg/mL of the primary antibody is placed on the 1:100 dilution plate). The antibody solution is loaded (50 μL per well). The antibody is incubated overnight at 4°C. The following day the plate is washed and the secondary biotin labeled goat anti mouse IgG antibody is diluted 1/1000 in BSA-TBSTZ and added (50 μL well) and incubated at room temperature for 2 hours. The plates are washed again and Strepavidin alkaline phosphatase is added (50 μL well). The plates are incubated at room temperature for another 2 hours, and then a final wash is performed. The last step is to add 100 μL per well of the alkaline phosphatase substrate, PMPP, dissolved at 1 mg/mL in the appropriate buffer. The color is allowed to develop for 30 minutes and the absorbance at 405 nM is read. A calibration curve is established by plotting the absorbance at 405 nM versus the known amount of vitellogenin in the standards. A plate reader is used in the last step to measure and calculate the data.

The monoclonal antibody used varies depending on the fish species. For fathead minnows, monoclonal antibody ND-28 is used. This antibody was raised against carp vitellogenin but cross-reacts with high affinity with purified fathead minnow vitellogenin. For largemouth bass, spotted bass, and red-eye bass monoclonal antibody ND-3G2 is used. This antibody was raised against striped bass but cross-reacts with high affinity against largemouth, spotted, and red-eye bass vitellogenin. Vitellogenin was purified from E_2-induced plasma for each species and used for standard curves.

A1.2.3 Notes and Comments on Variability

- Statistical tests are performed to determine the variability in the triplicate run of samples and standards. If an absorbance reading of one sample differs significantly from the other two in the triplicate it is thrown out. The coefficient of variation of the triplicate values should be less than 10%.

- A standard curve is constructed by taking the mean value of the triplicate samples at each concentration after the background has been subtracted and plotting these versus the known concentration. The curve is fitted to a quadratic equation.
- All samples must be in the range of the standard curve. If they either exceed or are below the range, they are re-diluted to a higher or lower concentration, respectively, and the assay is redone.
- Potential sources of variability in the assay come from dilution error, protein orientation in the wells, error in the preparation of stock solutions, improper handling and storage of standards and samples, and other sources. To minimize these sources, samples and standards are kept frozen at -20°C, positive controls are run on the assay to check for inter-assay variability and correct dilution and other standard operating procedures for the assay are strictly adhered to.
- A monoclonal antibody recognizes one site on the vitellogenin protein. A polyclonal antibody recognizes many. If a vitellogenin molecule has been degraded or fractured from freeze thawing a different result can be obtained in the assay especially if the assay uses a polyclonal antibody. Therefore, it is best to perform all the required dilutions at the same time and to run the assays concurrently.
- Points that lie on the extreme ends of the standard curve are considered less reliable than points in the middle of the standard curve.
- An intraassay test is run on the same sample by running the sample six times each in triplicate (for a total of eighteen wells) to ensure variability of protein orientation and dilution errors are acceptable. Interassay variation is determined by running a positive control at each dilution (in triplicate wells) in 6 different assays. Typically intraassay and interassay variation is less than 3% and 10% respectively for direct ELISA.
- The lower the dilution of the sample the more sensitive the assay.

A1.2.4 Standard Curves

To generate the points on the standard curve purified fathead minnow vitellogenin is diluted to the desired concentration with buffer containing diluted male fathead minnow plasma (diluted to the same concentration as the samples). Male plasma is always checked to be sure it is free of vitellogenin, prior to use. The absorbances obtained from the assay for these concentrations gives the numbers plotted for the curve. A quadratic regression of the absorbances versus concentrations yields the curve. Acceptable standard curves give a quadratic regression value of 0.95 or greater. Standard and unknown replicates must also pass the statistical test of coefficient of variation with a value of less than 10 percent. (Crowther, J.R., 1995; Denslow, N.D. et al, 1999).

APPENDIX A2

STEROID HORMONE AND PLASMA CHARACTERIZATION

A2.1 Steroid Hormones

Estradiol and testosterone were measured using kits that were manufactured by Oxford Biomedical Research (Oxford, Michigan) and distributed by Cedarlane Laboratories Ltd. (Hornby, Ontario).

The entire sample was processed. Deionized water (50 μL) was added to each vial so that there was adequate volume for analyses of plasma lactate, glucose, protein, and hormones. Preliminary test on fish plasma confirmed that glucose, lactate, and protein were not removed with an ether extract. Then the entire sample was extracted with 300 μL of ether. The solution was mixed for 10 seconds on a vortex mixer and the layers allowed to separate. The ether was removed and placed into a clean vial. The plasma was frozen for later analyses of serum glucose, protein, and lactate. The ether was evaporated off by placing the vials in a fume hood overnight. The vials were then sealed and frozen for later analyses of testosterone and estradiol.

Lactate and glucose were measured directly on the plasma. The plasma was diluted 100 times for analyses of protein levels. Estradiol and testosterone levels were measured directly on the ether extract exchanged into sample buffer.

The fish plasma was extracted with ether to remove both hormones (300 μL added to each vial containing the plasma sample plus 50 μL of deionized water). The solution was vortexed for ten seconds and allowed to separate for 30 min. The ether layer was placed into a clean 2 mL microfuge tube and allowed to evaporate at room temperature in a fume hood. The tube with residue was stored frozen (-20°C).

The samples were thawed and solubilized in 300 μL of extraction buffer (provided with the kits; the buffer was diluted five times with deionized water before use). Volumes of this preparation were taken for analysis of testosterone and estradiol. The samples were also stored frozen until analyzed.

Both methods were similar and required the same test solutions and volumes but with different conjugates and coated plates. Only one replicate or well was used for each sample and every tenth sample was analyzed in duplicate.

The ELISA of estradiol and testosterone in biological fluids was based on the competition between an enzyme conjugate and the hormone present in the sample for binding sites on an antibody coated microplate (each plate contained 12 strips with eight wells per strip). The sample was added followed by the enzyme conjugate.

Levels of 17-β-Estradiol (Product # EA78) were determined on 50 μL sample volumes. The samples were placed into the test wells followed by 50 μL of the enzyme conjugate diluted 50 times with EIA buffer provided with the kit. The solutions were mixed with gentle shaking on a MRX Microplate Reader (Dynatech Laboratories) and incubated for one hour at room temperature.

The well contents were dumped out by inverting the strips over a paper towel and blotting off the excess liquid. The wells were then rinsed three times with 200 μL volumes of a washing buffer (wash buffer provided with the kit diluted 10 times with deionized water). The well contents were dumped out between each rinse. The substrate, 3,3', 5,5' Tetramethylbenzidine (TMB), was then added to each well (150 μL) and the plates incubated for 30 min at room temperature.

Absorbances at 620 nM were obtained with a MRX Microplate Reader (Dynatech Laboratories). The absorbance readings were corrected for substrate background (blanks containing substrate). The percent of maximal binding was then obtained for the standards by dividing the corrected absorbance reading (B) by the absorbance of the blank (Bo) and multiplying by 100. The %B/Bo was plotted against the E2 concentration. The sample E2 concentrations were obtained by correcting for the different initial volumes.

The standard curve was prepared following the scheme outlined in the kit instructions. The stock solution was diluted 100 times with EIA buffer (20 μL in 980 μL buffer; 20 ng/mL). This working solution was then diluted ten times with EIA buffer to obtain 2 ng/mL and then again to obtain 0.2 ng/mL solutions of E2. Seven standards and one blank were prepared in duplicate. The test standards were 0.02, 0.04, 0.10, 0.20, 0.40, 1.0, and 2.0 ng/mL. The top three standards were prepared from the 2.0 ng/mL solution. The remaining standards were prepared from the 0.2 ng/mL solution.

Levels of testosterone (Product # EA70) were determined on 10 μL sample volumes, by a procedure that was the same as that described for E2. Sample testosterone concentrations were calculated by correcting for different initial sample volumes.

A2.2 Plasma Characterization

The plasma residue after addition of 50 μL of deionized water and extraction with ether was analyzed for lactate, glucose, and protein. Lactate and glucose were measured on the plasma sample without further dilution. The samples for protein were diluted 100 times before analysis.

Total lactate was determined enzymatically (Catalogue No. 735-10; Sigma-Aldrich). Lactic acid was oxidized to pyruvate and hydrogen peroxide. The hydrogen peroxide catalyzed the oxidative condensation of chromogen precursors to produce a colored dye with an absorbance maximum at 540 nm. The increase in absorbance at 540 nM was directly proportional to the plasma lactate concentration.

The plasma sample (10 μL) was dispensed to a 15 mL test tube followed by 2.0 mL of the lactate reagent. The solution was mixed and incubated for 1.5 h at room temperature. The absorbance at 540 nM was measured with a Novaspec II spectrophotometer. (Pharmacia Biotech). Lactate concentrations were obtained from a standard curve (control, 0.4, 0.8, and 1.2 mg/dL).

Total glucose was determined enzymatically with the Trinder reagent (Catalogue No. 315; Sigma-Aldrich). Glucose was first oxidized to gluconic acid and hydrogen peroxide. The hydrogen peroxide reacted with 4-aminoantipyrine and p-hydroxybenzene sulfonate to form a quinoneimine dye with an absorbance maximum at 505 nm. The increase in absorbance at 505 nM was directly proportional to the plasma glucose concentration.

The plasma sample (10 μL) was dispensed to a 15 mL test tube followed by 2.5 mL of the Trinder reagent. The solution was mixed and incubated for 10 min at room temperature. The absorbance at 505 nM was measured with a Novaspec II spectrophotometer. (Pharmacia Biotech). Glucose concentrations were obtained from a standard curve (control, 1.5, 3.0, 6.0, 12.0, and 24.0 mg/dL).

Levels of total plasma protein were measured with the reagent dye, Brilliant Blue G (Catalogue No. 610; Sigma-Aldrich). The dye reacts with protein in an acid-alcohol medium to form a blue-colored protein dye complex. The absorbance was measured at 595 nM and was directly proportional to the plasma protein concentration.

The plasma sample was diluted 100 times with deionized water (20 μL sample plus 980 μL of water). A volume (10 μL) was then dispensed to a 15 mL test tube followed by 2.5 mL of the protein dye reagent. The solution was mixed and incubated for 10 min at room temperature. The absorbance at 595 nM was measured with a Novaspec II spectrophotometer. (Pharmacia Biotech). Protein concentrations were obtained from a standard curve (control, 0.3, 0.6, 1.2, 2.4, and 4.8 mg/dL). All values were corrected for the initial volumes of plasma used.

APPENDIX A3

HISTOLOGY

A3.1 Histology Overview

Gonad and liver samples were collected from wild bass (largemouth, spotted, red-eye) in field samplings (spring & fall) in order to verify the sex of the fish and potentially identify any abnormal liver morphology. A sample of each tissue ($1cm^3$) was immediately preserved in 10 percent buffered formalin after collection and stored at room temperature. A microscope slide was prepared for each fish by the University of Florida -Veterinary Histology Core. Processing of the tissues was conducted as follows: the samples were washed with a series of ethanol and xylene washes, embedded in paraffin, sliced by microtome, and mounted on the microscope slides. The slides were then stained using hematoxylin and eosin (H&E) dyes and sealed with a cover slip.

Each slide was viewed using a compound microscope to visually verify the sex, reproductive stage, and liver morphology of each fish. A gonadal stage classification system was developed based on previous studies with *Fundulus heteroclitus* (Selman & Wallace, 1986) and *Micropterus salmoides* (Nelson, 2001). The classification system describes three maturational stages for male and female largemouth bass based on the predominant features. For Males stage 1 (spermatocytogenesis) is characterized by the presence of spermatogonia. These are normal sized cells containing a clearly defined central nucleus. Stage 2 (meiosis) is predominated by spermatocytes (1°&2°) and spermatids. These are much smaller than spermatogonia and the nucleus is not clearly distinguishable. Stage 3 (spermiogenesis) contains mature sperm. These are oval-shaped heads with long tails, both clearly visible in the microscope. Females were also divided into 3 stages of development. Stage 1 females contain primary phase follicle cells. These are small follicles that stain blue with the H & E stain. Stage 2 contain larger eggs with white, spherical cortical alveoli. Stage 3 females have enlarged eggs of varying diameters containing yolk. This last stage stains red with the H&E dye. Fish that were not yet sexually differentiated into male or females described above were denoted as undifferentiated (U).

Liver tissue was microscopically inspected for presence of aberrant cell morphology. Some cells were enlarged or hypertrophied (HYP) and others had enlarged spaces (SPC) between cells. Variations of cell morphology may be normal during the reproductive cycle. (Selman, K. and Wallace, R. 1986 and Nelson, M. 2001).

APPENDIX A4

YES ASSAY

A4.1 YES Assay Overview

This is the standard method for screening liquid samples for substances that exert an estrogenic effect with the recombinant yeast *Sacharomyces cerevisiae*. The method is based on the work of (Routledge et al, 1996). The DNA sequence of the human estrogen receptor (hER) was integrated into the yeast genome. A promoter sequence containing estrogen response elements (ERE) is linked to the lac-Z gene for ß-galactosidase and this is on a separate reporter plasmid. Estrogen binds to the estrogen receptor (expressed by hER) and this ligand complex then binds to the ERE elements on the reporter plasmid and turns on the expression of ß-galactosidase. The enzyme is synthesized and secreted into the medium and metabolizes the substrate chlorophenol red-ß-D-galactopyranoside (CPRG) turning the solution which is normally yellow to red. The recombinant yeast strain was obtained from J.P. Sumpter, Department of Biology and Biochemistry, Brunnel University, Oxbridge, Middlesex UB8 3PH, United Kingdom. Glaxo developed the strain and an agreement was signed with J.P. Sumpter not to commercialize the strain.

A4.2 Procedure

Cultures of yeast are stored frozen at –20°C in 1.5 mL sterile cyrovials (0.5 mL per vial). The yeast is grown up overnight – the day before testing the samples. The tests are conducted in 96 well microplates. All tests are set up with negative controls and positive controls. The endpoints are estimated relative to induction of ß-galactosidase activity in yeast exposed to 17 ß-estradiol (positive control). If there are concerns about bacterial contamination, then the sample is plated out at the end of the test (bacterial check). Bacterial contamination checks are done on plates prepared with trypticase soy agar. The yeast is grown on agar supplemented with yeast extract and tryptone peptone. The positive control is E2 prepared in methanol and stored at 4°C. The positive control is used to generate a standard curve. The standard curve is used to evaluate the samples tested. Some samples may be toxic to the yeast. Incubating the plates 2 or 3 days and checking for turbidity in the test wells can confirm toxicity. Turbidity is an indicator of growth.

The sample is toxic to the yeast if the wells remain clear after 2 to 3 days of incubation (viability check). Liquid samples and pore waters are tested as received. Methanol and water extracts are prepared on solid samples (4:1 solvent to solid by weight). The methanol extracts are tested at a final concentration of 2% (v/v in the assay medium). This is the no observed effect concentration or NOEC for methanol for the yeast. There is no need to filter sterilize or treat samples before testing. Endpoints are measured optically with a microplate reader. Turbidity is measured at a non-absorbing wavelength such as 630 nM or 750 nM. ß-galactosidase activity is measured at 540 nM, the wavelength for the maximum absorbance of phenol red but can also be measured at 570 nM.

The test conditions are summarized in this section.

Test Type:	static
Measurement:	turbidity and activity of ß-galactosidase
Temperature:	28±3°C (not fixed)
Light:	not applicable
Aeration:	not applicable
Vessel:	96 well microplate
Solution:	final test volume of 150 μL; sample volume of 100 μL
Duration:	1 or more days
Treatments:	negative control
	positive control
Replicates:	two for multiple dilution tests; six for single treatment tests
Organism:	exponentially growing recombinant yeast *Sacharomyces cerevisiae*
Loading:	3 mL/50 mL diluted 3 times = 2% of the 24 culture with A640 nM = 1
Water:	sample treatments prepared with minimal media in deionized water
Monitoring:	turbidity and red color

The endpoints can also be derived from absorbance measurements at 540 nM with a microplate reader. Replicate wells are pooled then read (averaging before measuring). With a plate reader, the averaging is done on measurements of individual wells. In both methods, the results are compared to the standard curve. Furthermore, both chlorophenol red and turbidity can be quantified (absorbance at 540 nM and 630 nM respectively).

The original method at the start of this project was to load samples directly to the plates. Test growth media was then added to the same plates resulting in a dilution of the sample. The same thing was done with the positive control standards. Later in the project the samples were concentrated with C18 columns and the elution's were added to the plates and allowed to evaporate off. The test growth media was then added to the plates. The standards were also added to the plates and allowed to evaporate with test growth media added. This allowed for the increased sensitivity of the assay. In the final version of the procedure blowing down the samples under nitrogen and then adding the samples to the plates further concentrated the elution samples.

APPENDIX A5

MRNA ASSAY DEVELOPMENT FOR FATHEAD MINNOWS

A5.1 mRNA Background

The action of estrogen or estrogen mimics is to induce the expression of genes by binding to soluble estrogen receptors and inducing the synthesis of mRNAs for specific genes. Two genes that are responsive in this manner are those that encode vitellogenin and the estrogen receptor (ER) itself. In the liver, the mRNAs for vitellogenin and ER are used as templates for the synthesis of vitellogenin protein (which is then secreted from the liver into the blood stream) and additional estrogen receptors (which stay within the hepatic tissue), respectively. The response at the mRNA level is quite fast, within 12-24 hours and reaches maximum response by 48 hours after a single exposure. With no further stimulation, the induced mRNAs are degraded and are not used for further protein synthesis. Thus, the response at the mRNA level is immediate and short term. This is in contrast to the response at the protein level. Proteins, which are made from the template mRNAs do not reach their peak concentration till 3 days after induction by a single exposure. Vitellogenin is particularly stable. Unlike the mRNA that encodes it, vitellogenin in males is not quickly degraded and remains in the blood for a longer time period. Depending on the level of exposure, vitellogenin can be monitored in the blood of males for weeks to months after an initial exposure. In females, vitellogenin is sequestered into developing oocytes. Thus, the response at the protein level is slower to be seen but it is of longer term.

Because of new technologies it is now possible to measure extremely low levels of mRNAs with great precision using real time quantitative polymerase chain reaction (Q-PCR) or the TaqMan assay. The TaqMan Assay is very similar in action to regular PCR, except that in addition to amplifying target DNA, it provides the means to calculate the amount of material present at each cycle by a fluorescent signal. This signal is compared to the fluorescent signal given off by a known concentration of cDNA from a standard curve. The signal comes from a Taq polymerase -dependent degradation of a third oligonucleotide (called a probe) that binds to the target cDNA somewhere between the forward and reverse primers.

This probe is a special oligonucleotide that is modified with two reporter dyes on the two ends. When the two reporter dyes are in close proximity to each other, the fluorescence of one is quenched by the other. Thus, intact probe molecules are not fluorescent. But when Taq polymerase which also has a 5' nuclease activity in addition to its 3' polymerase activity, makes its way down the chain, synthesizing complementary cDNA behind itself, it bumps into the probe, chews it up and releases the two fluorescent reporter dyes. This causes the dyes to separate from each other by enough distance that the quenching effect is gone. Thus the fluorescent signal that is measured is directly correlated with the amount of cDNA that has been amplified, doubling with every cycle. We have developed this assay to measure the steady state levels of both vitellogenin and ER mRNAs in fathead minnow.

A5.2 Procedure

In order to develop the TaqMan assay, it is necessary to first obtain and clone a piece of the mRNAs for vitellogenin and ER. These cloned fragments are then used to prepare standard curves to calibrate the amount of mRNA found in controls and test samples. The procedure will be divided into each of the steps that need to be done in order to perform the assay.

A5.2.1 Cloning of a Portion of the Vitellogenin and Estrogen Receptor mRNAs

A5.2.1.1 Preparation of Total RNA

The first step is to isolate the total RNA fraction from fathead minnow liver and then to copy the mRNAs into cDNAs using reverse transcriptase. Because it is difficult to obtain intact and discrete livers from female fathead minnows, we used the entire gut cavity. (In females the liver is found as little islands of tissue along the intestine). The gut cavity was frozen in liquid nitrogen and then homogenized in 1 mL of RNA STAT-60 (TEL TEST, Inc.) and prepared according to their protocol. The RNA was extracted with chloroform and centrifuged. The RNA portion remains in the upper aqueous layer, the DNA and proteins separate into a lower inter-phase and organic phase, respectively. The aqueous RNA is transferred into a new tube and re-extracted a second time with RNA STAT-60 to minimize DNA and protein contamination. Then the RNA is mixed with isopropanol to form a precipitate, which is then pelleted by centrifugation. The RNA pellet is washed with 0.75-1 mL of ethanol two times and is resuspended in 150 μL of RNA-secure (Ambion) to inactivate RNAses. 20 μL of RNA is then treated with 1 μl DNA-free (Ambion) to remove any remaining DNA contamination. The RNA is then checked for its quality by gel electrophoresis and it is quantified by reading its absorbance in a spectrophotomer at 260 nM and 280 nm. A ratio of absorbance 260/280 above 1.8 indicates a high quality RNA. The RNA is then converted to cDNA by using reverse transcriptase and incubating it in the presence of a specific buffer and nucleotides.

A5.2.1.2 Fathead Minnow Vitellogenin

The specific sequence for fathead minnow vitellogenin has been published and is available from the database. We used this sequence to design specific forward and reverse primers to allow us to use PCR as a method to amplify a 401 base pair (bp) fragment of the mRNA. The amplified piece of cDNA was visualized on an agarose gel, cut from the gel and extracted. It was then ligated into a prepared PGEM-T Easy Vector and subsequently transformed into E. coli. Upon bacterial replication the plasmid containing the vitellogenin fragment was amplified. The plasmid was then isolated from the bacterial cells and sent for DNA sequencing to confirm that we had indeed amplified the correct vitellogenin fragment. We also translated the DNA sequence into protein sequence to compare with existing proteins in the database. The results of both are shown in Figure A-1 Vitellogenin Cloning.

A5.2.1.3 Fathead Minnow ERα

The sequence for fathead minnow ERα was not known or available to us. We lined up all the available ER sequences from the databases for other species and found sections that were highly conserved. We designed forward and reverse primers to these sections and used PCR to amplify a 550 bp fragment. As with the vitellogenin segment described above, the ERα segment was gel purified, cloned into the PGEM-T Easy Vector, and transformed into bacteria. It was then purified from the bacteria and sent for DNA sequencing to confirm that we had amplified the correct fragment. We also translated the DNA sequence into protein sequence to compare with the databases. The results of both are shown in Figure A-2 ERα cloning.

A5.2.2 Development of Specific Primers for the TaqMan Assay

The sequences for the vitellogenin and ERα fragments were input into the Primer Express software (Applied Biosystems) to design appropriate primers for the TaqMan assay. These primers are closer together on the individual sequences than the originally designed primers. For good quantification by TaqMan each amplicon should be less than 150 bp. The location of the TaqMan primers on the two cloned segments is illustrated in Figure A-3.

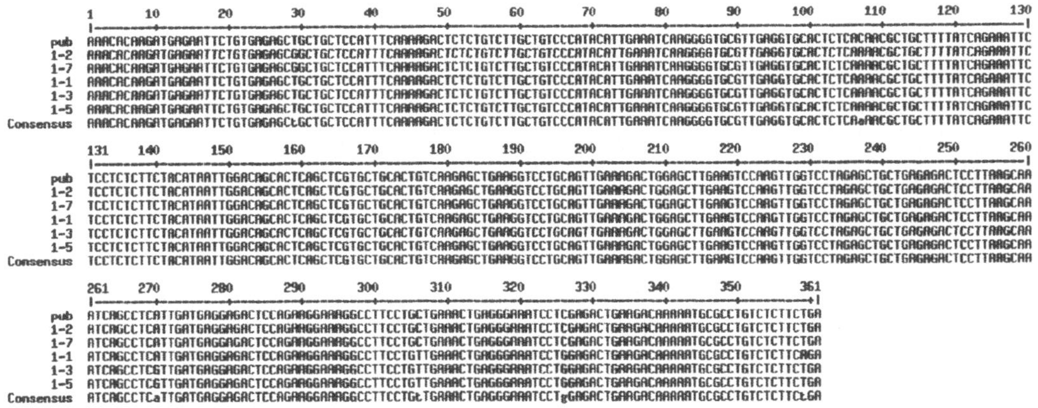

Cloned FHM VTG nucleotide sequences aligned with published FHM VTG

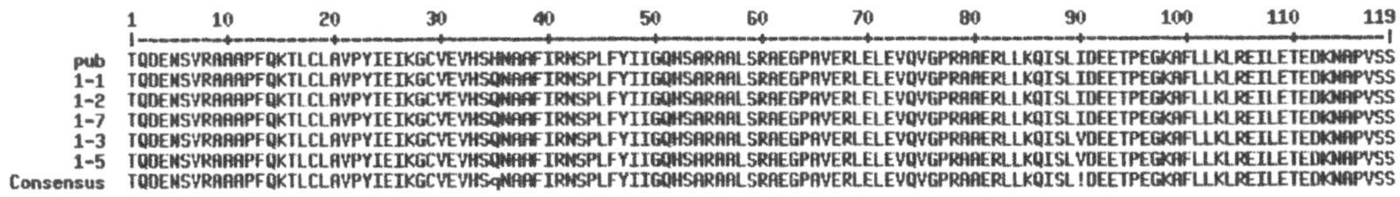

Cloned FHM VTG amino acid sequences aligned with published FHM VTG

Figure A-1. Vitellogenin Cloning

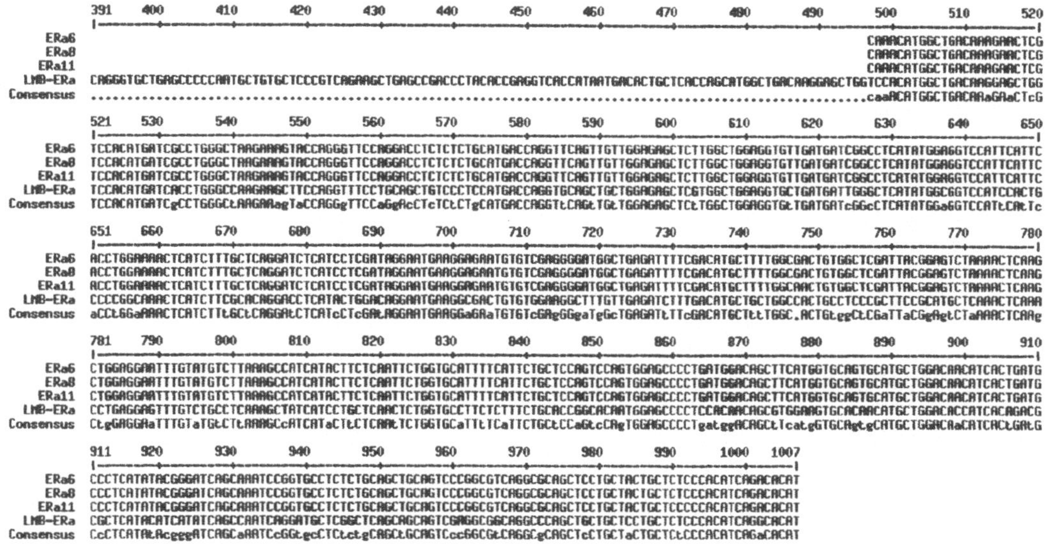

ERα cloned nucleotide sequences aligned with LMB published ERα sequence

ERα cloned amino acid sequences aligned with LMB published ERα sequence

Figure A-2. ERα Cloning.

A5.2.3 A Standard Curve and Optimization of Assay

Once TaqMan specific primers and probes were developed for both genes, it was necessary to optimize the concentration required as well as determine the amount of sample needed and the range of the standard curves. Figures A-4 and A-5 represent a typical amplification profile and a standard curve, respectively. Both assays were most efficient with a primer concentration of 300 nM and a probe concentration of 200 nM. The standard curve was based on known amounts of cDNA plasmid containing the gene of interest. For the vitellogenin standard curve, we used plasmid ranging from 2.68 –268 x 10^7 copies, while for the ER α curve, the plasmid varied from 2.56 –256 x 10^7 copies. To measure ERα, we used 100 pg of our reverse transcribed cDNA, while for vitellogenin; we used 5-100 pg of the reversed transcribed cDNA. The different amounts were necessitated because vitellogenin can be more highly induced (by a factor of 100) than ER.

Each sample was run in duplicate. ERα and vitellogenin values were normalized to 18 S rRNA (Applied Biosystems kit) to control for pipette error in preparation. Normalized data was averaged and then used to calculate the copy number per ug of total RNA.

Forward Primer

Reverse Primer

Fluorescent Probe

ERα Cloned Segment

CAAACATGGCTGACAAAGAACTCGTCCACATGATCGCCTGGGCTAAGAAAGTACCAGGGTTCCAGGACC

TCTCTCTGCATGACCAGGTTCAGTTGTTGGAGAGCTCTTGGCTGGAGGTGTTGATGATCGGCCTCATATGG

AGGTCCATTCATTCACCTGGAAAACTCATCTTTGCTCAGGATCTCATCCTCGATAGGAATGAAGGAGAAT

GTGTCGAGGGGAT**GGCTGAGATTTTCGACATGCTT**TTGGCGACTGTGGCTCGATTACGGA*GTCTAAAAC*

*TCAAGCTGGAGGAATTT*GTATGTCTTAAAGCCATCATACTTCTCAATTCTGGTGCATTTTCATTCTGCTCCA

GTCCAGTGGAGCCCCTGATGGACAGCTTCATGGTGCAGTGCATGCTGGACAACATCACTGATGCCCTCAT

ATACGGGATCAGCAAATCCGGTGCCTCTCTGCAGCTGCAGTCCCGGCGTCAGGCGCAGCTCCTGCTACTG

CTCTCCCACATCAGACACAT

VTG Cloned segment

AAACACAAGATGAGAATTCTGTGAGA**GCTGCTGCTCCATTTCAAAAG**ACTCTCTGTCTTGCTGTCCCAT

ACATTGAAATCAAGGGGT*GCGTTGAGGTGCACTCTCAA*AACGCTGCTTTTATCAGAAATTCTCCTCTCTTC

TACATAATTGGACAGCACTCAGCTCGTGCTGCACTGTCAAGAGCTGAAGGTCCTGCAGTTGAAAGACTGG

AGCTTGAAGTCCAAGTTGGTCCTAGAGCTGCTGAGAGACTCCTTAAGCAAATCAGCCTCATTGATGAGGA

GACTCCAGAAGGAAAGGCCTTCCTGTTGAAACTGAGGGAAATCCTGGAGACTGAAGACAAAAATGCGCC

TGTCTCTTCAGA

Figure A-3. Location of TaqMan Primers.

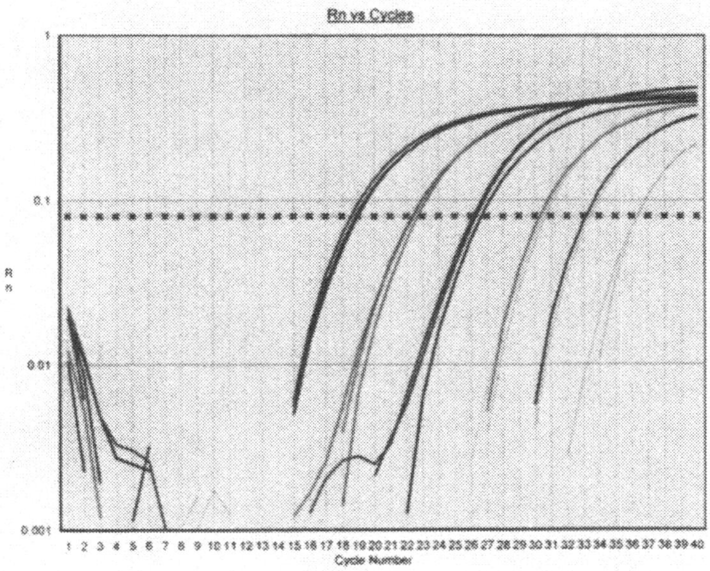

Plot of cycle number versus amount of fluorescent signal.

Figure A-4. **Amplification Plot**

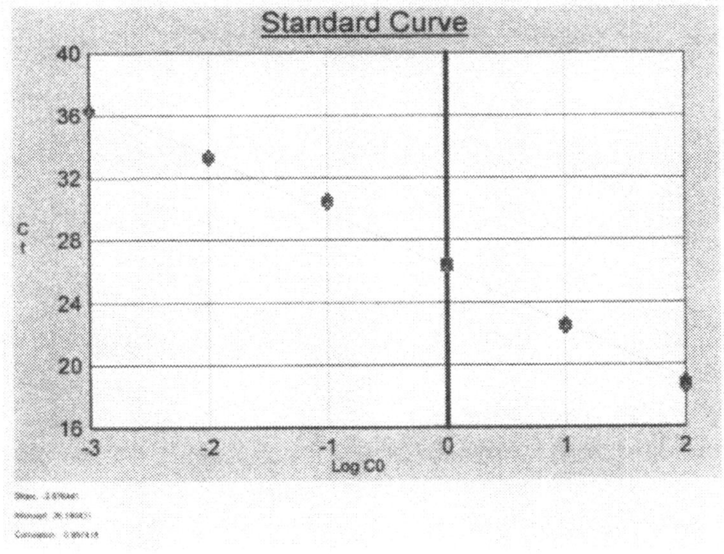

Based on data from the amplification plot of known copy numbers, a standard curve is created. From this curve it is possible to determine concentration of mRNA for a specific gene for unknown samples based on amplification data.

Figure A-5. Taqman Standard Curve

APPENDIX A6

HORMONES

A6.1 Hormone Analysis of Aqueous Samples

All reagent chemicals and products were purchased from Fisher Scientific at the highest possible purity. Steroidal compounds used as standards were purchased from Sigma Chemical, also at the highest possible purity. Aqueous solutions used deionized water produced using a Nanopure II system.

Identification and quantification of natural and synthetic steroidal compounds found in wastewater treatment plant effluents is accomplished through the use of solid phase extraction and GC/MS/MS. Samples are collected in 12-L fluorinated Nalgene jerricans. Samples are immediately brought to the laboratory and filtered through Millipore AP-40 glass fiber filters under pressure. The filtrate is collected and extracted under pressure through preconditioned Empore (3-M Corp, Minneapolis, Minn) C-18 solid phase extraction discs. The C-18 extraction discs are then rinsed with a 40% H_2O/60% MeOH solution to reduce organic matter loading. Steroidal compounds are then eluted from the C-18 discs with a 25% H_2O/75% MeOH solution. The eluent is then dried down and transferred to 1 mL volumetric flasks where it is resuspended in 200 μL acetonitrile in preparation for derivitization (as per Huang and Sedlak, 2000). Next, 50 μL of heptafluorobutyric anhydride (>98%; HFB) is added as the derivitizing agent. The volumetric flasks are sealed and placed in a 55°C oven for 1.5 hrs. The samples are then cooled and dried down under a gentle stream of nitrogen. The derivitized steroids are resuspended in 100 μL isooctane to which hexachlorobenzene (400 ppb) has been added as an internal standard. At this point, samples are ready for GC/MS/MS analysis.

Steroid derivatives are analyzed using a Finnigan GCQ ion-trap gas chromatograph-mass spectrometer system with an AS-2000 autosampler from Thermoquest Corp (San Jose, CA). The capillary column is a 30m * 0.25 mm ID * 0.25 μM film thickness RTX-5 column from Restek Corp (Bellefonte, PA). Splitless injections of 3.0 μL into a 250°C injection port are used. Helium was used as the carrier gas at a constant flowrate of 1.2 mL/min. The programmed temperature run consists of an initial 2.0 min hold at 80°C, followed by a 40°C/min ramp to 225°C with a 18 min hold, followed by a 40°C/min ramp to 290°C with a 1 min hold. Mass spectrometer conditions are electron impact ionization at 70 eV in a 230°C ion source with a 290°C transfer line from the gas chromatograph. Analytic conditions for the derivitized steroids and hexachlorobenzene (HCB) internal standard are as follows:

Compound	Retention Time (min)	Parent Ion	Product Ion (MS/MS only)	Collision Energy (eV)	L.O.D. (ng/L)	L.O.Q. (ng/L)
Mesterolone	21.30	414 (SIM)				
Hexachlorobenzene	6.76	284 (SIM)				
Testosterone	15.15	681	451, 466, 665	1.10	0.3-0.5	1.0-1.5
17β-Estradiol	15.58	664	451	1.20	0.3-0.5	1.0-1.5
Estrone	17.54	467	422, 448	1.00	0.5-0.7	1.5-2.0
Medroxyprogesterone	22.12	479	383	1.15	0.5-0.7	1.5-2.0

APPENDIX B1

LABORATORY STUDIES

APPENDIX B1

ESTROGENIC BIOMARKER RESPONSE IN SHORT-TERM SCREENING ASSAYS

B1.1 Introduction

Initial experiments were performed during the two-year study to evaluate how best to characterize effluents and field samples, and evaluate the presence and significance of biomarkers. This section presents the results from the data sets that provided initial guide to effluent screening. The following are summarized

- Comparisons of fathead minnow vitellogenin response in static and flow-through exposures to 17α -estradiol (E2),
- Laboratory handling and solution stability, and
- Potential for pooling plasma samples to increase sample volume.

B1.2 Overview of Laboratory Exposures

The studies conducted over the past 2 years evolved as each set of experiments were completed, and research by others provided insight into other factors to be addressed. Exposures and YES assays were done at the two laboratories participating in the study, HydroQual and Hydrosphere.

The fathead minnow is used extensively in laboratory-conducted bioassays. In the studies conducted larval and adult fish were evaluated. Larval fish were readily available. Typical bioassays for acute and chronic effects use larval fish that are newly hatched to 14 days old and on occasion up to 30 days old.

Much of the fathead minnow testing for endocrine disruption is done with adult fish. The bioassay market focuses on young fish produced under controlled conditions with adult fish raised under similar conditions being more difficult to obtain. It was discovered that the producers of larval fish rarely had adults. The only sources for adults were pond-raised. Due to the potential for pathogens coming in on these fish it was necessary to obtain a large pool of fish and to hold them at the testing facilities until needed.

Certain water quality criteria have to be met to successfully conduct these assays. These are easily met for larval fish. There are several methods readily available. For adult fish this can be more of a challenge. Fish larger than 3 grams were avoided because larger fish produce more ammonia, which is deleterious to overall fish health. The size of the fish is of less concern in flow-through studies; however, in static tests smaller fish are required to maintain water quality. The total weight of fish in any test tank for a static test was 1.35 g/L. The maximum concentration of ammonia for any test was 15-ppm (approximately 400-ppb unionized), and this maximum concentration did not appear to affect fish survival. Unionized ammonia is most toxic to fish and this form is more abundant at higher temperatures and pH. Feeding the fish minimally or not at all can control ammonia. Unionized ammonia begins to produce tissue damage in fish above 50 ppb (R. Francis-Floyd and C. Watson, 1996).

B1.3 Comparisons of Adult Fathead Minnow Vitellogenin Response in Static and Flow-through Exposures to 17b-Estradiol

B1.3.1 Purpose

A primary objective of the initial experiments was to optimize a fathead minnow exposure that could be potentially used to cost-effectively screen effluents in a laboratory. Therefore, the volume of effluent, number of renewals, and duration of exposure were critical. Initial assays provided a foundation to understand the magnitude and uncertainties in results.

B1.3.2 Overview

A series of studies were performed

1. At HydroQual, adult male and female fathead minnows were exposed to 1 nM E2 in a flow through system. Plasma vitellogenin was measured on specimens from the tanks on days 5, 10, and 20.
2. HydroQual also set up replicate tanks of the positive (1 nM E2) and negative controls for a 10-day exposure to compare with the flow-through experiments.
3. Hydrosphere and HydroQual set up additional static and static renewal tests varying the duration (7-10 days), number and timing of renewals, and concentration of the positive control. These allowed comparisons of responses between the laboratories, and sources of fathead minnows.

All flow-through results and results of static and static renewal assays are shown on Table B-1.

Table B-1. Control and Estradiol Exposure: Static and Flow-through Fathead Minnow Screening

Lab	Date	Test Exposure	Exposure Duration	Exposure Solution	Sex	GSI (%)	Vtg (mg/ml)	Length (cm)	Total Weight (g)	Gonadal Weight (g)	Plasma Length	RBC Length	Hematocrit (%)
HydroQual	4/16/00	Static Non-renewal	10 days	1nM E2	F		6.88	52	1.3		35	20	36
HydroQual	4/16/00	Static Non-renewal	10 days	1nM E2	F		13.79		1.4		20	10	33
HydroQual	4/16/00	Static Non-renewal	10 days	1nM E2	F		22.88	51	1.4		29	9	24
HydroQual	4/16/00	Static Non-renewal	10 days	1nM E2	F		23.03	50	1.4		29	8	22
HydroQual	4/16/00	Static Non-renewal	10 days	1nM E2	F		23.63	56	1.5		27	10	27
HydroQual	4/16/00	Static Non-renewal	10 days	1nM E2	F		24.64	69	2.1		33	13	28
HydroQual	4/16/00	Static Non-renewal	10 days	1nM E2	F		29.44	56	2.1		32	16	33
HydroQual	4/16/00	Static Non-renewal	10 days	1nM E2	F		49.27	51	1.8		20	15	43
HydroQual	4/16/00	Static Non-renewal	10 days	1nM E2	M		7.32	54	1.4		0		
HydroQual	4/16/00	Static Non-renewal	10 days	1nM E2	M		21.57	54	1.5		43	16	27
HydroQual	4/16/00	Static Non-renewal	10 days	1nM E2	M		25.90	58	2.1		34	13	28
HydroQual	4/16/00	Static Non-renewal	10 days	1nM E2	M		26.57	69	3		14	8	36
HydroQual	4/16/00	Static Non-renewal	10 days	Control	F		0.07	62	2.1		31	9	23
HydroQual	4/16/00	Static Non-renewal	10 days	Control	F		0.10	57	1.6		19	7	27
HydroQual	4/16/00	Static Non-renewal	10 days	Control	F		0.72	51	1.3		24	2	8
HydroQual	4/16/00	Static Non-renewal	10 days	Control	F		0.99	59	1.6		32	7	18
HydroQual	4/16/00	Static Non-renewal	10 days	Control	F		1.07	57	1.6		22	3	12
HydroQual	4/16/00	Static Non-renewal	10 days	Control	F		1.19	56	1.5		31	3	9
HydroQual	4/16/00	Static Non-renewal	10 days	Control	F		1.43	59	2		24	9	27
HydroQual	4/16/00	Static Non-renewal	10 days	Control	F		2.04	61	1.9		28	7	20
HydroQual	4/16/00	Static Non-renewal	10 days	Control	F		2.98	55	1.5		25	5	17
HydroQual	4/16/00	Static Non-renewal	10 days	Control	M		0.02	50	1.4		14	6	30
HydroQual	4/16/00	Static Non-renewal	10 days	Control	M		0.03	52	1.4		7	2	22
HydroQual	4/16/00	Static Non-renewal	10 days	Control	M		0.04	50	1.3		14	5	26
HydroQual	4/16/00	Static Non-renewal	10 days	Control	M		0.05	60	1.9		33	1	3
HydroQual	4/16/00	Static Non-renewal	10 days	Control	M		0.05	57	1.7		40	7	15
HydroQual	4/16/00	Static Non-renewal	10 days	Control	M		0.06	60	2.2		25	10	29
HydroQual	4/16/00	Static Non-renewal	10 days	Control	M		0.09	64	2.4		50	10	17
HydroQual	4/16/00	Static Non-renewal	10 days	Control	M		0.13	50	1.1		20	1	5
HydroQual	4/16/00	Static Non-renewal	10 days	MeOH Control, 1ppm	M		0.04	53	1.3		21	4	16
HydroQual	4/16/00	Static Non-renewal	10 days	MeOH Control, 1ppm	M		0.05	58	1.8		25	16	39
HydroQual	4/16/00	Static Non-renewal	10 days	MeOH Control, 1ppm	M		0.07	51	1.4		17	10	37
HydroQual	4/16/00	Static Non-renewal	10 days	MeOH Control, 1ppm	M		0.07	47	1		17	5	23

HydroQual	4/16/00	Static Non-renewal	10 days	MeOH Control, 1ppm	M		0.08	53	1.6		21	15	42
HydroQual	4/16/00	Static Non-renewal	10 days	MeOH Control, 1ppm	M		0.09	53	1.5		25	11	31
HydroQual	4/16/00	Static Non-renewal	10 days	MeOH Control, 1ppm	M		0.13	55	1.7		21	14	40
HydroQual	5/9/00	Flow-through	10 days	1nM E2	M		14.71						
HydroQual	5/9/00	Flow-through	10 days	1nM E2	M		41.44						
HydroQual	5/9/00	Flow-through	10 days	1nM E2	M		53.03						
HydroQual	5/9/00	Flow-through	10 days	1nM E2	M		58.05						
HydroQual	5/9/00	Flow-through	10 days	Control	M		0.02						
HydroQual	5/9/00	Flow-through	10 days	Control	M		0.05						
HydroQual	5/9/00	Flow-through	10 days	Control	M		0.06						
HydroQual	5/9/00	Flow-through	10 days	Control	M		0.001						
HydroQual	5/9/00	Flow-through	10 days	Control	M		0.001						
HydroQual	5/9/00	Flow-through	20 days	1nM E2	M		52.20						
HydroQual	5/9/00	Flow-through	20 days	1nM E2	M		144.50						
HydroQual	5/9/00	Flow-through	5 days	1nM E2	M		16.88						
HydroQual	5/9/00	Flow-through	5 days	1nM E2	M		31.07						
HydroQual	5/9/00	Flow-through	5 days	1nM E2	M		34.17						
HydroQual	5/9/00	Flow-through	5 days	1nM E2	M		37.43						
HydroQual	5/9/00	Flow-through	5 days	1nM E2	M		43.34						
HydroQual	5/9/00	Flow-through	5 days	1nM E2	M		43.88						
HydroQual	5/9/00	Static Renewal	2,8 days	1nM E2	M		2.77						
HydroQual	5/9/00	Static Renewal	2,8 days	1nM E2	M		4.40						
HydroQual	5/9/00	Static Renewal	4,6 days	1nM E2	M		26.90						
HydroQual	5/9/00	Static Renewal	4,6 days	1nM E2	M		30.70						
GLD-WF00039	6/13/00	Static Non-renewal	7 days	Control	F		2.04	50	1.48		13	2	13
GLD-WF00039	6/13/00	Static Non-renewal	7 days	Control	F			50	1.14		8	6	43
GLD-WF00039	6/13/00	Static Non-renewal	7 days	Control	F			52	1.39		17	14	45
GLD-WF00039	6/13/00	Static Non-renewal	7 days	Control	F			51	1.12		2	2	50
GLD-WF00039	6/13/00	Static Non-renewal	7 days	Control	M	1.023	0.04	54	1.62	0.0164	10	7	41
GLD-WF00039	6/13/00	Static Non-renewal	7 days	Control	M	0.669	0.001	58	1.73	0.0115	21	17	45
GLD-WF00039	6/13/00	Static Non-renewal	7 days	Control	M	0.493	0.001	55	1.55	0.0076	14	10	42
GLD-WF00039	6/13/00	Static Non-renewal	7 days	MeOH Control, 1ppm	F		8.34	58	1.72		19	17	47
GLD-WF00039	6/13/00	Static Non-renewal	7 days	MeOH Control, 1ppm	F			55	1.61		15	12	44
GLD-WF00039	6/13/00	Static Non-renewal	7 days	MeOH Control, 1ppm	F			50	1.2		13	9	41
GLD-WF00039	6/13/00	Static Non-renewal	7 days	MeOH Control, 1ppm	F			56	1.61		23	15	39
GLD-WF00039	6/13/00	Static Non-renewal	7 days	MeOH Control, 1ppm	M	1.250	0.01	59	2.09	0.0258	29	19	40

GLD-WF00039	6/13/00	Static Non-renewal	7 days	MeOH Control, 1ppm	M	1.209	0.01	61	2.26	0.027	31	24	44
GLD-WF00039	6/13/00	Static Non-renewal	7 days	MeOH Control, 1ppm	M	1.512	0.001	61	2.39	0.0356	29	27	48
GLD-WF00039	6/13/00	Static Non-renewal	7 days	MeOH Control, 1ppm	M	1.425	0.001	61	2.22	0.0312	24	21	47
GLD-WF00039	6/13/00	Static Non-renewal	9 days	0.1 nM E2	J		0.001						
GLD-WF00039	6/13/00	Static Non-renewal	9 days	0.1 nM E2	J		0.001						
GLD-WF00039	6/13/00	Static Non-renewal	9 days	0.1 nM E2	J		0.001						
GLD-WF00039	6/13/00	Static Non-renewal	9 days	0.1 nM E2	J		0.001						
GLD-WF00039	6/13/00	Static Non-renewal	9 days	1nM E2	J		0.001						
GLD-WF00039	6/13/00	Static Non-renewal	9 days	1nM E2	J		0.001						
GLD-WF00039	6/13/00	Static Non-renewal	9 days	1nM E2	J		0.001						
GLD-WF00039	6/13/00	Static Non-renewal	9 days	1nM E2	J		0.001						
GLD-WF00039	6/13/00	Static Non-renewal	9 days	Control	J		0.03						
GLD-WF00039	6/13/00	Static Non-renewal	9 days	Control	J		0.04						
GLD-WF00039	6/13/00	Static Non-renewal	9 days	Control	J		0.05						
GLD-WF00039	6/13/00	Static Non-renewal	9 days	Control	J		0.22						
GLD-WF00039	6/13/00	Static Renewal	7 days	1nM E2	F		92.70	52.75	1.4375		23.5	13.25	36.25
GLD-WF00039	6/13/00	Static Renewal	7 days	1nM E2	F		91.20	53.5	1.1925		17.25	14	45.25
GLD-WF00039	6/13/00	Static Renewal	7 days	1nM E2	M	0.271	17.10	55	1.22	0.0033	22	11	33
GLD-WF00039	6/13/00	Static Renewal	7 days	1nM E2	M	0.693	52.80	57	1.67	0.0115	17	16	48
GLD-WF00039	6/13/00	Static Renewal	7 days	1nM E2	M	1.031	67.30	58	2	0.0204	27	20	43
GLD-WF00039	6/13/00	Static Renewal	7 days	1nM E2	M	1.030	73.90	55	1.52	0.0155	22	15	41
GLD-WF00039	6/13/00	Static Renewal	7 days	1nM E2	M	0.829	85.10	55	1.69	0.0139	21	11	34
GLD-WF00039	6/13/00	Static Renewal	7 days	Control	F		14.07	54.75	1.52		10	8	45
GLD-WF00039	6/13/00	Static Renewal	7 days	Control	M	1.311	0.001	54	1.63	0.0211	20	17	46
GLD-WF00039	6/13/00	Static Renewal	7 days	Control	M	0.579	0.001	60	1.91	0.011	19	18	49
GLD-WF00039	6/13/00	Static Renewal	7 days	Control	M	1.005	0.001	62	1.97	0.0196	18	13	42
LE000624.1	6/24/00	Static Non-renewal	9 days	0.1 nM E2	F		18.03	61.67	2.31		13	12.33	48.33
LE000624.1	6/24/00	Static Non-renewal	9 days	0.1 nM E2	M	1.500	0.09	64	3.54	0.0523	49	14	22
LE000624.1	6/24/00	Static Non-renewal	9 days	0.1 nM E2	M	1.360	0.13	67	3.42	0.0459	46	16	26
LE000624.1	6/24/00	Static Non-renewal	9 days	0.1 nM E2	M	1.524	0.58	64	3.15	0.0473	47	16	25
LE000624.1	6/24/00	Static Non-renewal	9 days	1nM E2	F		19.05	68.67	2.89		27	13.67	34
LE000624.1	6/24/00	Static Non-renewal	9 days	1nM E2	M	0.934	16.70	59	2.15	0.0199	23	17	43
LE000624.1	6/24/00	Static Non-renewal	9 days	1nM E2	M	1.832	18.20	66	2.98	0.0536	22	27	55
LE000624.1	6/24/00	Static Non-renewal	9 days	Control	F		29.27	64.33	2.82		22	18	44
LE000624.1	6/24/00	Static Non-renewal	9 days	Control	M	1.656	0.001	77	5.52	0.0899	40	20	33
LE000624.1	6/24/00	Static Non-renewal	9 days	Control	M	2.292	0.001	66	3.91	0.0876	43	25	37
LE000624.1	6/24/00	Static Non-renewal	9 days	Control	M	1.105	0.02	64	2.7	0.0295	30	23	43
	8/22/00	Static Renewal	7 days	Control	M		0.001	55	1.42				
	8/22/00	Static Renewal	7 days	Control	M		0.026	51	1.27				
	8/22/00	Static Renewal	7 days	Control	M		0.001	55	1.16				
	8/22/00	Static Renewal	7 days	Control	M		0.001	57	1.54				

8/22/00	Static Renewal	7 days	Control	M	0.001	61	1.52
8/22/00	Static Renewal	7 days	0.1 nM E2	M	0.055	53	1.19
8/22/00	Static Renewal	7 days	0.1 nM E2	M	0.001	53	1.26
8/22/00	Static Renewal	7 days	0.1 nM E2	M	0.081	55	1.96
8/22/00	Static Renewal	7 days	0.1 nM E2	M	0.052	57	1.39
8/22/00	Static Renewal	7 days	0.1 nM E2	M	0.362	56	1.45
8/22/00	Static Renewal	7 days	0.1 nM E2	M	0.206	55	1.48
5/29/01	Static Renewal	7 days	0.01 nM E2	M	0.52		
5/29/01	Static Renewal	7 days	0.01 nM E2	M	0.01		
5/29/01	Static Renewal	7 days	0.01 nM E2	M	0.00		
5/29/01	Static Renewal	7 days	0.01 nM E2	M	0.03		
5/29/01	Static Renewal	7 days	0.01 nM E2	M	0.00		
5/29/01	Static Renewal	7 days	0.01 nM E2	M	0.0005		
5/29/01	Static Renewal	7 days	0.05 nM E2	M	0.16		
5/29/01	Static Renewal	7 days	0.05 nM E2	M	0.49		
5/29/01	Static Renewal	7 days	0.05 nM E2	M	0.03		
5/29/01	Static Renewal	7 days	0.05 nM E2	M	0.10		
5/29/01	Static Renewal	7 days	0.05 nM E2	M			
5/29/01	Static Renewal	7 days	0.05 nM E2	M	0.0005		
5/29/01	Static Renewal	7 days	0.1 nM E2	M	0.09		
5/29/01	Static Renewal	7 days	0.1 nM E2	M	0.01		
5/29/01	Static Renewal	7 days	0.1 nM E2	M	0.0005		
5/29/01	Static Renewal	7 days	0.1 nM E2	M	0.02		
5/29/01	Static Renewal	7 days	0.1 nM E2	M	0.02		
5/29/01	Static Renewal	7 days	0.1 nM E2	M	0.03		
5/29/01	Static Renewal	7 days	1 nM E2	M	41.55		
5/29/01	Static Renewal	7 days	1 nM E2	M	9.39		
5/29/01	Static Renewal	7 days	1 nM E2	M	20.47		
5/29/01	Static Renewal	7 days	1 nM E2	M	20.86		
5/29/01	Static Renewal	7 days	1 nM E2	M	28.16		
5/29/01	Static Renewal	7 days	1 nM E2	M	52.67		
5/29/01	Static Renewal	7 days	Control	M	0.41		
5/29/01	Static Renewal	7 days	Control	M	0.0005		
5/29/01	Static Renewal	7 days	Control	M	0.0005		
5/29/01	Static Renewal	7 days	Control	M	0.04		
5/29/01	Static Renewal	7 days	Control	M	0.0005		
5/29/01	Static Renewal	7 days	Control	M	0.0005		

a = Pooled plasma Vtg analyses.; U = Below the Detection Limit.; GSI = Gonadal Somatic Index; E2 =17B-Estradiol

B1.3.3 Results

B1.3.3.1 20-day Flow-through (April-May 2000)

The average plasma vitellogenin levels in male minnows exposed to 1 nM E2 for 5, 10, and 20 days averaged 34.5, 41.8, and 98.5mg/mL, respectively. The average vitellogenin concentration for the negative controls (i.e. fish in laboratory water with no added E2) was 0.026-mg/mL. (See Table B-1.)

During this flow through exposure, several observations were made. The fish became hard to sex at the 10-day exposure to 1 nM estradiol and males were harder to bleed than females and controls. There were no mortalities in the tests.

B1.3.3.2 10-day Static (April-May 2000)

The 10-day static tests run showed similar effects in that the males, ie. they were feminized and hard to bleed. No mortality occurred and it appeared that this static exposure with no renewals could be used from the standpoint of fish survival. The vitellogenin in the static exposures averaged 20.34mg/mL, approximately half the level after 10 days in the flow through system. A parallel set of static exposure for 9 days also showed average male plasma vitellogenin levels at 17.45-mg/mL.

Vitellogenin was present in male plasma in the negative controls for the initial exposures. However, vitellogenin levels in other batches of fatheads were lower, but frequently detectable. It became clear that time is needed to evaluate the baseline conditions of the fatheads, however, a difference in vitellogenin was readily distinguishable from the background levels.

B1.3.3.3 Renewals with Negative or Positive Controls

One disadvantage to static exposure experiments is that estradiol may degrade or be absorbed during the test time. Also water quality, as monitored by pH and dissolved oxygen, is affected by bioload, aeration, and fish feeding. Water quality is not an issue for flow-through systems because the test solution is constantly renewed. Therefore, shorter exposures to estradiol were evaluated to balance production of vitellogenin with test solution degradation.

Once exposed to an estrogenic compound, some time is required to produce the protein in the liver. The time lag between the mRNA signal and maximum protein concentration may be hours to days, but once produced, the vitellogenin is stable for some time in the plasma. Therefore, the exposure duration considers time subsequent to exposure to reach the maximum vitellogenin level. It also appeared from the early YES assays, that estrogenic response in the tank water containing fish biomass was much lower than influent solutions or tanks with no fish present. This is attributed to absorption, or interference in the assay. This was considered in evaluating static renewal exposures where the initial dose is not sustained with a constant solution concentration.

B1.3.3.4 Replacement with Negative Control Water

The volume of effluent needed for a laboratory screening assay needed to be optimized. A target exposure duration of 7-10 days was evaluated. If a static exposure (no renewals) deteriorated, renewal with laboratory control water was evaluated to allow adequate time to maximize vitellogenin even though E2 was no longer present.

Therefore, following exposure to the 1 nM E2 solution, the tank water was replaced with control water and exposure continued for a total of 10 days. Concurrent with the 10-day static and flow through exposures, minnows were exposed to the 1 nM E2 for two days, then the solution was renewed with laboratory negative control water and exposure continued for an additional 8 days for a total of 10 days. In a second tank, the renewal with negative control water was done on day four, then the test completed on day 10. The average male plasma vitellogenin after 10 days indicate that with continuous flow through exposure to 1 nM E2 vitellogenin levels reach 41.8 mg/mL; with no renewals 20.3 mg/mL; and with the initial 1 nM E2 solution renewed with clean water after 2 days vitellogenin was 3.59 mg/mL; and after 4 days, 28.8 mg/mL. This suggests additional activity was present in solution initial static exposure, because replacement with control water after 2 days showed significant, but reduced response. However, when exposed in positive control water for 4 days, the results were comparable to the static non-renewal.

B1.3.3.5 Renewal with 1 nM E2 Solution

In another exposure, the test was renewed with the 17α-estradiol solution on day 4 and the test was ended on day 7. The vitellogenin in this exposure was 59.24 mg/mL. Well over twice that of the corresponding static non-renewal exposure. This is similar to expectations in the flow through system, where renewals increase the total exposure to the estrogenic compound. A static test that includes renewals significantly increases the estrogenic response.

B1.3.3.6 Nine-day Static Comparing Adults and Juvenile Fathead Minnows at 1 and 0.1 nM E2 (June 2000)

The initial exposures were conducted using a positive control of 1 nM 17α-estradiol solution. Subsequently exposures were performed at lower concentration likely for effluents. Nine day exposures to 0.1 nM of 17 α -estradiol (E2), male minnows produced vitellogenin levels statistically different from the negative control (i.e. method blank) in a static test showing a 30-fold increase in resulting vitellogenin concentration.

Literature values indicate that significant reductions in male fathead minnow gonadosomatic index GSI were not observed in exposure concentrations up to 100 ng/L, but were observed at 320 and 1,000 ng/L (Panter et al, 1998). A concentration of 1 nM of E2 is equivalent to 272 ng/L of estradiol. In the 9-day exposures to 1 nM E2 solutions, there was a statistically significant difference between the positive and negative control test solutions. However, at lower exposures the differences were not significant.

In addition to the evaluation of exposure duration and response for adults, potential use of juvenile fathead minnows were evaluated. These were considered potentially advantageous because of their small size requiring smaller effluent volumes. 25-day old juvenile minnows were exposed to 0.1 nM and 1nM E2 for nine-days in a static test without renewal. Whole animal homogenates were analyzed for vitellogenin. Vitellogenin did not increase for juvenile fish with the exposures used. After reviewing other studies, it appears that frequent renewals and longer exposure times may be required to see induction in juveniles at low concentrations of estrogen.

While emphasis was placed on males, some observations of the results for females were noted. For example, in the initial 10-day exposures, the females weighed approximately 1.3-2 gms and the vitellogenin levels in the controls were 0.1 to 3 mg/mL. The females used in the 9 day static weighed about 3 gm and showed an average vitellogenin level of 29.3 mg/mL in the controls. These baseline conditions may influence the responses in these assays. For the initial exposures a large number of male and female adults were shipped and held two days likely increasing vitellogenin in male plasma.

B1.3.3.7 Static Renewals (January 2001)

Following the series of exposures performed in April and May, an approach to screening effluents in the laboratory was implemented that included a 7-day exposure with effluent renewal after 3 days. This was calibrated using one renewal, based on the significant response from the positive control of 0.1 nM E2, and the assumption that if estrogenicity of effluents were below this value, at least one renewal would enhance the response. In September 2000, the 0.1 nM E2 control did not show a positive response (See Table B-3 Effluent Screening) prompting additional evolution of responses at low concentrations.

As the assays proceeded, parallel exposures to lower concentrations of E2 and additional renewals were evaluated to obtain data on the sensitivity of the assay. It was hypothesized that this would enhance production of vitellogenin biomarker due to the increased overall dose, as was shown with the initial tests during protocol development with high E2 concentrations. This would indicate whether additional renewals would enhance the sensitivity of the screening, and also better clarify the detection limits of this assay.

Seven-day exposures to controls and E2 concentrations of 0.01, 0.05, 0.1, and 1 nM with renewal at day 3 were also performed. For this set of exposures the vitellogenin levels were variable for the controls and lower concentrations. Only the exposures at 1 nM in this case were significantly different from controls.

B1.4 Laboratory Handling

B1.4.1 Potential Effects From Plastics/Cleaners

In establishing boundaries for the initial sample handling and laboratory exposures, experiments were conducted to evaluate the effects of readily available plastic containers and laboratory cleaners on production of vitellogenin. Potential effects of three types of plastic container material were evaluated by placing container material (cut sections) in the test tank during the entire length of the test. 10 mg/L concentrations of each of 3 cleaning solutions were used to simulate the worst-case scenario for cleaner contamination. Increases in vitellogenin production were noted for these worst-case container and cleanser exposure scenarios (Table B-2).

Table B-2. Plastics and Cleaners: 10-day Static Non-renewal Exposure of Fathead Minnows

Exposure Solution	Vitellogenin (mg / mL) Males	Females
Control	0.02	0.07
Control	0.03	0.10
Control	0.04	1.07
Control	0.05	2.04
Control	0.06	0.72
Control	0.09	1.19
Control	0.05	2.98
Control	0.13	0.99
Control		1.43
Plastic A	0.08	5.14
Plastic A	0.60	
Plastic B		20.16
Plastic C	101.06	117.83
Plastic C	79.12	
Cleaning Agent A	4.07	50.39
Cleaning Agent A	1.12	
Cleaning Agent B	1.07	17.45
Cleaning Agent B	0.65	
Cleaning Agent C	0.62	2.49
Cleaning Agent C	0.54	
Note: Female vitellogenin measurements for plastics and cleaning agents were made on pooled plasma samples		

It was determined that some precautions are needed in handling solutions, and cleaning/rinsing of equipment. Because effluents may have low estrogenicity, small changes attributed to these other sources required careful monitoring of control solutions.

B1.4.2 Potential for Pooling Plasma Samples to Increase Sample Volume

A typical 2- to 3-gram fathead minnow can produce blood plasma sample volumes of 4- to 200- uL. At least 10 uL of blood plasma is required for the Vitellogenin analysis, and if other measurements (e.g. hormone levels) were of interest, sufficient plasma would likely not be present. Additional experiments were conducted by pooling plasma from 3 fish and the pooled results were compared to individual fish results. The pooled sample results were generally within one standard deviation of the average of the levels measured for the individual fish. While this can lose some information on variability, many of the vitellogenin results showed fairly consistent results. The proposed EPA protocol includes taking the average of measurements on individual fish in a single tank, for each of the four exposure replicates. These averages provide the range of response for comparing controls and samples. Our data suggest that pooled samples may provide similar levels of information.

Experiments were conducted to determine if pooling fathead minnow samples would give a substantially different result versus running each sample individually on a direct vitellogenin enzyme-linked immunosorbent assay (ELISA). The samples were diluted individually then 5 µL were pooled with two other samples giving a 15-µL pool. The pool was then run as if it were a separate sample. These data (Table B-4) indicate that when values tend to be clustered, this provides a tool that could be applied in some situations. It was used in some cases for female plasma samples that were used only for supporting information.

Table B-3. Evaluation of Pooled Plasma Vitellogenin Sample Concentrations (mg/mL) Compared to Average of Levels in Individual Fish in 10-day Static Exposures

	1	2	3	4	5	6	7	8	9
Gender	F	F	M	F	F	F	M	M	F
Solution	E2	E2	-	-	-	-	-	-	-
Individual Vitellogenin Measurements									
1	6.88	23.63	>0.09	0.07	0.10	0.72	>0.13	0.07	24.64
2	29.44	22.88	0.06	1.07	2.98	0.99	>0.09	0.08	23.03
3	13.79	49.27	0.05	2.04	1.19	1.43	>0.13	0.07	
Average	16.70	31.93	0.07	1.06	1.42	1.05	>0.12	0.07	23.83
Pooled Plasma Measurement	23.27	37.26	0.07	1.64	1.25	0.89	>0.11	0.07	38.13

E2 = 1 nM estradiol. ;- = negative control.

B1.5 Conclusions

Based on the initial laboratory tests, effluent screening proceeded with a seven day exposure to adult male fathead minnows using a 0.1 nM E2 positive control solution, with effluent or E2 solution renewals on day 3.

The following highlight observations from laboratory exposures:

1. Adult fathead minnows produce a more detectable increase in vitellogenin more rapidly than exposures to juveniles.
2. Suppliers provide adults generally from ponds, and these can vary in baseline conditions.
3. Survival was not a problem in the 7-10 day exposures.
4. The 0.1 nM E2 solution produced positive responses in exposures through August, but not in September. This concentration is near reported threshold levels.

5. Vitellogenin was a much more sensitive biomarker in the 7-day exposure than GSI. No statistically significant differences in GSI were noted for these short-term exposures, even with 1 nM E2 control solution.

APPENDIX B2

EFFLUENT SCREENING

Once a fathead minnow screens program was developed, it was applied to a number of effluents. This section highlights the screening program results.

B2.1 Protocol

Individual grab samples were collected, packed with ice and shipped via an overnight night service to the laboratory. Once received at the laboratory the samples were logged in, checked for arrival temperature, sub-sampled for hardness and alkalinity and transferred to a cold room for storage at 4°C until needed to begin exposures. Samples were warmed to 20°C prior to exposure initiation. Samples were tested for the presence of total residual chlorine. Samples that exceeded 0.15 mg chlorine/L were dechlorinated with sodium thiosulfate. Samples were checked to ensure that water quality did not exceed acceptable parameters for dissolved oxygen (4.0 mg/L to saturation) and pH (6 to 9). Conductivity was also checked.

Samples for the YES Assay were collected from the fathead minnow exposure vessels just after loading with sample but prior to stocking with minnows. These samples were stored in 1.5 ml cryovials and stored at –17°C until used for the YES Assay.

B2.2 Effluent Sources

One objective was to understand the frequency and level of estrogenic responses for a variety of effluent types, predominantly domestic wastewaters. It is important to note that these samples represent a single grab sample and may not reflect the variability of the plant effluent. This screening process provided a broad rapid tool that could be applied by a facility for several samples or composites for a more detailed plant specific evaluation.

Evaluating samples for estrogenic compounds is not currently part of the NPDES or other monitoring programs. Therefore, each of the participating facilities were assured that no information would be provided in the report that would attribute individual results to a specific facility. Eighteen facilities agreed to provide effluent samples for experiments included in this study. Facilities were selected to represent a variety of communities in Florida (for example, facilities represent both large urban areas as well as smaller, more rural communities), a variety of treatment plants (simple primary treatment for an industrial WWTP), and willingness to participate in the study.

Characteristics of the eighteen facilities included the following:

- Fourteen of the eighteen facilities were municipal WWTPs, which included only domestic and light commercial users in the influent stream;
- Influent to one utility facilities was dominantly industrial with two primary industrial contributors (a chemical plant and a paper mill),

- Two were effluents from private facilities (food manufacturers),
- Capacity, as measured in million gallons per day (mgd) averaged from less than 1 to more than 45 mgd. Lower capacity WWTPs included both a small, rural community as well as one industrial WWTP;
- All the WWTP included primary treatment using sand filtration to remove solids,
- About half of the WWTPs include tertiary treatment systems which focus on nutrient removal for nitrogen and phosphorus,
- In general, all the WWTPs reach maximum flow during storm water events; storm water is incorporated into the WWTP influent stream; and
- About half of the WWTPs chlorinate the discharge. A single WWTP uses UV disinfection. One facility specified that sodium thiosulfate was used to dechlorinate prior to surface water discharge, which may also be done for other facilities that chlorinate in order to meet permit limits for chlorine.

B2.3 Results

In June 2000, two effluent samples were used in addition to standards, to guide the evaluation of effluent responses (Table B-4). No significant responses were noted during these experiments, encouraging additional focus on lower E2 concentration range (0.1 nM) for positive controls.

The primary exposures were performed in August 2000, on effluents from six WWTPs and one industrial sample. These data are shown on Table B-5 and illustrated on Figure B-1. A single effluent (Sample A) produced an estrogenic response in excess of the 0.1 nM E2 standard. This facility was constructed in 1998 and has an average daily capacity of 0.250-million gallons per day. The influent stream is comprised of domestic users and light commercial (for example, domestic waste from shopping centers). The waste stream is aerated upon arrival, followed by primary and secondary treatment, as well as tertiary treatment for removal of nitrogen and phosphorus. The effluent is chlorinated. The 5-day Biological Oxygen Demand (BOD) removal efficiency is greater than 99 percent.

Two other facilities showed a slight elevation of vitellogenin levels, but these were perhaps not significant. One of these (Sample B) had characteristics similar to Sample A, described above. The second, (Sample G) was a utility that receives discharges from a paper mill and chemical plant as the predominant flow. This plant was constructed in 1974 and has an average daily capacity of 23 mgd. Each of the industrial contributors has an NPDES permit which requires effluent from the plant to be treated to remove solids and BOD. This WWTP has primary treatment and secondary treatment (96-percent BOD removal). The effluent is not chlorinated or disinfected. Because the plant receives only industrial and no domestic waste, the BOD and fecal coliform are monitored, but the effluent is not disinfected before it is released from the plant.

Table B-4. Data from the 7-day Static Non-renewal and Renewal Exposure of Fathead Minnows to Two Effluent Samples

Exposure Solution	Test Exposure	Length (mm)	Total Weight (g)	Gonadal Weight (g)	GSI	Plasma Length (mm)	RBC Length (mm)	Hematocrit (%)	Vtg. (mg V_{tg}/ mL)
Males (June 13, 200)									
Control	Static Non-renewal	58	1.73	0.0115	0.669	21	17	45	<0.0005
		54	1.62	0.0164	1.023	10	7	41	0.04
		55	1.55	0.0076	0.493	14	10	42	<0.0005
		57	1.61	NA	NA	NA	NA	NA	NA
Effluent A, 100%	Static Non-renewal	54	1.5	0.0191	1.290	27	17	39	<0.0005
		55	1.74	0.0095	0.549	25	20	44	<0.0005
		52	1.31	0.0146	1.127	15	11	42	<0.0005
Effluent B, 100%	Static Non-renewal	61	2.32	0.0248	1.081	20	19	49	<0.0005
		57	2.18	0.0446	2.089	32	26	45	0.01
		52	1.4	0.0131	0.945	13	9	41	0.01
Control	Static Renewal at day 3 with control water	54	1.63	0.0211	1.311	20	17	46	<0.0005
		60	1.91	0.011	0.579	19	18	49	<0.0005
		62	1.97	0.0196	1.005	18	13	42	<0.0005
Effluent A, 100%	Static Renewal at day 3 with control water	55	1.88	0.0169	0.907	24	21	47	0.01
		59	2.08	0.0276	1.345	26	24	48	<0.0005
Effluent B, 100%	Static Renewal at day 3 with control water	63	2.2	0.0272	1.252	19	14	42	<0.0005
		52	1.26	0.0103	0.824	13	10	43	<0.0005
		57	1.83	0.0196	1.083	17	15	47	<0.0005
Effluent B, 100%	Static Renewal at day 3 with effluent	57	1.68	0.013	0.780	22	19	46	0.04
		60	2.18	0.0353	1.646	20	19	49	<0.0005
		61	2.16	0.0167	0.779	24	16	40	0.01
		52	1.61	0.0042	0.262	12	12	50	<0.0005

Females

Exposure Solution	Test Exposure	Length (mm)	Total Weight (g)		Plasma Length (mm)	RBC Length (mm)	
Control	Static Non-renewal	50	1.48		13	2	
		50	1.14		8	6	
		52	1.39		17	14	
		51	1.12		2	2	
Effluent A, 100%	Static Non-renewal	53	1.41		10	8	
		52	1.16		12	9	
		51	1.27		16	11	
Effluent B, 100%	Static Non-renewal	50	1.12		15	10	
		53	1.43		14	13	
		54	1.55		21	15	
		57	1.59		16	14	
Control	Static Renewal at day	57	1.92		6	6	

Exposure Solution	Test Exposure	Length (mm)	Total Weight (g)		Plasma Length (mm)	RBC Length (mm)	
	3 with control water	53	1.27		4	3	
		53	1.31		13	9	
		56	1.58		17	14	
Effluent A, 100%	Static Renewal at day 3 with control water	54	1.61		12	12	
		52	1.17		12	9	
		53	1.38		14	12	
		55	1.51		8	6	
Effluent B, 100%	Static Renewal at day 3 with control water	54	1.64		9	7	
		52	1.39		8	5	
		63	2.04		22	17	
		58	1.9		14	9	
Effluent B, 100%	Static Renewal at day 3 with effluent	54	1.45		15	12	
		50	1.26		9	4	
		57	1.81		18	15	
		50	1.29		12	12	
Effluent B, 100%	Static Renewal at day 3 with effluent	63	1.81		15	8	
		54	1.11		13	6	
		54	1.38		14	14	
		52	1.3		13	9	

Males: Hydrosphere Job Number: GLD-WF00039, FM ID Number 918 .Females: Hydrosphere Job Number: GLD-WF00039, FM ID Number 918. Renewals on Day 3 With Control Water and Effluent.

Table B-5. Effluent Screening: 7-day Male Fathead Minnow Static Renewal Exposures (August, 2000)

Exposure Solution	Effluent type	Sample No.	Total Wt (gms)	Length (mm)	Vtg. (mg / mL)		Average Vtg
Control (-)	Control	1	1.42	55	0.001	<	0.005
Control (-)	Control	2	1.27	51	0.026		
Control (-)	Control	4	1.16	55	0.001	<	
Control (-)	Control	5	1.54	57	0.001	<	
Control (-)	Control	6	1.52	61	0.001	<	
0.1 nM E2	Control	7	1.19	53	0.055		0.126
0.1 nM E2	Control	8	1.26	53	0.001	<	
0.1 nM E2	Control	9	1.96	55	0.081		
0.1 nM E2	Control	10	1.39	57	0.052		
0.1 nM E2	Control	11	1.45	56	0.362		
0.1 nM E2	Control	12	1.48	55	0.206		
A	Municipal WWTP	13	1.36	54	3.156		2.027
A	Municipal WWTP	14	1.59	56	1.958		
A	Municipal WWTP	15	2.03	64	0.967		
B	Municipal WWTP	19	1.32	56	0.026		0.011

Exposure Solution	Effluent type	Sample No.	Total Wt (gms)	Length (mm)	Vtg. (mg / mL)		Average Vtg
B	Municipal WWTP	20	1.82	60	0.001	<	
B	Municipal WWTP	22	1.66	56	0.017		
B	Municipal WWTP	23	1.45	56	0.001	<	
C	Municipal WWTP	24	1.26	51	0.001	<	<0.001
C	Municipal WWTP	25	1.27	56	0.001	<	
C	Municipal WWTP	26	1.50	55	0.001	<	
C	Municipal WWTP	27	1.38	55	0.001	<	
C	Municipal WWTP	28	1.43	56	0.001	<	
D	Municipal WWTP	29	1.65	58	0.001	<	<0.001
D	Municipal WWTP	30	1.98	60	0.001	<	
D	Municipal WWTP	31	1.28	54	0.001	<	
D	Municipal WWTP	32	2.01	62	0.001	<	
D	Municipal WWTP	33	2.08	61	0.001	<	
D	Municipal WWTP	34	1.43	56	0.001	<	
E	Municipal WWTP	35	1.35	53	0.001	<	<0.001
E	Municipal WWTP	36	1.43	55	0.001	<	
E	Municipal WWTP	37	1.58	58	0.001	<	
E	Municipal WWTP	38	1.33	55	0.001	<	
E	Municipal WWTP	39	1.20	54	0.001	<	
E	Municipal WWTP	40	1.49	57	0.001	<	
F	Municipal WWTP	41	1.13	54	0.033		0.007
F	Municipal WWTP	42	1.22	53	0.001	<	
F	Municipal WWTP	43	1.30	55	0.001	<	
F	Municipal WWTP	44	1.50	56	0.001	<	
F	Municipal WWTP	45	1.03	50	0.011		
F	Municipal WWTP	46	1.44	56	0.001	<	
G	95% Ind (P&P)	47	1.35	55	0.081		0.014
G	95% Ind (P&P)	48	1.88	60	0.001	<	
G	95% Ind (P&P)	49	2.12	62	0.001	<	
G	95% Ind (P&P)	50	1.60	58	0.001	<	
G	95% Ind (P&P)	51	1.19	53	0.001	<	
G	95% Ind (P&P)	52	1.84	59	0.001	<	

Fathead Minnow, Pimephales promelas Test Exposure: 7-day Static with renewal at 72-hr. Sample 3 was not clearly sexed, presumed to be female. (Vtg measured 0.806)
No samples could be collected from samples 16, 17, 18 and 21.Laboratory Test Event ID# GLD-WF 00056
< Less than detection limits .

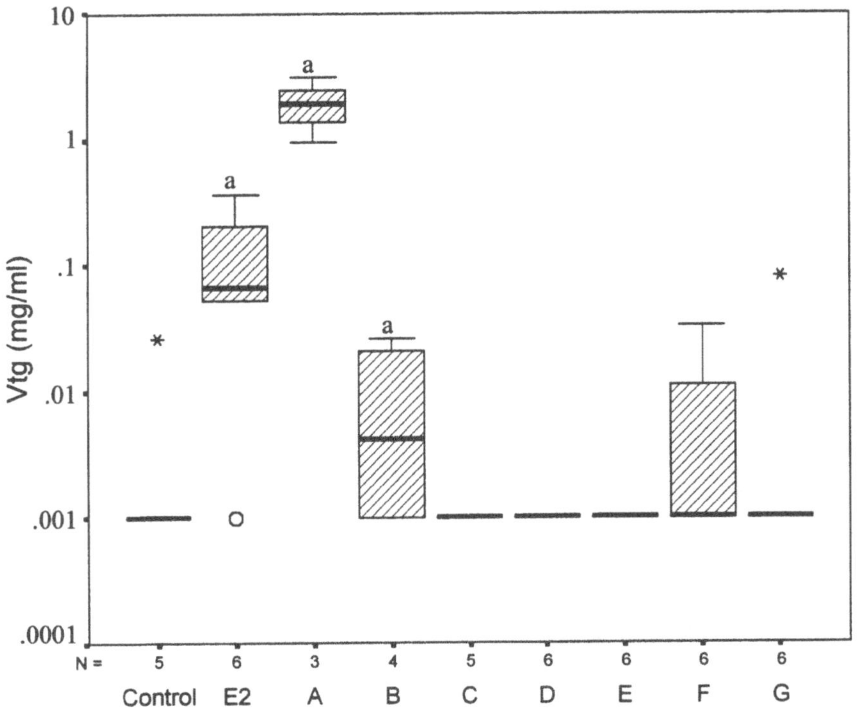

Note: Static exposure with renewal on day 3. Positive control 0.1nME2
[a] Samples significantly greater than negative control at 0.05 level.

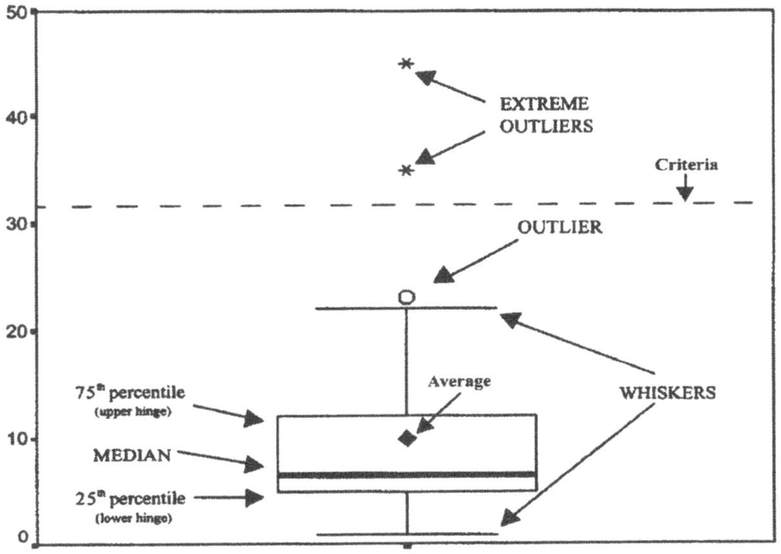

Box plots are used to partition data into four groups, or quartiles, each representing 25% of the measurements (McLave and Sincich, 2000). The data are presented graphically with the lower quartile (25th percentile), median (50th percentile, corresponding to bold line on figure), and upper quartile (75th percentile) values forming a box with "hinges" drawn at the upper and lower quartiles. The distance between the upper and lower quartile is defined as the inter-quartile range. Extending beyond the upper and lower hinges are "whiskers," which represent the furthest points within 1.5 inter-quartile ranges beyond the upper and lower quartiles. Points that extend beyond the whiskers to within three inter-quartile ranges beyond the hinges are known as outliers, indicated by open circles. Those outliers that extend beyond three inter-quartile ranges are extreme outliers, indicated by asterisks.

Figure B-1. Effluent Screening Evaluation: Plasma Vitellogenin in 7-day Static Renewal Fathead Minnow Exposure (August 2000).

At the end of September effluents from five WWTPs and three industrial effluents were screened. See Table B-6. Fathead minnow plasma samples from all effluent exposures and the positive control showed vitellogenin below detectable levels or only at trace concentrations. This was the first event where the 0.1 nM E2 did not show a positive response.

Table B-6. Effluent Screening: 7-day Male Fathead Minnow Static Renewal Exposures (September, 2000)

Exposure Solution	Effluent type	Fish Sample Number	Total Wt (g)	Length (mm)	Vtg (mg / mL)		Average Vtg
Control (-)	Control	1	1.78	62	0.001	<	0.002
Control (-)	Control	2	2.10	62	0.001	<	
Control (-)	Control	3	3.20	71	0.007		
Control (-)	Control	4	1.98	62	0.001	<	
Control (-)	Control	5	2.02	62	0.001	<	
Control (-)	Control	6	2.26	65	0.001	<	
0.1 nM E2	Control	7	2.39	64	0.001	<	<0.001
0.1 nM E2	Control	8	2.99	68	0.001	<	
0.1 nM E2	Control	9	2.89	69	0.001	<	
0.1 nM E2	Control	10	2.53	67	0.001	<	
0.1 nM E2	Control	11	1.81	60	0.001	<	
0.1 nM E2	Control	12	2.45	67	0.001	<	
A	Municipal WWTP	13	1.78	60	0.001	<	0.003
A	Municipal WWTP	14	2.25	65	0.010		
A	Municipal WWTP	15	2.35	66	0.001	<	
A	Municipal WWTP	16	1.62	60	0.001	<	
A	Municipal WWTP	17	2.18	64	0.001	<	
A	Municipal WWTP	18	3.38	74	0.001	<	
B	Municipal WWTP	19	1.77	61	0.001	<	<0.001
B	Municipal WWTP	20	2.31	61	0.001	<	
B	Municipal WWTP	21	2.47	69	0.001	<	
B	Municipal WWTP	22	2.32	66	0.001	<	
B	Municipal WWTP	23	2.61	66	0.001	<	
B	Municipal WWTP	24	2.87	69	0.001	<	
C	Municipal WWTP	25	2.12	62	0.001	<	<0.001
C	Municipal WWTP	26	2.21	65	0.001	<	
C	Municipal WWTP	27	2.60	67	0.001	<	
C	Municipal WWTP	28	2.26	63	0.001	<	
C	Municipal WWTP	29	2.07	62	0.001	<	
C	Municipal WWTP	30	4.07	76	0.001	<	
D	Municipal WWTP	31	1.88	60	0.001	<	<0.001
D	Municipal WWTP	32	2.36	65	0.001	<	
D	Municipal WWTP	33	1.71	61	0.001	<	
D	Municipal WWTP	34	2.07	62	0.001	<	
D	Municipal WWTP	35	2.58	67	0.001	<	
D	Municipal WWTP	36	3.72	74	0.001	<	
E	Industrial	37	2.00	62	0.001	<	<0.001

Exposure Solution	Effluent type	Fish Sample Number	Total Wt (g)	Length (mm)	Vtg (mg / mL)	Average Vtg
E	Industrial	38	2.66	68	0.001 <	
E	Industrial	39	2.34	65	0.001 <	
E	Industrial	40	1.87	63	0.001 <	
E	Industrial	41	2.15	61	0.001 <	
E	Industrial	42	2.10	65	0.001 <	
F	Industrial	43	2.73	66	0.001 <	<0.001
F	Industrial	44	2.04	63	0.001 <	
F	Industrial	45	2.97	69	0.001 <	
F	Industrial	46	2.24	63	0.001 <	
F	Industrial	47	1.95	62	0.001 <	
F	Industrial	48	2.64	67	0.001 <	
G	Industrial	49	2.06	63	0.001 <	<0.001
G	Industrial	50	2.31	64	0.001 <	
G	Industrial	51	2.43	67	0.001 <	
G	Industrial	52	2.45	65	0.001 <	
G	Industrial	53	2.00	63	0.001 <	
G	Industrial	54	2.50	65	0.001 <	
H	Municipal WWTP	55	2.54	66	0.001 <	<0.001
H	Municipal WWTP	56	2.21	64	0.001 <	
H	Municipal WWTP	57	3.47	72	0.001 <	
H	Municipal WWTP	58	1.81	62	0.001 <	
H	Municipal WWTP	59	2.87	70	0.001 <	
H	Municipal WWTP	60	2.92	69	0.001 <	

< = Below detection limits; WWTP=Waste Water Treatment Plant; Fathead Minnow, *Pimephales promelas*
Test Exposure: 7-day Static with renewal at 72-hours, 9/27/2000 - 10/3/2000.

Because of these negative results, a WWTP with influent from domestic and light commercial sites, shown during previous screening to have a positive response, was studied further in January 2001. See Table B-7. The WWTP supplied a 5-gallon sample. The sample was split in two, and one volume was spiked with 0.2 nM E2. A positive control was also spiked with 0.2 nM E2.

This effluent was previously positive, and this also suggests that effluent characteristics may vary over time. Differences in fish length, weight, and hematocrit are within the statistical variation for normal adult fathead minnows.

The preliminary conclusion, consistent with the literature, is that vitellogenin levels increase with additional doses, providing the concentration is above a selected threshold concentration. Below the threshold, no increases in vitellogenin may be observed even for constant flow through conditions. The threshold for fathead minnows appears to be near the 0.1 nM concentration previously used for our exposures. As this study progressed it was hypothesized that the adult fathead minnows obtained in late September may be less responsive in the fall/winter than during times of active spawning.

Table B-7. Effluent Screening: 7-day Static Renewal Exposure of Male Fathead Minnows (January 2001)

Treatments	Specimen			Hematocrit (%)	Vitellogen (mg/ml)
	Number	Length (mm)	Weight (g)		
Negative Control,	1	73	3.87	35	<0.0005
DTF	2	77	4	49	<0.0005
	3	75	4.17	35	<0.0005
	4	72	3.3	57	0.002
	5	79	4.52	57	<0.0005
	6	70	2.94	54	0.005
Positive Control,	7	74	3.62	36	<0.0005
DTF + E2	8	77	4.57	50	<0.0005
(0.02 nM)	9	69	3.19	44	<0.0005
	10	75	3.55	44	<0.0005
	11	74	4.46	46	<0.0005
	12	67	2.38	55	<0.0005
Effluent	13	66	2.45	47	<0.0005
Sample A	14	71	3.41	35	<0.0005
	15	76	4.47	45	<0.0005
	16	71	3.12	45	<0.0005
	17	71	3.76	50	<0.0005
	18	NS	NS	NS	NS
Effluent	19	85	5.66	43	<0.0005
Sample A	20	73	3.83	36	<0.0005
+ E2 (0.02 nM)	21	75	4.73	48	<0.0005
	22	69	3.42	45	<0.0005
	23	74	3.63	49	<0.0005
	24	69	3.21	38	<0.0005

NS = Not sampled
Hydrosphere Job Number GLD-WF 00098

B2.4 Comparison of Fathead Minnows and Juvenile Trout in Effluent Flow-through Experiment

Our team provided support to a large WWTP facility interested in evaluating potential endocrine disruptors in their effluent. Their laboratory was equipped to perform longer-term flow-through experiments, with both fathead minnows and rainbow trout. We provided supplies, discussed protocols for obtaining plasma for vitellogenin analyses, exposed trout to high levels of estradiol to develop necessary additional trout vitellogenin for standards, and provided results.

Vitellogenin was not detected or was present at only trace levels in any of the male fathead minnow samples (Table B-8) following 21 and 45-day exposures to effluent in a flow through system.

This suggests the estrogenic potential of this effluent is below the 0.1 nM E2 threshold response observed in our static renewal assay.

Table B-8. WWTP Effluent Flow-Through Experiment: Fathead Minnow Plasma Vitellogenin

Fish Sample No.	Treatment	Sex	Vitellogenin (mg/mL)
Data Set 1 - ~ 21-day Exposure			
1	Control	M	<0.0005
2	Control	M	<0.0005
3	Control	M	<0.0005
4	Control	M	<0.0005
5	Control	M	<0.0005
6	Control	M	<0.0005
7	Final Effluent	M	<0.0005
8	Final Effluent	M	<0.0005
9	Final Effluent	M	<0.0005
10	Final Effluent	M	<0.0005
11	Final Effluent	M	<0.0005
12	Final Effluent	M	<0.0005
13	Stock Tank	F	**1.870**
14	Stock Tank	F	**1.907**
15	Stock Tank	M	<0.0005
Data Set 1 - ~ 45-day Exposure			
1	Final Effluent	M	<0.0005
2	Final Effluent	?	<0.0005
3	Final Effluent	M	<0.0005
4	Final Effluent	M	<0.0005
5	Final Effluent	M	0.057
6	Final Effluent	M	<0.0005
7	Final Effluent	M	0.003
8	Final Effluent	M	<0.0005
9	Stock Tank	F	<0.0005

Vitellogenin- determined by direct ELISA (UF/ICBR Dr. N. Denslow lab).

(Miles-Richardson et al, 1999) published a study of fathead minnow reproductive effects and induction of plasma vitellogenin. They indicated that the concentration of E2 that causes a statistically significant increase in plasma vitellogenin in adult male fathead minnows is approximately 0.1 nM E2. A threshold has been described that indicates that below this lower level no response (vitellogenin formation) would be expected. According to this study, this threshold concentration corresponds to the threshold for causing histological changes in the gonads.

Eighty-one rainbow trout plasma samples were submitted, forty of these from control exposures and forty-one from effluent exposures. Because these were juveniles and not sexed, individual samples were analyzed (13 control, 12 treated). Results are summarized in Table B-9 and illustrated on Figure B-2. Prior to analysis, several fish were exposed to a 1 nM concentration of estradiol (Static renewal at 4 days, exposure duration of 7 days). Trout vitellogenin was purified from the plasma from these samples and used for developing the standard.

Table B-9. Effluent Screening for Estrogenic Activity Trout Plasma Vitellogenin Sample Date September 09, 2000

Controls			Effluent Exposure		
Ref #	Tank-Sample	Vitellogenin (mg/ml)	Ref #	Tank-Sample	Vitellogenin (mg/ml)
1	C2-1	0.005	43	T2-3	0.173
4	C2-4	<0.0005	46	T2-6	0.093
7	C2-7	<0.0005	49	T2-9	0.282
10	C2-10	<0.0005	52	T2-12	0.108
13	C3-1	0.013	55	T2-15	0.173
16	C3-4	0.033	58	T3-3	0.084
19	C3-7	0.174	61	T3-6	0.297
22	C3-10	0.016	64	T3-9	0.045
25	C3-13	0.035	67	T3-12	0.556
28	C4-3	0.014	71	T4-4	0.238
34	C4-9	0.006	76	T4-9	0.233
37	C4-12	<0.0005	78	T4-11	0.464
40	C4-15	0.033			

Vitellogenin- determined by direct ELISA (UF/ICBR Dr. N. Denslow lab)

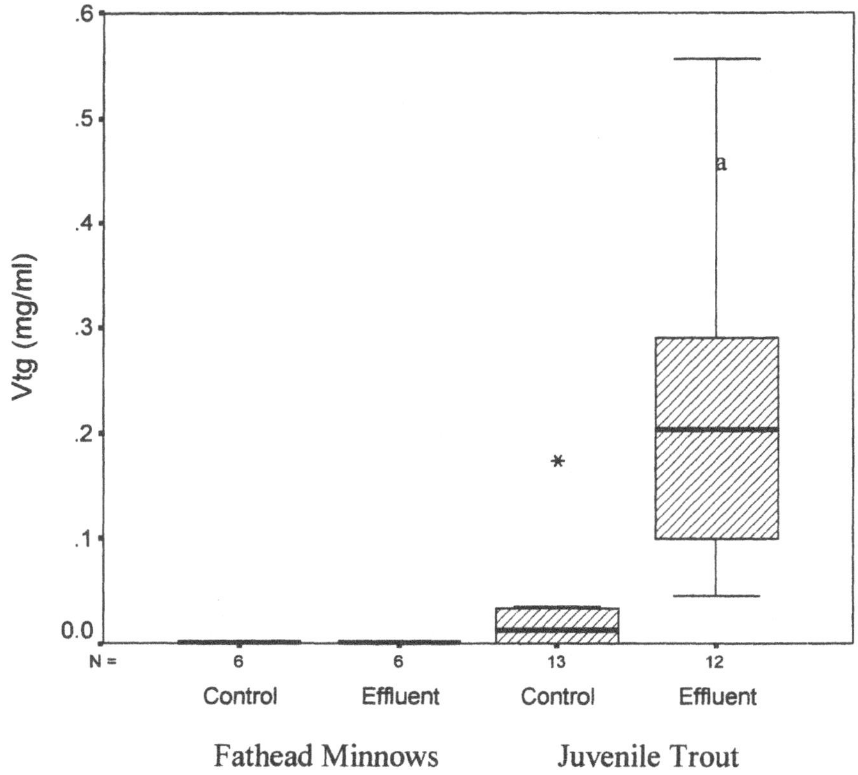

Figure B-2. Flow-through Effluent Exposure – Comparison of Response in Fathead Minnows and Juvenile Trout.

The results from the trout samples suggest some elevation of vitellogenin in the effluent exposed fish as compared to the controls. These results were compared to the published study by (Thorpe et al, 2000) for vitellogenin induction in juvenile rainbow trout. Some differences in the absolute reported values may occur when comparing measurements from different laboratories. Because these are not standard methods, differences may arise from several factors like purification of the vitellogenin used as the standard, procedural differences in protocols used in the assays, etc. Therefore, some uncertainty in the absolute concentrations may be present.

The control samples show considerable variability in the background vitellogenin levels and higher values than those reported in the article by (Thorpe et al, 2000). It is possible that differences in the fish stock used contribute to this effect. Levels in controls were reported at an average of approximately 25 ug/mL vitellogenin. (Thorpe et al, 2000). controls and lowest concentrations all showed levels below 10, frequently nearer to 1.

The response in the effluent exposed samples showed an average near 225 ug/mL plasma vitellogenin. This corresponds to the range of about 24 ng/L nominal exposure of E2. Because there appears to be some background estrogenic effect in the trout used in these exposures, the level of vitellogenin observed in the treated samples may overstate the estrogenicity of the effluent.

The E2 concentration dependent increase in vitellogenin is not linear. Increasing E2 concentrations from 10 to 100 ng/L produces 2 to 3 orders of magnitude increase in vitellogenin. Therefore, there may be an enhanced additive effect with trace levels of estrogenic exposure when there is an initial residual level of vitellogenin.

The rainbow trout data, taken together with the results of the fathead minnow results and literature studies, suggest that estrogenicity of the effluent may be equivalent to E2 concentrations between 5 and 25 ng/L. The fathead minnow results suggest the effluent is below the threshold level (assumed to be below 0.1 nM or 27 ng/L of E2). As indicated previously, based on our results, we believe it may be lower than this upper bound value since our exposures were generally significant at 27 ng/L for a simple 7-day static renewal assay.

The trout data suggest some estrogenicity is present in the effluent. Since the trout are more sensitive, concentrations above 5 ng/L E2 would be expected to produce an induction of vitellogenin. The plasma vitellogenin concentrations in the treated samples may overstate the effluent estrogenicity, because of the initial levels of vitellogenin indicated in the controls. Once fish are sensitized to estrogen, responses may be enhanced and can be additive.

APPENDIX B3

SEDIMENT SCREENING

B3.1 Introduction

When potentially discharging effluents into a receiving stream, the sediment characteristics that may contribute to or inhibit responses are potential issues in the evaluation of impacts in receiving streams. The primary objective of the sediment screening was to evaluate if sediment placed in tanks with the fathead minnows would produce an estrogenic response as measured by elevation in the vtg concentrations in the male fathead minnows.

A total of nine samples were included in the sediment evaluation. At that time, the field study was proposed for the St. Johns River. Therefore, the selection of sediments included two locations in the St Johns, with samples provided by John Higman of the St. Johns River Water Management District. Other sediments locations included river sediment, downstream from various point sources (e.g. waste water treatment plant, meat rendering), near stormwater outfalls, and adjacent to various land uses (agricultural, golf courses, landfill).

B3.2 Fathead Minnow Sediment Screening

The primary goal was to evaluate potential estrogenic responses, similar to the protocol being used to screen effluents. Between 0.5 and 1.0 cm of sediment was uniformly added to the bottom of each exposure tank, thoroughly mixed, and allowed to settle overnight before the addition of fish. Negative controls included tanks with laboratory control water, one with control sediment and one with only water. Positive controls included 1 nM E2 as previously used in developing the assay for effluent screening, with the addition of control sediment to one of the positive control tanks.

At test termination, plasma was collected from fish and transported to the University of Florida for vitellogenin ELISA. Survival and vtg results are summarized in Table B-10 with complete data in Table B-11. The two mortalities observed during the 7-day test occurred in positive control tanks – one each in the positive control with sediment and without sediment. The positive controls were run in duplicate, and two of the sediment samples (rendering plant and trout hatchery samples) were set up in replicate tanks.

Table B-10. Male Fathead Minnow 7-day Sediment Exposures: Summary of Survival and Vitellogenin Data

Treatment	Average Plasma Vitellogenin (mg/mL)	Fish Survival (%)
Negative Controls		
No sediment present	0.006	100 (6/6)
Control (clean) sediment	0.002	100 (6/6)
Positive controls – 1.0 nM 17b-Estradiol		

No sediment (rep A)	112	83 (5/6)
No sediment (rep B)	105	100 (6/6)
With control sediment (rep A)	39.1	100 (6/6)
With control sediment (rep B)	26.5	83 (5/6)
Sediment Samples		
Rendering plant (rep A)	0.12	100 (6/6)
Rendering plant (rep B)	3.81	100 (6/6)
Ditch – residential/golf	0.003	100 (6/6)
Ditch – agriculture/landfill/golf	0.004	100 (6/6)
St. Johns watershed #1	0.085	100 (6/6)
St. Johns watershed #2	0.023	100 (6/6)
Stormwater outfall	0.42	100 (6/6)
Trout hatchery (rep A)	4.45	100 (6/6)
Trout hatchery (rep B)	26.9	100 (6/6)
Wastewater treatment plant	3.78	100 (6/6)

Table B-11. Plasma Vitellogenin Results for Sediment Screening Male Fathead Minnow, *Pimephales promelas*
Test Exposure: 7-day Static

Positive controls - 1.0 nM 17b-Estradiol				Sediment Samples		
No sediment (rep A)		PC	9	Ditch - agriculture/landfill/golf		
	140	PC	9		0.004	
	92	PC	9		0.007	
	83	PC	9		0.008	
	139	PC	9		0.001	<
	105				0.001	<
					0.001	<
No sediment (rep B)		PC	16	Ditch - residential/golf		
	71	PC	16		0.001	<
	69	PC	16		0.001	<
	141	PC	16		0.001	<
	116	PC	16		0.001	<
	127				0.001	<
					0.014	
With control sediment (rep A)		PCSed	13	Rendering plant (rep A)		
	35.9	PCSed	13		0.013	
	49.4	PCSed	13		0.001	<

Positive controls - 1.0 nM 17b-Estradiol					Sediment Samples		
	25.5	PCSed	13			0.361	
	39.4	PCSed	13			0.001	<
	31.1	PCSed	13			0.23	
	53.1						
With control sediment (rep B)		PCSed	15		Rendering plant (rep B)		
	21.7	PCSed	15			0.007	
	27.8	PCSed	15			0.256	
	46.9	PCSed	15			0.022	
	19.0	PCSed	15			0.015	
	17.3					22.46	
						0.083	
Negative Controls							
No sediment present		NC	1		St. Johns watershed #2		
	0.001	NC	1	<		0.009	
	0.01	NC	1			0.015	
	0.012	NC	1			0.042	
	0.001	NC	1	<		0.029	
	0.007	NC	1			0.005	
	0.006					0.036	
Control (clean) sediment		NCSed	5		St. Johns watershed #1		
	0.001	NCSed	5	<		0.001	<
	0.005	NCSed	5			0.022	
	0.001	NCSed	5	<		0.29	
	0.001	NCSed	5	<		0.01	
	0.001	NCSed	5	<		0.001	<
	0.001			<		0.017	

< below detection limits; (All results in mg/mL)

Several control samples were included in the evaluation. This included 1 nM E2 exposures with and without control sediment present. These data indicated that the presence of sediments reduced the estrogenic response, with a 3- to 4-fold higher concentration of vitellogenin without sediment present, compared to positive controls with sediment. Therefore, sediments can reduce the response presumably by sorbing the estrogenic compounds. (See Figure B-3).

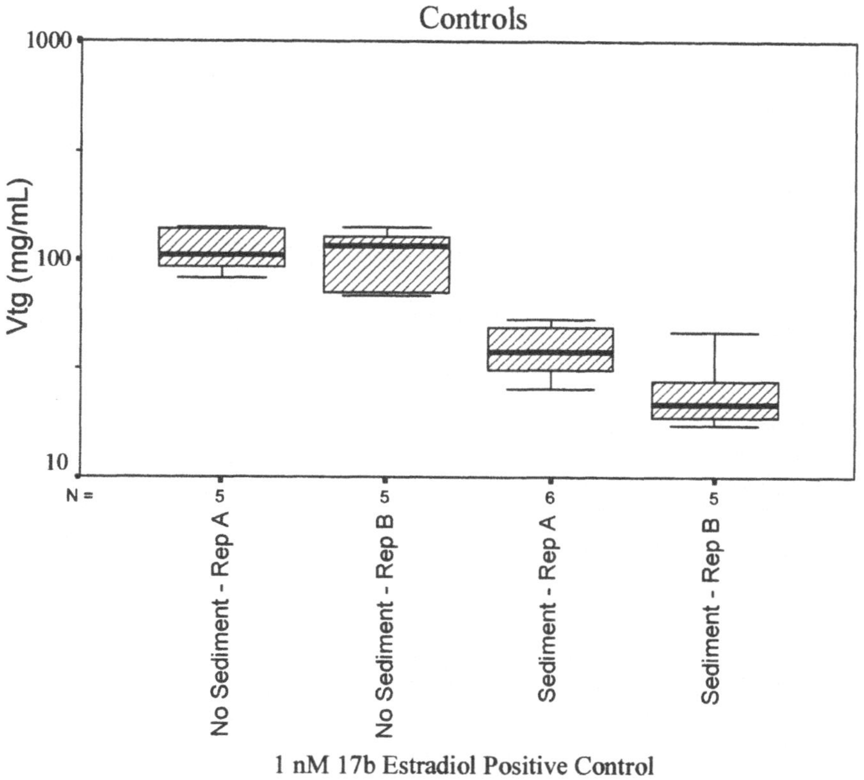

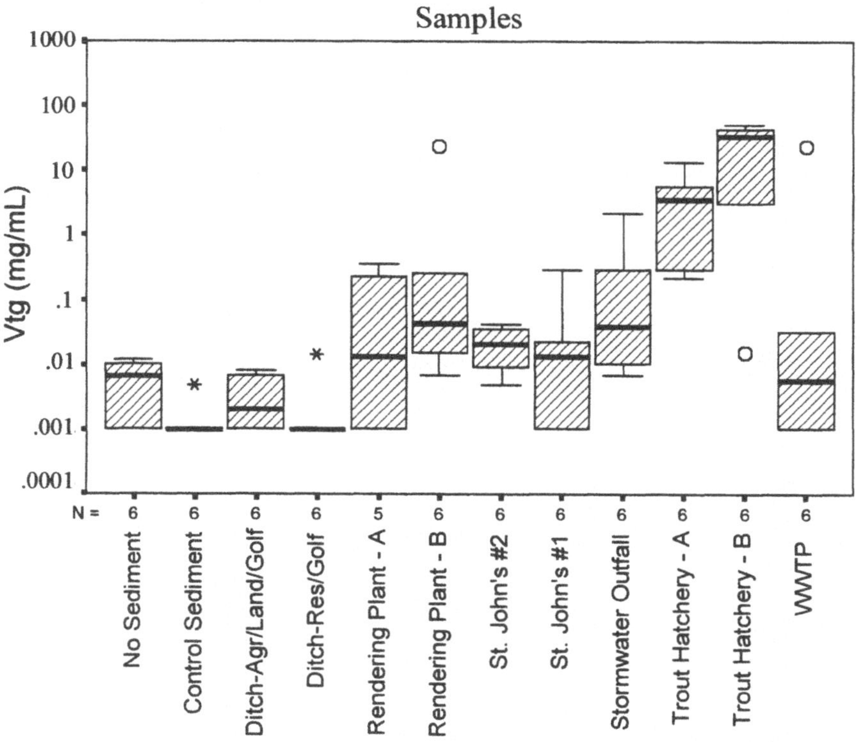

Figure B-3. Vitellogenin Results for Sediment Screening.

B3.3 Pore Water Testing

To place these data in context, several more standard measurements were made including porewater characteristics (Table B-12), and toxicity tests (Table B-13). These data provide some additional context for potential results from the in vivo exposures. These are summarized below.

Table B-12. Physical and Chemical Characteristics of Porewater Collected from Test Sediments

Sample HQ20000958-	Description	pH	Conductance (µS/cm)	Color	Odor	Ammonia (mg/L)
1	St. John's 1	8.2	820	Clear	Organic	0.6
2	St. John's 2	7.7	1360	Clear	Organic	0.6
3	Ditch, residential, golf course	7.8	350	Clear	Organic	0.2
4	Urban lake	7.6	400	Brown	Organic	10
5	Rendering plant effluent	7.3	1680	Brown	Organic	0.7
6	Wastewater treatment plant	7.6	1210	Grey	Organic	61
7	Stormwater outfall	7.3	950	Clear	Chemical	11
8	Trout farm	7.6	1090	Clear	Organic	0.2
9	Ditch, agricultural, landfill, golf	7.6	500	Clear	Organic	1

Table B-13. Summary of Sediment Porewater Toxicity Characterization

Sample HQ20000958		Fathead Minnow Survival	Algal Growth (IC_{50})	Microtox (% of controls)	THB ($x10^4$ cells/mL)	Hyallela azteca Survival
1	(St. John's River)	100%	Non- toxic	>91	2.2	ND
2	(St. John's River)	100%	Non-toxic	>91	1.7	ND
3	(Urban lake)	ND	36% sediment porewater	>91	2.1	2%
4	(Canal-residential/ golf course)	100%	Non-toxic	>91	3.3	ND
5	(Rendering plant)	100%	Non-toxic	>91	0.7	ND
6	(Wastewater treatment plant)	100%	Non-toxic	35	110,000	ND
7	(Stormwater outfall)	100%	Non-toxic	>91	1,100	ND
8	(Trout farm)	100%	Non-toxic	2	11,000	ND
9	(Canal-agriculture, landfill, golf)	100%	Non-toxic	>91	0.049	ND

ND = No data, test was not conducted.
Non-toxic = IC50 value > maximum exposure concentration (100-percent test sediment)

B3.3.1 Yeast Estrogen Screening (YES) Assay

The YES assay was conducted with two replicates per treatment level, with the exception of St. John's samples 1 and 2, which were tested with only one replicate per treatment level. After 4 days of incubation absorbants weere measured. Endpoints measured included cytotoxicity (LC_{50}, corresponding to concentrations lethal to 50 percent of the test yeast population, measured at A620) and estrogenic induction potential (corresponding to a doubling of β-galactosidase activity compared to controls). The induction potential was measured as the increase in β-galactosidase activity (A540) corrected for cell viability.

None of the test sediments elicited a cytotoxic response ($LC_{50} > 67$-percent porewater), nor did any of the test sediments produce an estrogenic induction potential ($EC_{200} > 67$-percent porewater). Positive control treatments (E2) were not cytotoxic at concentrations up to 667 pM, while the estrogenic potential (EC_{200}) was estimated at 333 pM.

Therefore, the estrogenic effect observed in the fathead minnow exposures were not replicated in the YES assay with the procedure used.

B3.3.2 72-hour Algal Growth Inhibition

Algal growth inhibition tests were conducted in accordance with the protocol, Growth Inhibition Test Using the Freshwater Alga (*Selenastrum capricornutum)*, Environment Canada, EPS 1/RM/25 (1992, amended 1997). Serial dilutions of each test sediment were performed to achieve treatments of 0.8, 1.6, 3.1, 6.3, 12.5, 25, 50, and 100-percent test sediment. Tests were conducted in 96 well microplates (220 μL in 5 wells per treatment; 3 plates per sample) with 6 day-old algal cells (*Raphidocelis subcapitata*, formerly *Selenastrum capricornutum*) at an average inoculum density of 9732 cells/mL. After 72 hours, final cell densities were expressed as a percent of the control cell densities. All test sediments, with the exception of sediment HQ20000958-3, were non-toxic to *Raphidocelis subcapitata* using the endpoint of 50-percent inhibition compared to control.

B3.3.3 Microtox

Sediment toxicity was investigated using measurements of bacterial luminescence (Microtox). Three replicates per treatment were measured for luminescence at 0, 5, and 15 minutes. Resulting 15-minute averages were compared to control response, with all samples exhibiting values greater than the controls except HQ200000958-6 (wastewater treatment plant) and HQ200000958-8 (trout farm).

B3.3.4 14-day *Hyallela azteca* Growth and Survival

Insufficient quantities of test sediment HQ20000958-4 (urban lake) were collected for the fathead minnow exposure, therefore the *Hyallela azteca* 14-day growth and survival test was substituted. The test method protocol followed Environment Canada's Test for Survival and Growth in Sediment Using the Freshwater Amphipod (*Hyalella azteca)* (1997). Five replicates of the control and treatment sediment with 10 organisms per replicate were exposed for 14 days. At test termination, mortality in the control was 16 percent (8/50), while sediment exposure in the treatment group was lethal to 98 percent of the test organisms.

B3.3.5 Sediment Total Heterotrophic Bacteria Counts

Porewater from test sediments was diluted with growth medium (factor of ten) in a 96 well microplate. Five replicates per sediment sample were analyzed, with counts conducted after three days. Scores are the lowest dilution of the sample in each replicate exhibiting growth (turbidity). From the scores in each replicate, the most probable number of total heterotrophic bacteria was calculated.

APPENDIX B4

SHORT-TERM REPRODUCTIVE STUDY

B4.1 Introduction

The purpose of this study was to perform a 21-day flow-through bioassay test using adult fathead minnows (*Pimephales promelas*). This study focused on evaluating the ecological significance of estrogenic compounds in a single representative fish test species. Adult fathead minnows were chosen as a representative test species, because there is an extensive body of scientific knowledge on the life cycle and habitat requirements since fathead minnows are used in the standard EPA bioassay tests for chronic and acute toxicity. More importantly, the short-term screening for endocrine disruption will use the fathead minnow model to evaluate potential adverse reproductive effects.

A flow-through system was designed to dose two endocrine disrupting compounds in exposures to adult fathead minnows (Pimephales *promelas).* The system was adapted from methodology outlined in the EPA manual "Methods for Measuring the Acute Toxicity of Effluents and Receiving Waters to Freshwater and Marine Organisms" (EPA/600/4-90/027F, August 1993). Solutions of the estrogen ICI 182,780[1] were used in the exposures. The exposures ran for 21-days. Egg production was monitored throughout the study. At the end of the study, blood samples were collected and analyzed for vitellogenin.

B4.2 Flow-Through Design and Materials

The flow-through had to be designed so that there was enough flow to maintain adequate water quality to support spawning minnows yet slow enough to not consume large quantities of test material. The volumes of various components of the system were designed with this in mind. The goal was to produce a system that would provide continuous levels of exposure over the course of the study.

The water used in this study for controls and exposure solutions was municipal water that was conditioned with three laboratory grade carbon cylinders to remove impurities and chlorine. This water has been used successfully over the years in a variety of aquaculture settings. A 200-gallon fiberglass head tank was used to supply the water to the system. The level of water in the head tank was maintained with a float valve. The carbon removes dissolved oxygen so the tank was vigorously aerated. The water in the head tank was also heated to maintain test temperature during the study.

---

[1] 17ß-Estradiol (E_2) and the anti-estrogen 7a-[9-[(4,4,5,5,5,-Pentafluoropentyl)sulphinyl]nonyl]-estra-1,3,5(10)-triene-3,17b-diol

A Cole-Parmer Programmable System IV Controller was used to regulate flow through the system in conjunction with a series of stainless steel solenoids and a float switch. This controller has been used successfully in field flow-through studies. The controller was set with two cycles. During the first cycle the dosing chambers were filled with water from the head tank. A float switch activated the second cycle when all of the dosing chambers were filled. The controller timed the second part of the cycle during which the contents of the dosing chambers dispensed their contents to a series of mixing chambers and then to the exposure vessels. Then the cycle started again.

Masterflex® peristaltic pumps were used to dose the test compounds into the first mixing chambers. The pumps were checked daily for any leaks. The amount of tubing used was minimized and was run straight to Teflon lined tubing. Exposure vessels were 10-gallon aquaria. At the back of the aquaria was a standpipe used to maintain the exposure volume. Test solutions entered the aquaria at the front or opposite the standpipe. The tanks were supplemental aerated with laboratory air.

The rest was primarily the delivery part of the system and was constructed of mainly glass tubing, stainless steel tubing, glass aquaria and glass solution vessels. Due to their absorptive nature regarding the test compounds, plastics were avoided as much as possible. However, the use of plastic tubing was unavoidable and kept to a minimum. Where possible Teflon-lined tubing was used. Figure B4-1 is a picture of the running system.

B4.2.1 Conditioning System and Minnows

The flow-through system was designed and built months before exposures began. At the time of completion the system was cycled with the water to be used in the study. The system was then stocked with adult male and female minnows. The ratio was two males to four females in each exposure vessel. The minnows were fed commercial frozen brine shrimp. PVC arches were used as a media for egg deposition. The minnows were allowed to acclimate for several months in the system before the exposures to E2 and ICI began. Egg production was monitored during this time to determine when the minnows were producing consistently. Minnows were swapped around during this time to establish successfully spawning populations in each of the aquaria.

B4.2.2 Study Design

The test groups were as follows: a solvent control, 0.5nM E2, 0.5nM ICI and a combination of 0.5nM E2 and 0.5nM ICI. These compounds do not readily go into aqueous solution so a solvent carrier was used. In this study tri-ethylene glycol (TEG) was used. The test solutions were prepared as follows: The dry compounds were first dissolved in absolute ethanol to make the ethanol stock. TEG is somewhat viscous so you can't easily dissolve dry compounds in it. An aliquot of the ethanol stock was dissolved in TEG to make the primary stock. A secondary stock was made from the primary to get the solvent concentration down. The secondary stock solution was pumped into the system over a 24-hour period. This solution was made fresh on a daily basis.

Figure B-4. Apparatus for 21-day flow-Through System for Fathead Minnow Exposures to Measure Biomarker and Reproductive Endpoints.

The test solutions enter the system through the mixing chambers. From here they are flushed every 15-minutes into the exposure vessels. The overall flow through the exposure vessels was set at 2.5-L/hour. The exposure vessels hold 10-L of solution. This makes the 24-hour 90-percent replacement rate 2.7. The exposures were conducted with four replicates per treatment, which gives a total of eight males and sixteen females per exposure. Two arches were in each aquarium at all times. The exposures were conducted for a period of 21-days.

During the exposure period, aquariums were checked on a daily basis. The overall condition of the adults was checked. Arches were checked for the presence of eggs. Arches with eggs were collected and isolated in bins to determine fertilization and hatching success. At the end of the study the adults were sacrificed. Wet weight and total lengths were determined. Blood was collected, processed and sent off for vitellogenin analysis. Gonadal tissue was weighed.

B4.3 Results

All data collected during the exposures are shown in Table B-14 (males), Table B-15 (females), and Table B-16 (reproductive endpoints). The endpoints included gonadosomatic index (GSI) and vitellogenin production, reproductive endpoints (spawning, egg production and fertilization). Vitellogenin and GSI measurements are illustrated on Figure B-5, and cumulative fecundity on Figure B-6.

- Male fathead minnows exposed to E2 and the E2+ICI combination produced significant amounts of vitellogenin. The GSI's were all similar, with the E2+ICI combination showing the greatest difference from the control. The reduction was not significant of the 0.05 level (see Table B-17).
- The female minnows also produced significant amounts of vitellogenin in the E2 and E2+ICI combination. The GSI value for the E2+ICI combination was significantly less than the control (Table B-15).
- Egg production was significantly reduced in the ICI and E2+ICI combination. Egg production was reduced but not significantly in the E2 test group. Fertilization was not adversely effected in any of the exposure groups (Table B-16).

Table B-14. Short-term Fathead Minnow Reproductive Study: 21-day Flow-through Study Comparing Exposures to Estrogen, Anti-estrogen and Mixture in Male Fathead Minnows

Exposure Solution (Conc.)	Sex	Length (mm)	Total Weight (g)	Gonadal Weight (g)	GSI (%)	Hematocrit (%)	Wtg. (mg V_{tg}/ mL)
Control	M	79	3.2	0.0268	0.8375	31.6	0.011
Control	M	75	5.26	0.0513	0.9753	47.7	<0.0005
Control	M	80	6.05	0.0401	0.6628	35.5	<0.0005
Control	M	73	4.05	0.0231	0.5704	35.1	<0.0005
Control	M	85	7.82	0.0889	1.137	38.2	<0.0005
Control	M	80	5.59	0.0697	1.247	40.6	<0.0005
Control	M	73	4.68	0.0523	1.118	36.9	0.012
Control	M	78	5.19	0.0431	0.8304	40.7	<0.0005
ICI	M	75	4.51	0.0215	0.4767	44	<0.0005
ICI	M	74	5.54	0.0431	0.778	44.6	<0.0005
ICI	M	79	6.78	0.0611	0.9012	50	<0.0005
ICI	M	80	4.87	0.0411	0.8439	42	<0.0005
ICI	M	80	7.81	0.1001	1.282	55.1	0.014
ICI	M	79	6.55	0.0908	1.386	52.9	<0.0005
ICI	M	74	6.03	0.0818	1.357	52.1	0.383
ICI	M	85	6.29	0.0461	0.7329	43.7	<0.0005
ICI & E2	M	81	4.66	0.0401	0.8605	32.2	20.473
ICI & E2	M	79	4.29	0.0112	0.2611	34.8	0.487
ICI & E2	M	83	7.29	0.0121	0.166	31.6	59.746
ICI & E2	M	77	4.72	0.0435	0.9216	37.5	45.103
ICI & E2	M	84	6.01	0.0728	1.211	34.6	57.904
ICI & E2	M	77	4.49	0.0417	0.9287	36.2	62.250
ICI & E2	M	73	5.15	0.0412	0.8	36.1	10.676
ICI & E2	M	78	5.19	0.0603	1.162	38.8	58.099
E2	M	86	7.65	0.1316	1.72	40.9	62.984
E2	M	77	5.1	0.0442	0.867	40.2	45.241
E2	M	84	7.52	0.0901	1.198	31.3	48.214
E2	M	79	3.29			36.8	1.083
E2	M	82	6.48	0.0462	0.713	45.5	96.664
E2	M	80	6.17	0.0651	1.055	40.6	49.601
E2	M	79	6.16	0.0671	1.089	37.3	12.538
E2	M	82	5.91	0.0617	1.044	36	5.322

Control - triethylene glycol solvent .

E2 -17β- estradiol

- Gonadal Somatic

GSI Index

ICI - 182,789(anti-estrogen) 7a-[9-[(4,4,5,5,5,-Pentafluoropentyl)sulphinyl]nonyl]-estra-1,3,5(10)-triene-3,17β-diol

ICI and E2 concentrations 0.5 nM

Table B-15. Results of 21-Day Flow-Through: Female Fathead Minnow Endpoints

Exposure Solution	Sex	Length (mm)	Total Weight (grams)	Gonadal Weight (grams)	GSI Index (%)	Hematocrit (%)	Vtg. (mg / mL)
Control	F	70	3.98	0.3019	7.585	34.7	3.239
Control	F	70	3.04	0.1908	6.276	30	4.656
Control	F	70	3.29	0.3467	10.538	28.9	2.176
Control	F	73	3.22	0.2629	8.165	38.5	7.375
Control	F	65	3.33	0.3099	9.306	40	2.360
Control	F	65	2.51	0.1706	6.797	27.5	9.606
Control	F	62	2.23	0.1378	6.179	41.2	18.681
Control	F	67	2.9	0.375	12.93	42.4	14.017
Control	F	67	3.52	0.5516	15.67	39.3	8.124
Control	F	67	3.36	0.5524	16.44	34.8	2.006
Control	F	70	3.94	0.6015	15.27	34.8	11.523
Control	F	67	3.35	0.3212	9.588	38.4	9.695
Control	F	61	2.21	0.15	6.787	50.9	4.940
Control	F	67	3.03	0.2831	9.343	34.7	1.326
Control	F	67	3.08	0.1331	4.321	33.8	4.545
ICI	F	64	2.74	0.1319	4.814	54	4.643
ICI	F	67	3.25	0.3025	9.308	41	16.504
ICI	F	63	2.37	0.2631	9.498	43.2	2.133
ICI	F	67	2.82	0.361	12.8	43.9	8.871
ICI	F	64	2.43	0.2012	8.28	36.4	5.594
ICI	F	71	3.43	0.3921	11.43	42.3	16.532
ICI	F	65	2.71	0.2432	8.974	31.4	1.724
ICI	F	69	3.36	0.33909	11.63	41.3	12.278
ICI	F	74	3.54	0.1811	5.116	42.4	6.175
ICI	F	77	4.15	0.0202	0.4843	44.4	<0.0005
ICI	F	65	2.67	0.1909	7.15	47.8	4.719
ICI	F	76	4.49	0.1528	3.403	46.1	2.879
ICI	F	69	3.04	0.2054	6.757	35.7	5.312
ICI	F	60	2.2	0.1111	5.05	34	4.528
ICI	F	70	3.93	0.2712	6.901	47.5	0.184
ICI	F	62	2.62	0.1721	6.569	37.7	4.309
ICI & E2	F	68	3.11	0.3113	10.01	40	28.010
ICI & E2	F	65	3.19	0.1044	3.26	34	79.042
ICI & E2	F	65	3.1	0.1631	5.26	36.3	76.384
ICI & E2	F	76	4.12	0.0612	1.485	43	145.838
ICI & E2	F	70	3.62	0.4924	13.6	41.9	28.371
ICI & E2	F	67	2.69	0.0108	0.4015	30.8	13.357
ICI & E2	F	70	2.86	0.2631	9.199	37.1	83.497
ICI & E2	F	67	3.14	0.4817	15.34	46.6	78.339
ICI & E2	F	62	2.42	0.0601	2.483	33.3	57.708
ICI & E2	F	68	2.96	0.1381	4.666	43.1	43.166
ICI & E2	F	68	3.16	0.1792	5.67	39.4	44.536
ICI & E2	F	64	2.08	0.0701	3.37	25.7	33.527

Exposure Solution	Sex	Length (mm)	Total Weight (grams)	Gonadal Weight (grams)	GSI Index (%)	Hematocrit (%)	Vtg. (mg / mL)
ICI & E2	F	70	3.3	0.1121	3.397	30.4	84.929
ICI & E2	F	68	2.65	0.1231	4.645	34.1	66.327
ICI & E2	F	70	2.8	0.3222	11.51	37.07	10.219
ICI & E2	F	67	2.62	0.2616	9.985	43.8	30.763
E2	F	72	3.22	0.0918	2.851	35.3	89.035
E2	F	78	4.63	0.5016	10.83	36.7	26.956
E2	F	64	2.61	0.3928	15.05	37	33.421
E2	F	69	3.25	0.4234	13.03	51.7	34.120
E2	F	67	2.64	0.2131	8.072	30.8	56.688
E2	F	64	2.49	0.1461	5.867	31.1	104.096
E2	F	67	3.13	0.1871	5.978	30.3	82.873
E2	F	62	2.34	0.2781	11.88	56.3	44.709
E2	F	64	2.76	0.4652	16.86		27.761
E2	F	64	2.48	0.1171	4.722	34.6	71.353
E2	F	64	2.82	0.4718	16.73	42.4	45.647
E2	F	67	2.69	0.288	10.71	33.8	124.437
E2	F	66	3.29	0.401	12.19	38.5	62.925
E2	F	80	4.67			44.6	10.575
E2	F	68	2.71	0.1671	6.166	35.2	38.186

ICI - 182,780

Control - triethylene glycol solvent control

ICI and E2 concentrations 0.5 nM

Table B-16. Results of 21-day Flow-Through: Male Reproductive Endpoints

CONTROL		
Tank #	Spawn #	Total #. Eggs
1	1	202
1	2	201
1	3	208
1	4	205
1	5	24
1	6	118
1	7	31
1	8	100
1	1	195
1	2	194
1	3	186
1	4	156
1	5	20
1	6	28
1	7	30
1	8	96
2	1	163
2	2	257
2	3	54
2	4	196
2	5	152
2	6	303
2	1	150
2	2	313
2	3	2
2	4	190
2	5	111
2	6	265
3	1	296
3	2	211
3	3	325
3	4	57
3	5	26
3	6	67
3	7	122
3	1	295
3	2	200
3	3	316
3	4	57

ICI		
Tank #	Spawn #	Total # Eggs
5	1	101
5	2	26
5	3	35
5	4	63
5	5	56
5	6	10
5	1	96
5	2	26
5	3	35
5	4	61
5	5	55
5	6	10
6	1	84
6	2	266
6	3	340
6	4	188
6	5	53
6	6	172
6	7	53
6	8	109
6	9	156
6	10	102
6	1	81
6	2	242
6	3	326
6	4	148
6	5	50
6	6	165
6	7	51
6	8	100
6	9	118
6	10	89
7	1	36
7	2	56
7	3	20
7	1	30
7	2	50
7	3	20
8	1	86

ICI & E2		
Tank #	Spawn #	Total # Eggs
9	1	56
9	2	123
9	3	94
9	4	57
9	1	50
9	2	96
9	3	84
9	4	50
10	1	38
10	2	306
10	3	286
10	4	28
10	5	67
10	6	10
10	1	38
10	2	126
10	3	257
10	4	27
10	5	60
10	6	0
11	1	156
11	2	285
11	3	146
11	4	30
11	1	152
11	2	279
11	3	112
11	4	18
12	1	56
12	2	286
12	3	159
12	4	36
12	5	23
12	1	53
12	2	170
12	3	122
12	4	35
12	5	5

17ß-estradiol (E2)		
Tank #	Spawn #	Total # Eggs
13	1	56
13	2	23
13	3	100
13	4	201
13	5	156
13	6	89
13	7	53
13	8	109
13	9	203
13	10	98
13	1	55
13	2	23
13	3	98
13	4	196
13	5	102
13	6	84
13	7	53
13	8	101
13	9	200
13	10	56
14	1	111
14	2	138
14	3	56
14	4	112
14	5	48
14	6	68
14	7	203
14	8	158
14	9	23
14	1	73
14	2	99
14	3	54
14	4	110
14	5	22
14	6	50
14	7	176
14	8	110
14	9	15
15	1	260

CONTROL		
Tank #	Spawn #	Total #. Eggs
3	5	26
3	6	60
3	7	101
4	1	65
4	2	496
4	3	302
4	4	147
4	5	132
4	6	141
4	7	54
4	1	60
4	2	425
4	3	300
4	4	112
4	5	115
4	6	111
4	7	53

ICI		
Tank #	Spawn #	Total # Eggs
8	2	33
8	3	208
8	4	56
8	5	96
8	6	33
8	1	70
8	2	31
8	3	156
8	4	50
8	5	95
8	6	33

ICI & E2		
Tank #	Spawn #	Total # Eggs

17ß-estradiol (E2)		
Tank #	Spawn #	Total # Eggs
15	2	110
15	3	57
15	4	86
15	1	254
15	2	108
15	3	57
15	4	76
16	1	246
16	2	26
16	3	158
16	4	265
16	5	176
16	6	207
16	1	230
16	2	25
16	3	156
16	4	233
16	5	56
16	6	156

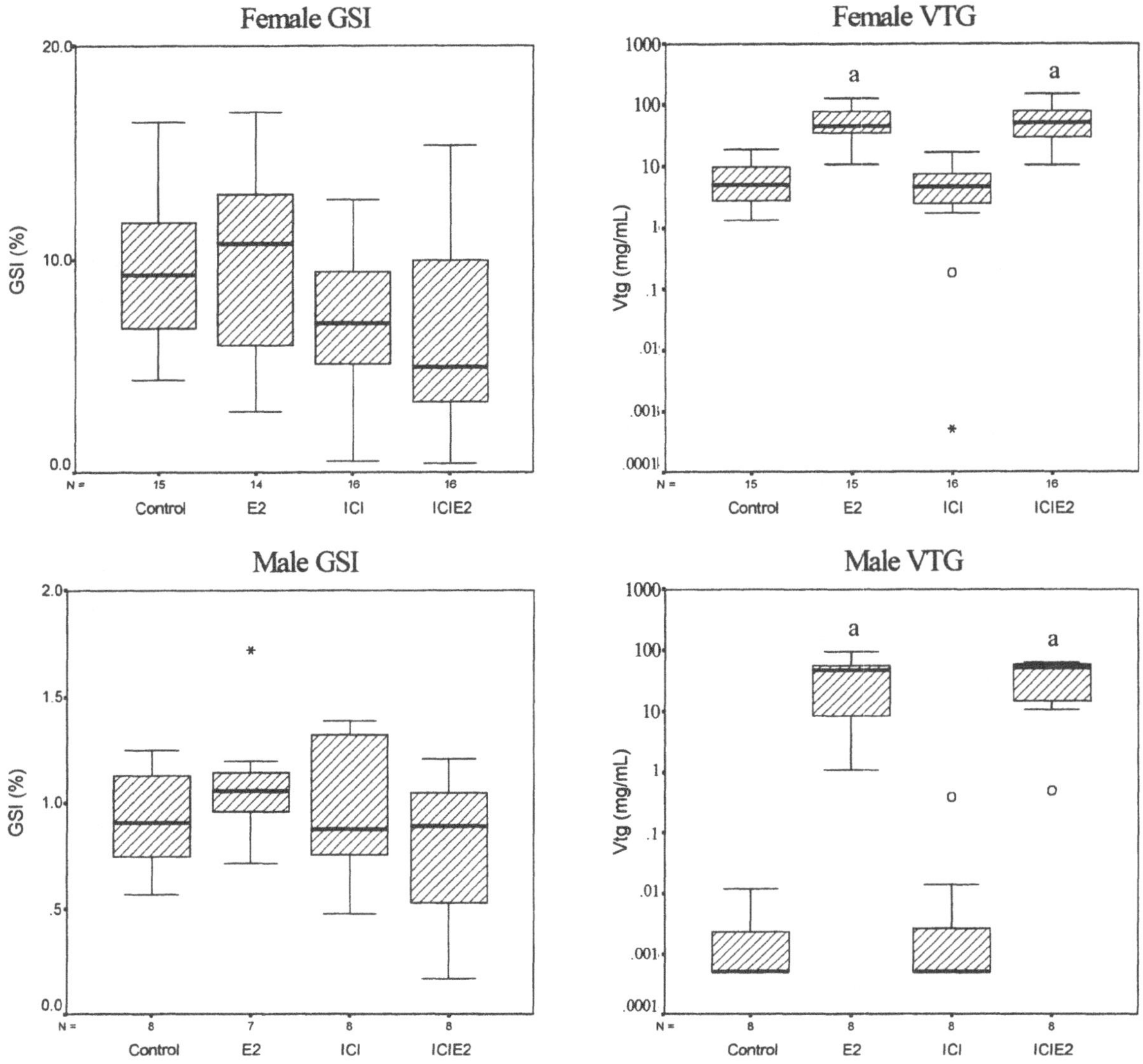

ICI - 182,780

Control - triethylene glycol solvent control

ICI and E2 concentrations 0.5 nM

Figure B-5. Short-term Fathead Minnow Reproductive Study: 21-day Flow-through Study Comparing Plasma Vitellogenin (Vtg) Levels and Gonadal Somatic Index (GSI) for Control, Estrogen (E2), Anti-estrogen (ICI) and Mixture.

Table B-17. Summary Statistics for 21-day Flow-through Study

Male Fathead Minnows

Treatment	GSI	Vitellogenin (mg/ml)
TEG, Control	0.92	0.0029
0.5 nM ICI	0.96	0.0496
0.5 nM E2	0.96	39.34*
0.5 nM ICI + E2	0.78	40.20*

Female Fathead Minnows

Treatment	GSI	Vitellogenin (mg/ml)
TEG, Control	9.67	6.95
0.5 nM ICI	7.38	6.02
0.5 nM E2	10.06	60.15*
0.5 nM ICI + E2	6.51*	56.50*

Fathead Minnows Reproduction

Treatment	# of Spawns	# of Eggs / Spawn	% Fertilization
TEG, Control	28	41.6	85.6
0.5 nM ICI	25	24.4*	89.5
0.5 nM E2	29	31.0	84.2
0.5 nM ICI + E2	19	29.5*	77.3

* Denotes a significant difference between the control and the sample for this endpoint (alpha = 0.05).

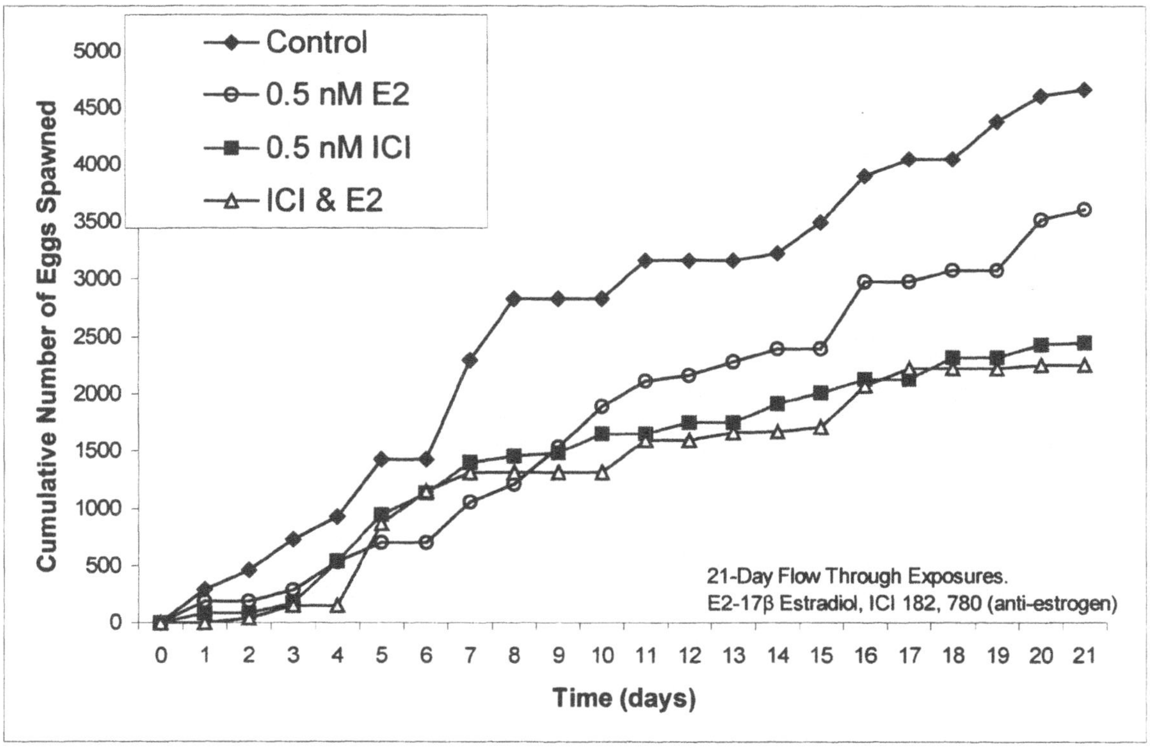

Figure B-6. Effects of E2 and ICI on the Cumulative Fecundity of Fathead Minnows.

APPENDIX B5

mRNA MEASUREMENTS FROM STATIC FATHEAD MINNOW EXPOSURES

This experiment was set up to measure the effects of exposing female fathead minnows to a strong estrogen (E2), a strong anti-estrogen (ICI 182,780), a strong androgen (5-androstan-17-methyl-17-ol-3-one) and a potent anti-androgen (flutamide) on vitellogenin and estrogen receptor expression, both at mRNA and protein levels. Since there is a certain extent of cross-talk between receptors, these exposures were performed to evaluate if the expression of vitellogenin mRNA in reproductively active females could be reduced. The extent of change in laboratory studies could help interpret the results from the field exposures.

B5.1. Exposure Conditions

Female, adult fathead minnows were exposed to 4 different compounds at 1nM static concentrations; E2, ICI #182-780 (anti-estrogen), 5-androstan-17-methyl-17-ol-3-one (androgen), flutamide (anti-androgen), and solvent control, tri-ethylene glycol (TEG). Fish were exposed to each compound in separate tanks (20 fish per tank) and 4 fish were collected at each time point; 2-, 4-, 6-, and 24-hours; and 7 days after exposure. Samples processed at each time point included: blood from tail for plasma collection to measure plasma vitellogenin protein concentrations and interior cavity of the mid torso, frozen in liquid nitrogen, for RNA analysis. Tissue samples were stored at -80°C until processed for RNA. Plasma vitellogenin protein concentrations were determined by the fathead minnow vitellogenin ELISA. Unknown sample concentrations were calculated from a standard curve based on the known amount of fathead minnow vitellogenin protein, as described in Section A-1.

B5.2 Results and Discussion for In vivo Exposure Experiment

Effects at the level of plasma vitellogenin were only seen after 24 hours of treatment. (See Table B-17.) Only E2 increased plasma levels of vitellogenin significantly greater than controls in a time dependent manner starting at 27.75 mg/ml after 24 hours and increasing to 133.23 mg/ml after 7 days of exposure to a static concentration (Fig B-7). After 7 days of treatment it appears that flutamide may increase plasma vitellogenin over background. But, because only 4 fish were included at every time point and fathead minnows are not perfectly synchronized, it is difficult to evaluate whether this increase is significant. This should be repeated with more concentrations of flutamide to see if there is a dose response. It appears that plasma levels are reduced by ICI 182,780, but again because of the low number of fish, it is impossible to determine if this is significant.

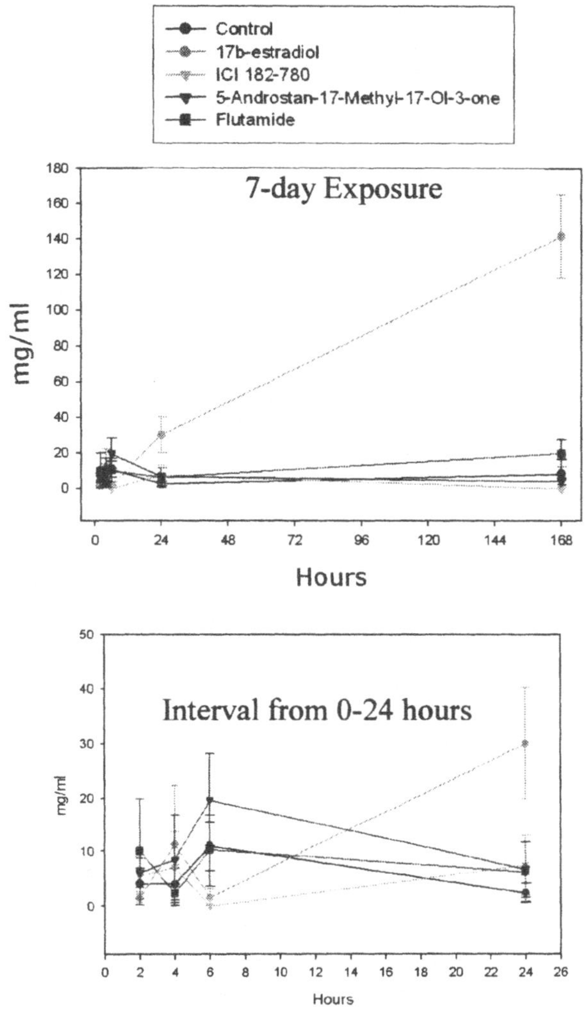

Figure B-7. Plasma Vitellogenin Protein Concentration for Static Exposures to Hormone Active Compounds

Using mRNA as the biomarker, we see some differential gene expression patterns for ERα and vitellogenin (Figure B-8). Both were induced by E2, but ERα appears to be induced more quickly than vitellogenin mRNA. ERα mRNA induction peaks at 24 hours and is down after 7 days, while vitellogenin mRNA continues to increase in this time frame. Since these were static exposures, we expected that E2 would be metabolized during this time frame, as is suggested by the response of the ERα mRNA. There are some suggestions in the literature that vitellogenin mRNA is stabilized by a factor that itself maybe induced by E2. Thus, vitellogenin mRNA continues to accumulate in the seven-day time frame, suggesting that the plasma vitellogenin seen at 7 days was there due to continued synthesis from mRNAs. Flutamide appears to have a positive effect on vitellogenin mRNA synthesis by 7 days, increasing vitellogenin mRNA levels from 0 copies/ug total RNA (control levels) to 1×10^9 copies after 7 days. This suggests that the increase seen in plasma vitellogenin is really due to *de novo* synthesis of vitellogenin. Exposure to the ICI compound appears to reduce the level of vitellogenin mRNA copy number below control levels after 7 days. The effect of this anti-estrogen on E2-induced transcription should be verified in a dose dependent manner.

Table B-18. Vitellogenin Concentrations - ELISA Data from Fathead Minnow Plasma

	Control	E2	ICI	Androgen	Flutamide
2 hours	4.916	2.843	0.077	5.162	39.302
	< 0.0005	<0.0005	14.193	12.869	0.476
	11.104	3.844	<0.0005	0.002	0.436
	< 0.0005	0.014	8.162	6.394	< 0.0005
4 hours	0.209	44.663	1.677	< 0.0005	0.011
	< 0.0005	0.157	0.004	0.412	7.171
	12.244	<0.0005	26.820	< 0.0005	0.048
	3.689	<0.0005	0.001	33.368	2.348
6 hours	10.554	6.459	<0.0005	27.466	0.161
	< 0.0005	0.001	0.059	38.778	13.176
	22.104	0.002	0.240	<0.0005	0.0005
	11.248	0.009	0.063	12.179	27.606
24 hours	2.582		0.010	< 0.0005	<0.0005
	0.006	46.486	0.0005	21.507	
	7.211	36.363	6.574	0.074	< 0.0005
	< 0.0005	7.235	23.412	5.173	18.720
7 days	33.351	206.240	0.003	0.169	21.697
	< 0.0005	139.969	<0.0005	7.591	0.771
	< 0.0005	94.830	<0.0005	7.279	19.692
	<0.0005	125.232		1.083	37.357

B5.3 RNA Isolation and Reverse Transcription

Total RNA was used to measure the amount of mRNA for both the ERα and vitellogenin genes in each fish following exposure. To isolate total RNA, the entire gut cavity from each fish (approximately 200-500mg) was homogenized in 1ml of RNA STAT-60 (TEL TEST Inc.) as described above.

Good quality, intact RNA was recovered from each sample as evidenced by RNA gel electrophoresis and spectrophotometer readings (Figure A-6). 1 ug of total RNA was reverse-transcribed, for each sample, to get corresponding cDNA using random hexamers. Each cDNA sample was run on the Taqman to measure ERα and vitellogenin mRNA, and 18S rRNA for normalization. mRNA values were normalized, calculated and graphed.

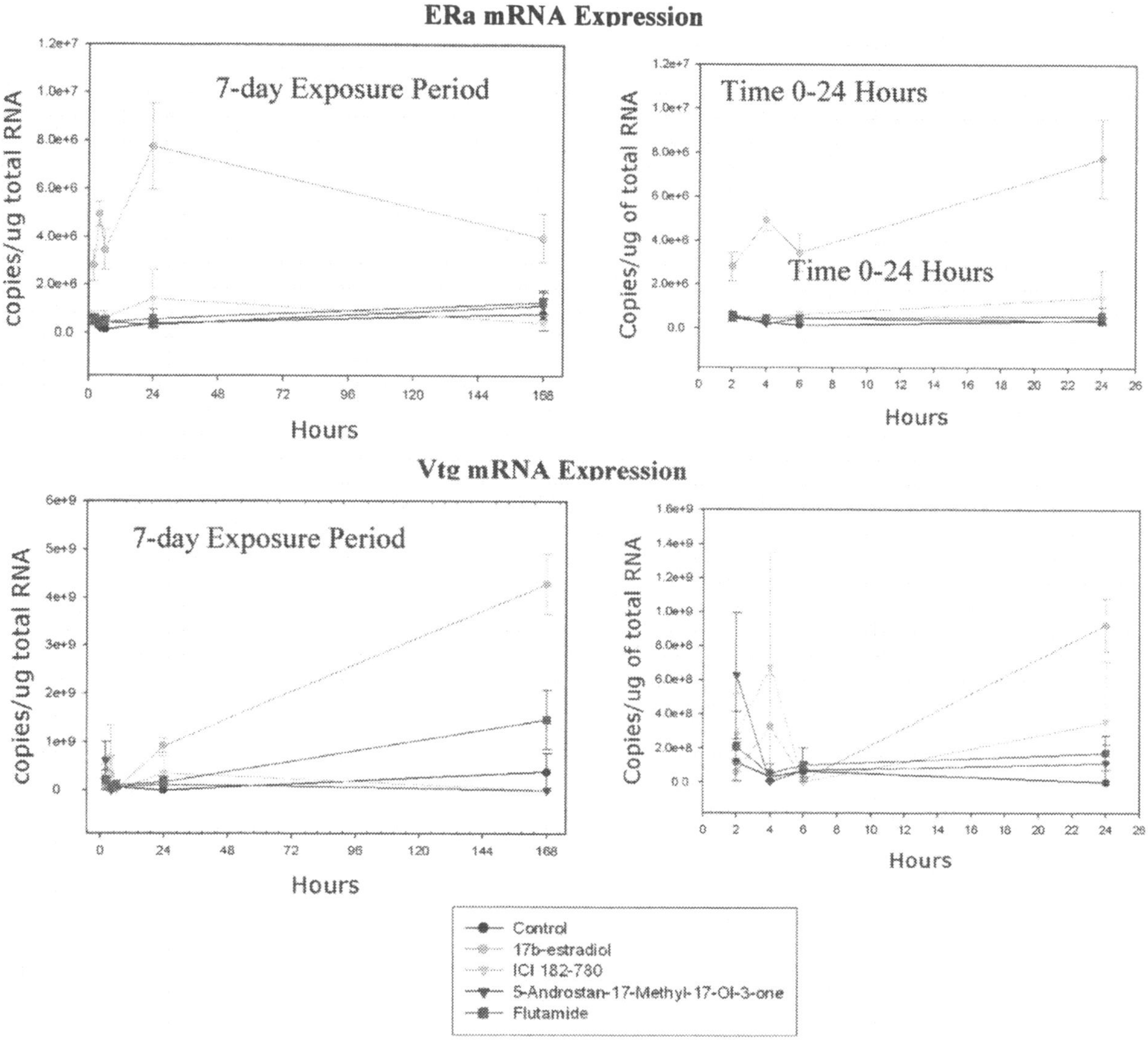

Figure B-8. mRNA Expression for Static Exposures to Hormone Active Compounds.

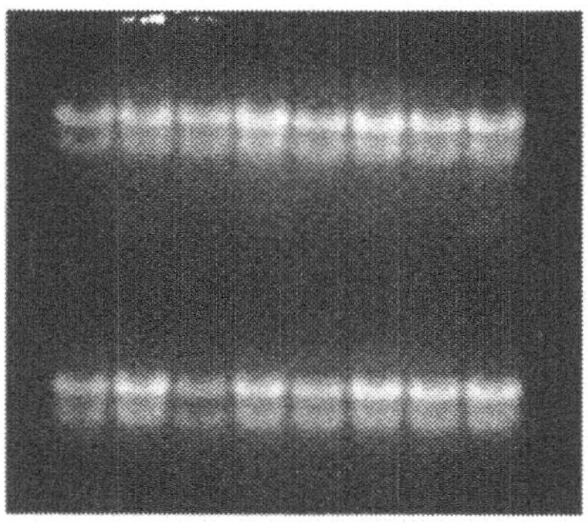

Figure B-9. Total RNA Isolation Check Gel. Presence of two intact bands by electrophoresis, representing the large and small ribosomes, indicates quality RNA (non-degraded).

APPENDIX C

FIELD STUDIES

C.1 Overview of the Field Study

This appendix documents the selection of the watersheds and stations used for this field sampling effort, the interpretation of these results, the field sampling protocol and sampling analyses, and results. The interpretation of these results, and implications for design and implementation of field studies to detect potential endocrine mediated effects are presented in Section 4, Field Studies. The metro-Atlanta region was selected for the study area due to the large urban/mixed use growth and proximity to the personnel performing the sampling.

C.1.1 Study Area

In the development of the initial work plan, the objective was to select sampling locations representing distinct land uses (forested, agricultural, urban/mixed use) where bass could be collected to allow comparison with ongoing University of Florida field studies on largemouth bass. The St. Johns River in Florida, which was originally proposed, was rejected because it transitions to brackish water as it discharges into the Atlantic Ocean. Therefore, by extending the sampling up to the Atlanta region provided an extensive network of freshwater streams with bass. The highly urbanized areas, near Atlanta have been studied by others and these provide a context for the limited data collection/biomarker measurements supported in this study.

Consequently, two rivers in central and north Georgia were selected for the biomarker study (Figure C-1). These include the Chattahoochee River, which flows through metro Atlanta, and the Conasauga River, which is located north of Atlanta near the Tennessee border.

The Chattahoochee River was chosen due to a number of wastewater treatment facilities that discharge into the Chattahoochee River. This river has a relatively small watershed compared to the size city it feeds and the land use is pre-dominantly urban/mixed use within the basin. The Conasauga River was chosen because it progresses from mostly forested land use to agricultural to urban/mixed use.

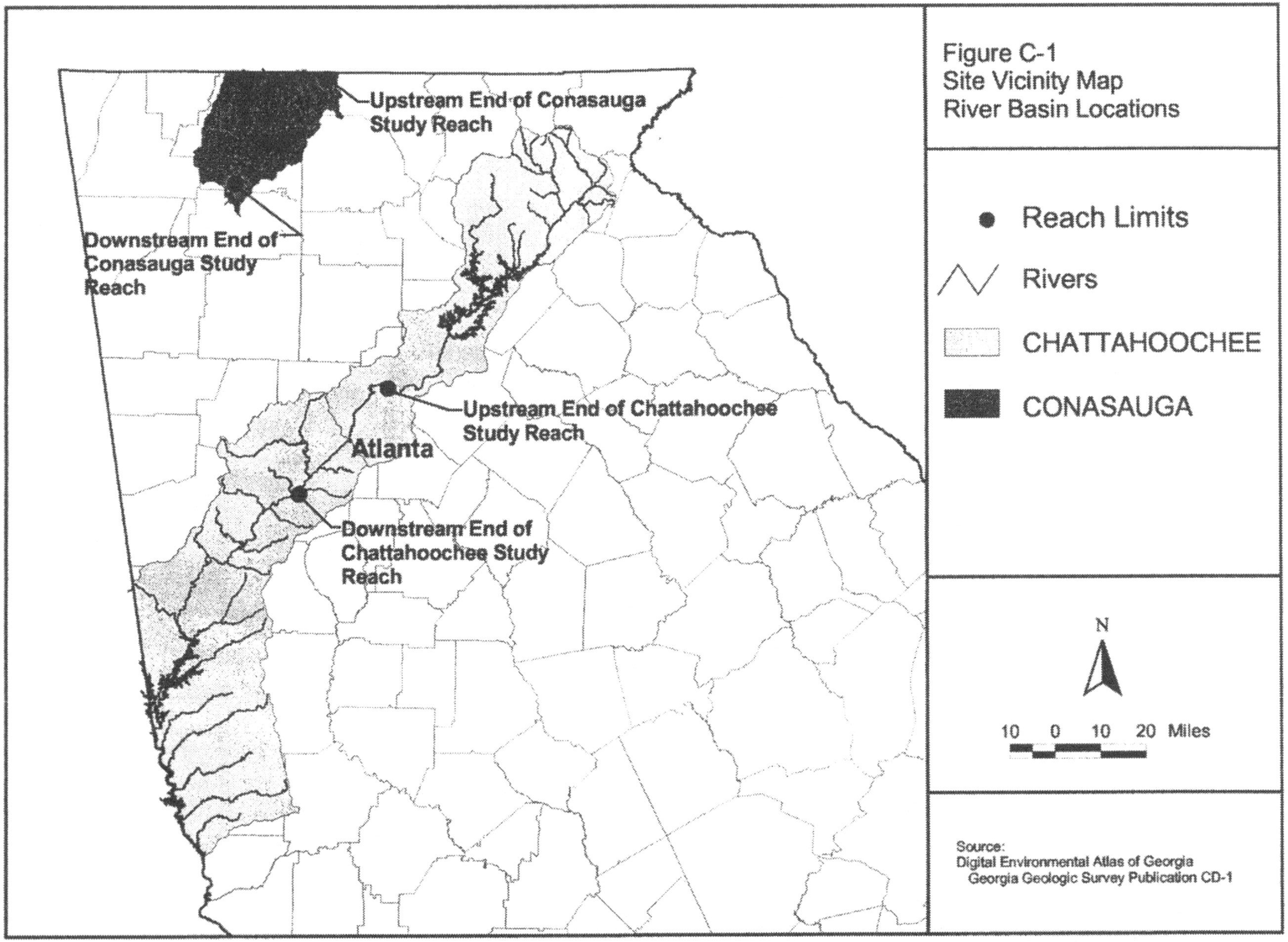

Figure C-1
Site Vicinity Map
River Basin Locations

APPENDIX C1

CHATTAHOOCHEE RIVER

C1.1 Overview of Chattahoochee River Field Study

The Chattahoochee River was only sampled within the urban/mixed use metro-Atlanta area. The Chattahoochee River is dammed in north metro-Atlanta, which forms Lake Sidney Lanier, a major drinking water supply and hydropower reservoir for the area. The dam releases are from the hypolimnion, and therefore the river downstream from the reservoir is too cold for bass to survive until the river reaches Bull Sluice Lake, a shallow non-detention time lake, approximately 35 miles downstream from Lake Lanier (Freeman, 2001) (Figure C-1). The most upstream fish sampling location is in Bull Sluice Lake.

The Chattahoochee River begins as a cold-water mountain stream in the Blue Ridge north of Helen, Georgia. The river flows 430 miles to the confluence with the Flint River in South Georgia to form Lake Seminole and then the Apalachicola River, which then flows through Florida and into the Gulf of Mexico. The river drains 8,770 square miles and is the most heavily used water resource in Georgia (drinking water, irrigation, industry, etc.) (United States Geological Survey, 2000). Metro Atlanta is located approximately 100 miles downstream from the headwaters. There are thirteen dams on the main stem of the river, which have dramatically altered the minimum and maximum flows in the system. Mean annual discharge upstream of Roswell, Georgia (USGS Station 02335450) is 1301 cfs from 1980 to present (United States Geological Survey, 2002a), and mean annual discharge near Fairburn, Georgia (USGS Station 02337170) is 3912 cfs from 1967 to present (United States Geological Survey, 2002b).

Major land uses within the watershed vary depending on the location along the river. Upstream of Lake Lanier and Atlanta is mostly forested and rural towns. Downstream of Lake Lanier and through Atlanta is heavily urban/mixed use with dense residential development and industry (in pockets south of the confluence with Peachtree Creek). Downstream of Atlanta, agriculture and rural towns dominate until Columbus, Georgia (downstream of the study area) (Atlanta Regional Commission, 1998 and Alhadeff et al, 2000). Surface water withdrawals are primarily for drinking water and industrial use.

The Chattahoochee River receives multiple point source discharges, including domestic and industrial wastewater inputs. Within the portion of river studied (Figures C-2A and C-2B), the cumulative current permitted discharge from the ten wastewater treatment facilities within the basin is 347 million gallons per day (Mgal/d) (Burke, 2000).

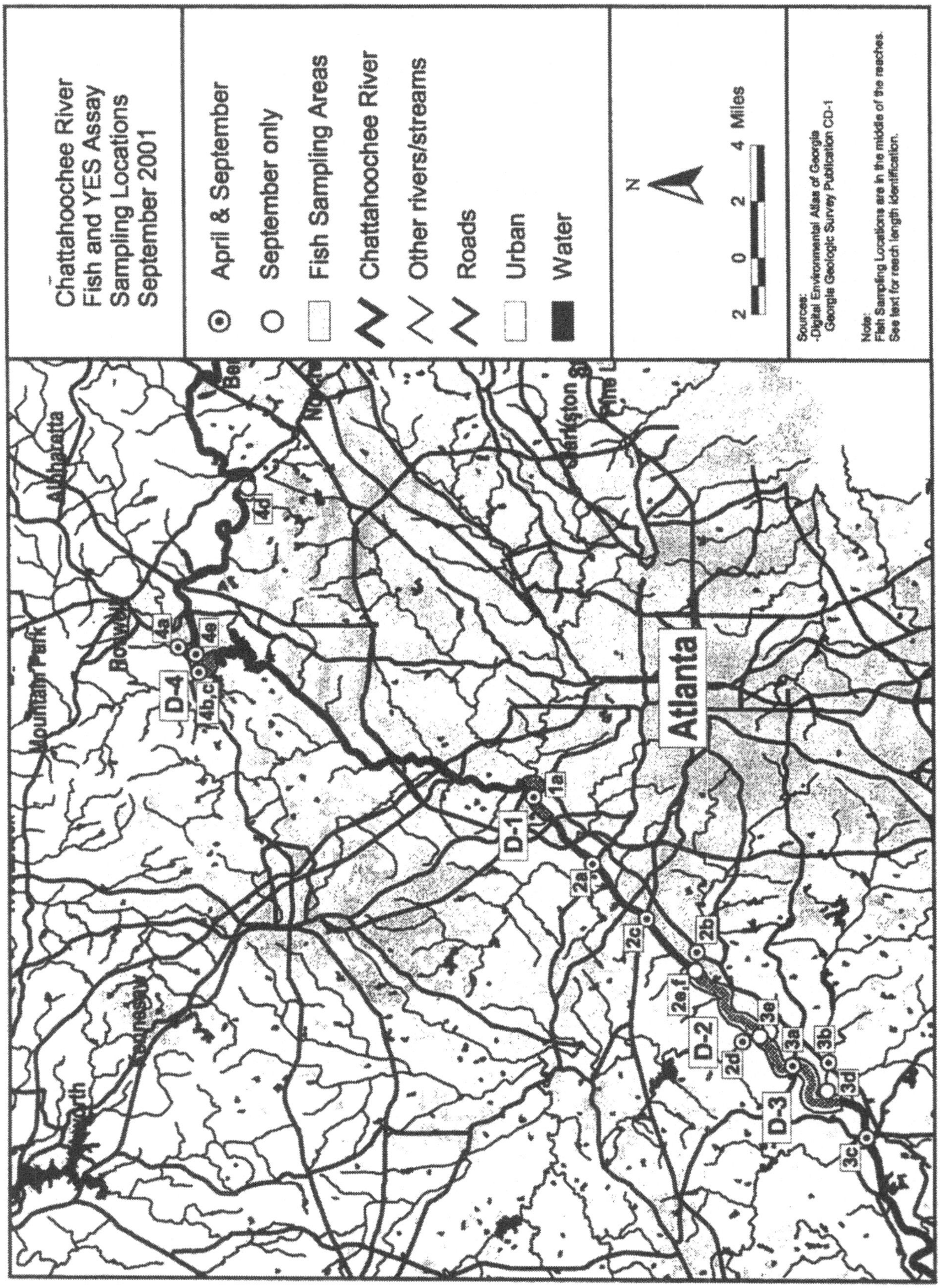

Water Environment Research Foundation

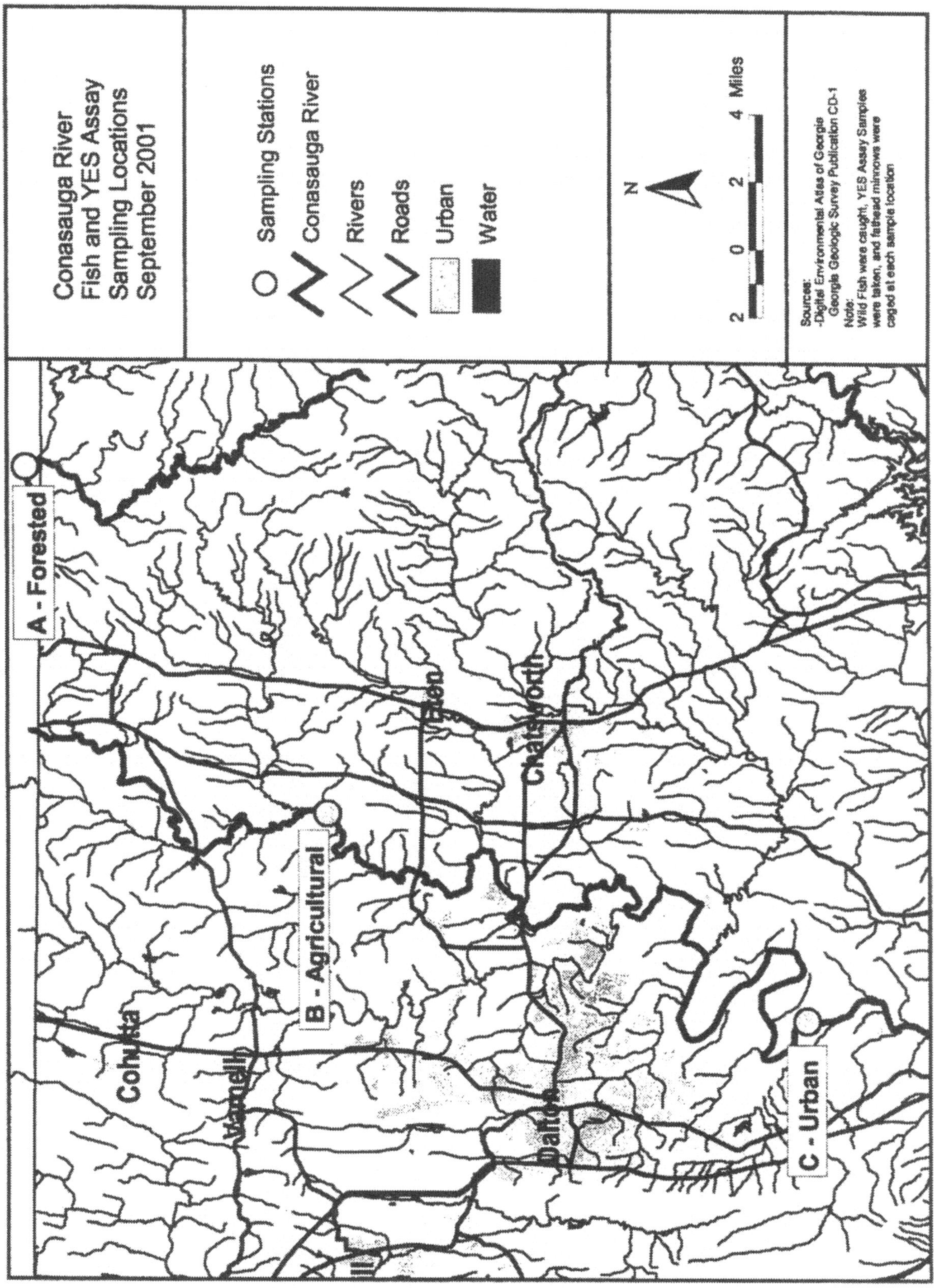

Figure C-1B. Chattahoochee River Fish and YES Assay Sampling Locations, Sept. 2001

Several regional watershed assessments have been developed in the Atlanta region over the past several years. Assessments or portions of assessments reviewed included:

- Watershed Management Guidance Document. Metro Atlanta Urban Watersheds Initiative (City of Atlanta, 1998);
- West Watershed Impacts Assessment Report (City of Atlanta, 1998);
- Gwinnett County Watershed Protection Plan (Gwinnett County, 2000);
- Fulton County Sandy Springs Watershed Assessment. Summary of Watershed Characterization Activities (Fulton County, 2000);
- Fulton County – Big Creek Water Resources Management Plan. Watershed Characterization Findings to Date. Fact Sheet (Fulton County, 2000);
- Characterization Summary for the Johns Creek Study Area (Fulton County, 2001);
- Gothard's Creek Watershed Assessment Study: Volume I – Watershed Characterization and Volume II – Watershed Management Plan. (Douglasville-Douglas County Water and Sewer Authority, 1999);
- Final Report on the Anneewakee Creek Watershed Assessment Study. (Douglasville-Douglas County Water and Sewer Authority, 2000); and
- Big Creek Watershed Study. Master Plan (Atlanta Regional Commission, 2000)

The Georgia Environmental Protection Division (GaEPD) requires any party that is requesting an increase in wastewater discharge under a NPDES permit to develop a watershed assessment for the service area. The rationale is that if new sewer capacity is built, new development will occur, and the potential for increased degradation to the streams increases. The watershed assessment is an attempt to plan for the increased stress on the streams. From a review of these watershed assessments, the major problems in the metro Atlanta region that drains into the area picked for this study on the Chattahoochee River are fecal coliform and sediment. The likely causes are urban/mixed use growth, which causes increased erosion during construction, and larger impervious areas. Increasing the percentage of impervious area in a basin increases the peak flows from storm events and carries more pollutants to the streams. In addition, metro-Atlanta has pockets of combined sewers, which periodically overflow during large storm events, and leaking sanitary sewer systems throughout the watershed. From data provided in the studies listed above, the overall water quality of the Chattahoochee River basin is shown in Table C-1.

Table C-1. Chattahoochee River Water Quality

Watershed Study	Anneewakee	Gothards	Johns Creek	Sandy Springs[3]	Metro Atlanta Urban Watershed Initiative	Big Creek Watershed Study
Area of Study	Douglas County	Douglas County	Fulton County	Fulton County	Fulton County - Peachtree Creek	Fulton County
Drains Into	Just downstream of Station 3	Chattahoochee at divide between Stations 2 and 3	Approx. 10 miles upstream of Station 4	Near Station 4	Just upstream of Station 1	Just upstream of Station 4
Sampling Period	1998	1998	1999	1999	1996-1997	1994-1996
Water Temp. (C)	22.1	24.2	-	-	-	-
BOD (mg/L)	<4	3	-	-	10.2[5]	1.8
Conductivity (uhmos/cm)	95.7	146.7[2]	-	-	-	-
DO (mg/L)	7.5	5.1	-	-	7.3	-
Ammonia (mg/L)	<0.2	0.1	-	-	0.5[6]	0.07
Nitrite/Nitrate (mg/L)	0.87	-	-	-	0.45	0.7
Nitrite (mg/L)	-	0.026	-	-	-	-
Nitrate (mg/L)	-	0.5	-	-	-	-
TKN	-	0.24	-	-	0.9[7]	0.48
Total Nitrogen (mg/L)	-	-	1.7	-	-	-
Total Phosphorus (mg/L)	0.07	0.02	0.17	0.12	0.3[8]	0.27
pH	6.6	6.9	-	-	-	-
TSS (mg/L)	7.9	6.6	111	109[3]	144[9]	82.7
Turbidity NTU	9.5	15	-	-	16.4[10]	-
Fecal Coliform[1]	209-1737	100-498	-	-	620-23,000	397-6096
Zinc (ug/L)	8.2	50	-	-	97.2[11]	-
Copper (ug/L)	0.85	-	-	-	21[12]	-
Lead (ug/L)	0.39	-	-	-	18.8[13]	-
Cadmium (ug/L)	-	-	-	-	2.0[14]	-

The United States Geological Survey (USGS) conducts several studies in the Chattahoochee River basin including the NAWQA, National Reconnaissance Program, among others. The objectives of NAWQA, which began in 1991, are to describe the water quality conditions of the streams and ground water of identified watersheds and to understand how this water quality is changing over time and how natural and human factors play a role. The Chattahoochee River is studied as part of the Apalachicola-Chattahoochee-Flint River Basin study where 147 surface water sites and 161 groundwater sites have been established for water quality, water quantity, sediment, and biological sampling (Frick et al, 1995). Overall, the water quality concerns mostly originate in urban/mixed use areas, primarily the metro Atlanta region (United States Geologic Survey, 2000).

The National Reconnaissance Program began in 1999 to develop a baseline database on the occurrence of emerging contaminants in water resources. The study incorporates 36 states with stations in 142 streams, 55 wells, and 7 effluent discharges. Big Creek, (our Station #4 on the Chattahoochee River), is part of the uses program. The program targets 94 chemicals, including 15 reproductive and steroidal hormones (United States Geological Survey, 2001). Preliminary results from the Big Creek study indicated the detection of 16 pharmaceuticals in wastewater streams and 3 pharmaceuticals in finished drinking water (Moll et al, 2001). Not all 94 chemicals were tested for in the Big Creek system. The steroidal hormones tests included 17α-Estradiol, cholesterol, and 3b-Coprostanol (none were found in the samples) (Frick, 2001). The final report for this research has not been published by the USGS, so data cannot be presented at this time.

A large percentage of the Chattahoochee River and tributaries within the region of this study (downstream of Lake Lanier to most downstream study reach near Anneewakee Creek) are identified on the State 303(d)/305(b) list (Georgia Departmental of Natural Resources Environmental Protection Division, 2000) for water quality impairment. Approximately 118 miles of stream or 31 percent of the stream miles listed on the 305(b)/303(d) list are partially supporting their designated use for constituents including fecal coliform, lead, cadmium, copper, dissolved oxygen, and fish consumption guidance. The evaluated cause(s), as described by the GaEPD, are urban/mixed use runoff, combined sewer overflow, and nonpoint source pollution. Approximately 223 miles or 60 percent of the stream miles listed on the 305(b)/303(d) list do not support their designated use for constituents including fecal coliform, lead, copper, zinc, toxicity, dissolved oxygen, fish consumption guidance, and temperature (in the Chattahoochee River). The evaluated cause(s), as described by the GaEPD, are urban/mixed use runoff, combined sewer overflow, industrial facilities, residuals from industrial facilities, nonpoint source pollution, municipal facility, and dam release (from Lake Lanier). The GaEPD is in the process of evaluating all streams on the 303(d) list.

APPENDIX C2

Conasauga River

C2.1 Overview of the Conasauga River Field Study

The Chattahoochee River provided a number of sampling locations representing urban/mixed use locations, however, alternate locations were identified in the Conasauga River that included other land uses. The Conasauga River was selected for its linear land use progression. The headwaters are located in mostly the Cohutta Wilderness Area. The river then flows through a heavy agriculture section and finally reaches the city of Dalton (Figures C-3A and C2-3B). Potential sources include urban/mixed use runoff, septic systems, permitted point discharges, and pesticides/herbicides/animal waste from agricultural operations.

The Conasauga River flows in a northerly direction for approximately 13 miles from its beginning near Blue Ridge, Georgia. It then flows in a westerly direction 13 miles where it bends, after emerging from the mountains within the Chattahoochee National Forest, and flows in a southerly direction for 62 miles. There it joins the Coosawattee River near Resaca, Georgia, and forms the Oostanaula River. The entire system is part of the Coosa River valley that flows into Alabama (United States Geological Survey, 1981).

The Conasauga River drains an area of 727 square miles (465,280 acres). About 127 square miles (81,280 acres) are in Tennessee and the remaining area (384,000 acres) is in Georgia. Mean annual discharge at Eton, Georgia (USGS Station 02384500), just downstream of the agriculture station, is 856 cfs from 1981 to present (United States Geological Survey, 2002c), and mean annual discharge at Tifton, Georgia (USGS Station 0238700), just downstream of the urban/mixed use station, is 1216 cfs from 1936 to 2000 (United States Geological Survey, 2002d). The largest tributaries are Coahulla Creek with a drainage area of 178 square miles and Jacks River, which drains 88 square miles (United States Geological Survey, 1981).

Major land uses in the watershed include undeveloped forest in the headwaters, agriculture and small rural towns in the mid-section, and urban/mixed use areas (Chatsworth and Dalton) further downstream. The main industry in the watershed is carpet manufacturing. Surface water withdrawals are primarily for drinking water and carpet manufacturing.

Due to there being federally protected species in the river (three fish and nine mussels), several studies (some conducted by Golder Associates Inc. for Dalton Utilities) have been conducted within the watershed. The studies, geared toward studying the biology of the river, suggest that the water quality concerns are primarily from nonpoint source runoff from agricultural operations and development. In addition, bank erosion is evident throughout the lower reaches of the river. Overall, the biotic diversity is high and the river is extremely productive.

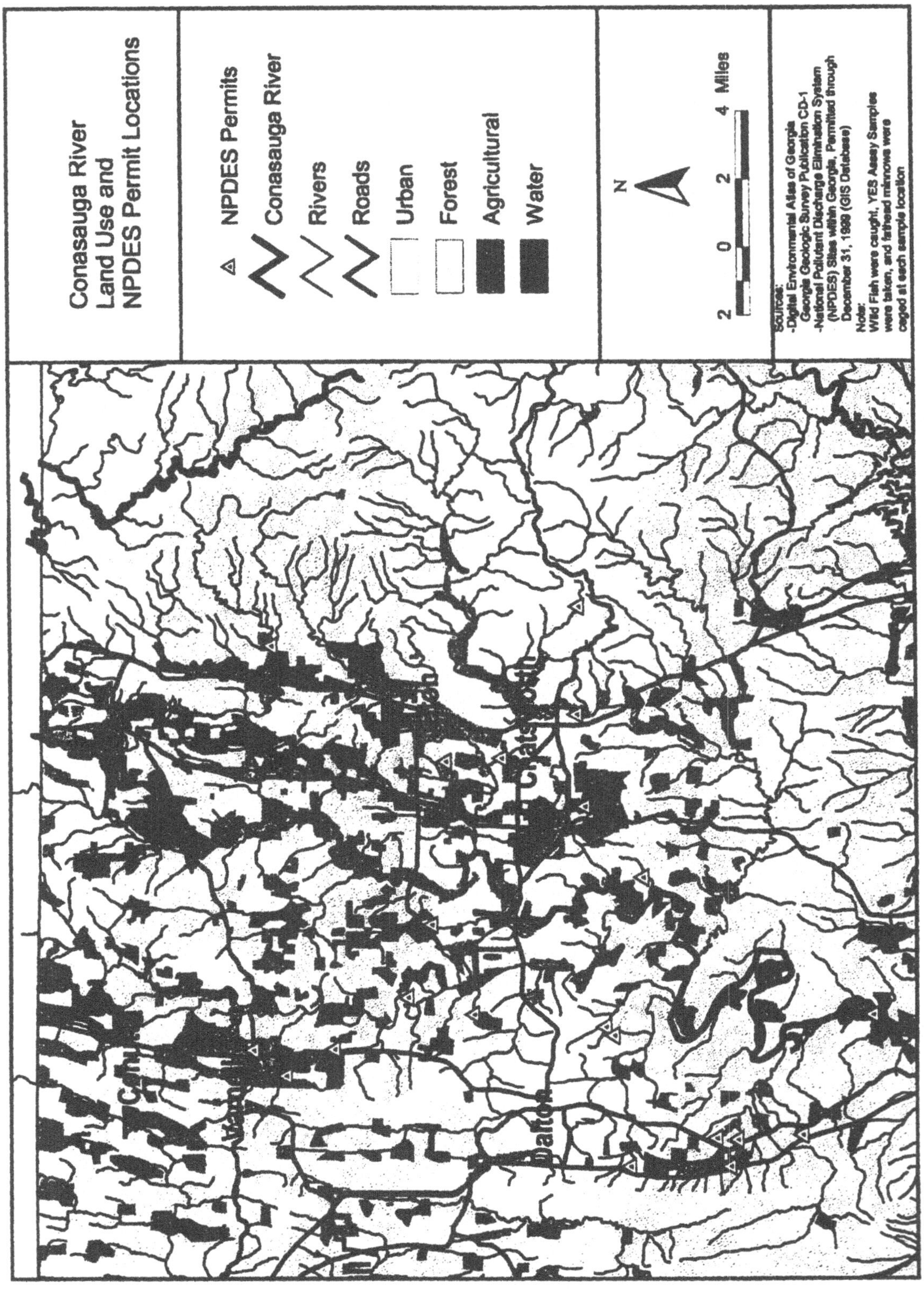

Figure C-2A. Conasauga River Land Use and NPDES Permit Locations.

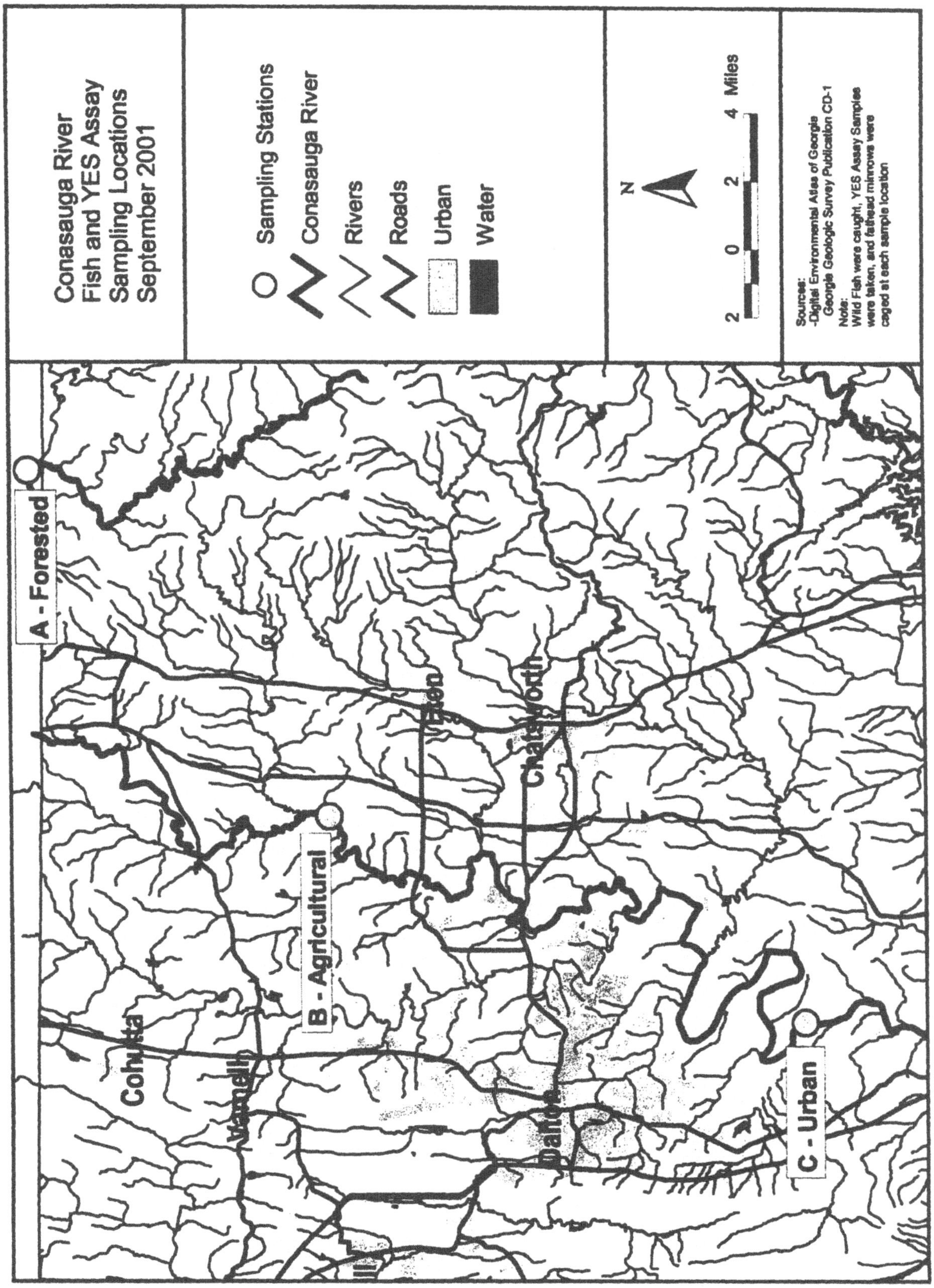

Figure C-2B. Conasauga River Fish and YES Assay Sampling Locations, Sept. 2001.

From data collected by Golder Associates in the studies referred to in the paragraph above, the overall water quality of the Conasauga River in a reach in the middle of the overall study area is shown in Table C-2 and Table C-3.

Table C-2. Conasauga Water Quality Data Table

	Season*	
Parameter	Spring	Fall
Air Temperature (C)	24.9	18
Water Temperature (C)	20.7	16.3
Biologicaal Oxygen Demand (mg/L)	0.53	1.05
Conductivity (umhos/cm)	100.1	115.8
Dissolved Oxygen (mg/L)	8.27	8.75
Ammonia (ug/L)	38.7	0
Nitrate (ug/L)	299.1	145.3
Total Phosphorus (ug/L)	50	25.8
pH	7.4	7.5
Total Suspended Solids (mg/L)	22.8	5.1
Turbidity (NTU)	15.5	9.2

* Spring samples were taken in either May or June from 1995 - 1997.

A Source Water Assessment Plan (SWAP) was completed earlier in 2001 to determine the potential risks to the regional drinking water supply for the Conasauga River basin. The only concern was the potential for spills from the interstate and railroad corridors. Otherwise, the risk to polluting the drinking water (Conasauga River) was considered low (Brown and Caldwell, Golder Associates Inc., Jordan, Jones and Goulding, 2000).

The Conasauga River and tributaries within the watershed are identified on the State 305(b)/303(d) list (Georgia Department of Natural Resources Environmental Protection Division, 2000) and show impairment for constituents including metals and fecal coliform. The evaluated cause(s), as described by the GaEPD, are urban/mixed use runoff, nonpoint source pollution, and municipal facilities. The GaEPD is in the process of evaluating all streams on the 303(d) list.

Table C-3. Water Quality Impairment in the Conasauga River Basin Within the Region of this Study (Georgia Department of Natural Resources Environmental Protection Division, 2000)

Partially Supporting Designated Uses

Basin/Stream	Location	Water Use/ Classification	Criterion Violated	Evaluated Cause(s)
Coahulla Creek	Below 728 Road to Mill Creek	Fishing	Fecal Coliform	Urban Runoff
Holly Creek	Downstream Chatsworth WPCP (Murray County)	Fishing	Cadmium	Nonpoint Source Pollution
Holly Creek	Rock Creek to Conasauga River (Murray County)	Fishing	Fecal Coliform	Nonpoint Source Pollution
Jacks River	West/South Forks to Stateline (Fannin County)	Wild and Scenic	Zinc	Nonpoint Source Pollution
Swamp Creek	Headwaters to confluence with Conasauga River	Fishing	Fish Consumption Guidance	Nonpoint Source Pollution

Not Supporting Designated Uses

Basin/Stream	Location	Water Use/ Classification	Criterion Violated	Evaluated Cause(s)
Conasauga River	Holly Creek to Oostanaula River (Murray/Gordon County)	Fishing	Fecal Coliform	Municipal Facility, Urban Runoff

The GaEPD determines the level of impairment in each water body in Georgia and also determines the source of impairment. The GaEPD or USGS are the agencies that collect the data used to make this determination. The amount and type of data collected is variable and sometimes very little.

APPENDIX C3

FIELD SAMPLING SUMMARY

C3.1 Summary of Field Sampling Stations

Sampling stations were selected based on the following criteria:

- Boat ramp accessibility for launching the electrofishing boat,
- Separate reaches with distinct point and non-point discharges, and
- Knowledge of being able to catch the targeted amount of fish, 20 per station or reach.

In addition, locations, which would most likely not attract vandals, were selected on the Conasauga River to decrease the likelihood of losing a fathead minnow trap.

Field surveys were conducted from April 30 – May 10, 2001 (Spring survey) and from September 17 – 28, 2001 (Fall survey) on the Chattahoochee and Conasauga Rivers. Seven stations were sampled, four in the Chattahoochee River and three in the Conasauga River (Figures C-2A, C-2B, C-3A, and C-3B):

- Forested Station (Conasauga River) – in the Cohutta wilderness, furthest upstream station,
- Agricultural Station (Conasauga River) – surrounded by primarily row crop and cattle production land uses, upstream of urban/mixed use land use,
- Urban Station (Conasauga River) – mixed use area, downstream of Dalton and Chatsworth,
- Station #1 (Chattahoochee River) – downstream of the confluence of Peachtree Creek with the Chattahoochee River,
- Station #2 (Chattahoochee River) – downstream of the confluence of Utoy Creek and upstream of the confluence of Sweetwater Creek,
- Station #3 (Chattahoochee River) – downstream of Sweetwater Creek to the confluence of Camp Creek, and
- Station #4 (Chattahoochee River) – downstream of confluence of Big Creek to the confluence of Willeo Creek

Table C-4 provides a summary of the data collected during the spring and fall sampling events at all seven sampling locations.

Table C-4. Summary of Field Sampling During Spring and Fall 2001 Field Sampling Events

Sampling Event	Station	ID	Caged Fathead Minnows (#Survived/#Processed/ Total Caged)	Wild Fish Collection Method	Species Collected (number)	YES Assay Samples*	Chemical Testing
Spring	Forested Station	A	Yes (39/20/40)	Angling	Redeye Bass (19)	1	
					Spotted Bass (1)		
	Agricultural Station	B	Yes (30/20/40)	Angling	Spotted Bass (19)	2	
					Largemouth Bass (1)		
	Urban Station	C	Yes (19/19/40)	Electrofishing	Spotted Bass (19)	2	
					Largemouth Bass (1)		
	Station #1	D1	No	Electrofishing	Largemouth Bass (15)	1	
					Spotted Bass (5)		
	Station #2	D2	No	Electrofishing	Spotted Bass (12)	3	
					Largemouth Bass (7)		
	Station #3	D3	No	Electrofishing	Spotted Bass (14)	4	
					Largemouth Bass (6)		
	Station #4	D4	No	Electrofishing	Spotted Bass (17)	3	
Fall	Forested Station	A	Yes (30/26/30)	Angling	Redeye Bass (19)	1	1
					Spotted Bass (2)		
	Agricultural Station	B	Yes (15/15/30)	Angling	Spotted Bass (10)	1	
					Redeye Bass (2)		
	Urban Station	C	Yes (39/24/40)	Electrofishing	Spotted Bass (21)	1	1
	Station #1	D1	No	Electrofishing	Largemouth Bass (8)	1	1
					Spotted Bass (12)		
					Unknown Species of Bass (2)		
	Station #2	D2	No	Electrofishing	Spotted Bass (14)	5	4
					Largemouth Bass (4)		
	Station #3	D3	No	Electrofishing	Spotted Bass (7)	5	4
					Largemouth Bass (10)		
	Station #4	D4	No	Electrofishing	Largemouth Bass (21)	3	3

* Equals the number of samples taken at or upstream of each station and up to the next upstream station.

ID is designation used in Figures.

APPENDIX C4

FATHEAD MINNOWS

C4.1 Overview of Fathead Minnows Field Study

Fathead minnows are commonly reported for use in screening for estrogenic effects. They can readily be sexed and were the model used in many of the laboratory tests. To facilitate comparison with laboratory data collected during Phase 1, laboratory fathead minnows were exposed in the Conasauga River sampling areas for 2 weeks prior to the collection of wild fish.

The fathead minnows used for the field study were obtained from Kurtz Fish Hatchery in Pennsylvania. They were adult specimens and were easily sexed by observation by an experienced biologist. All specimens were at sexual maturity and the mean weight was 2.5 g (females = 2.3 g, males = 2.9 g, unknown sex = 1.3 g). The fathead minnows were treated for external parasites with formalin and fed frozen brine shrimp to increase the size of the specimens before beginning the caged tests. Laboratory water quality holding conditions were 21 +_ 1°C, a pH of 7.5-8.0 S.U., dissolved oxygen greater than 95 percent, and a conductivity of 300 umho/cm. The photoperiod was 16 hours of light and 8 hours of darkness. Fish were randomly selected and placed in plastic bags for transportation to the Conasauga River. Ten males and ten females were placed in each bag.

Six fathead minnow cages (2 at each station) were deployed on April 12, 2001 for the spring survey and on September 5, 2001 for the fall survey at the three sampling stations on the Conasauga River. Two cages with 20 minnows each were attached approximately mid-depth along the riverbank. Twenty fathead minnows per station were targeted for analysis; the additional minnow cage at each station was deployed as a contingency against vandalism, mortality, or loss. Minnows were exposed for two weeks and retrieved and processed during the wild fish sampling.

Up to twenty survivors from each station were processed in the field. The fish were anesthetized with MS-222 (50-100 ppm), the tail removed with a razor blade, and blood collected into a heparinized hematocrit tube from the caudal vein. Plasma was separated from the blood by centrifugation and stored in labeled microcentrifuge tubes on dry ice for transport to the University of Florida. At the University, samples were stored at -80°C until processed.

One of the methods under development in this study included measurement of mRNA. Ideally, the liver would be targeted for the biomarker measurement, but due to the small liver size, the entire body cavity (internal organs, liver, & gonad) was removed, finely cut, and preserved in RNAlater™ fixative for RNA analysis. The samples were stored on wet or gel ice at 0 to 4°C for transport to the University of Florida. At the University, samples were stored at – 20°C until processed. These results are presented in the section entitled MRNA Measurements.

A sample identifiacation, sex, weight, and any observable health problems were recorded for each specimen. Sex was verified by visual examination of the gonads.

While in spring there was toxicity at the urban/mixed use station, differences in vitellogenin levels at these three locations were not significant, α=0.05 however, in the fall the vitellogenin levels in females were lower than the forested station at the α =0.1 level.

All data collected for the fathead minnows caged in the Conasauga River for two weeks are presented in Table C-5. The vitellogenin concentrations are illustrated in Figure C-4.

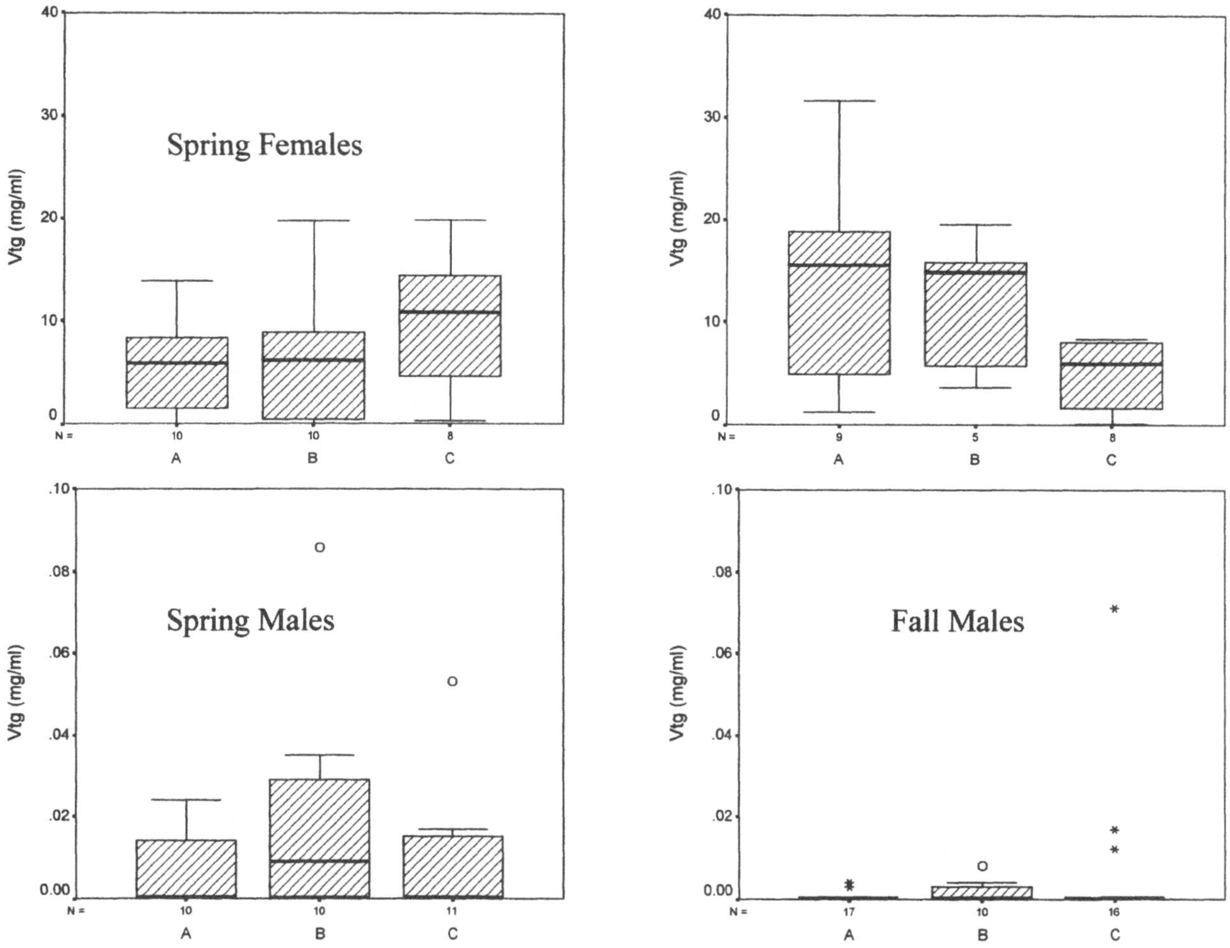

[1] A = Forested; B = Agricultural; C = Urban/Mixed use. [2] For the station locations see Figures C-3.

Figure C-4. Plasma Vitellogenin Levels: Caged Fathead Minnows Exposed Two Weeks at Three Stations of the Conasauga River.

Table C-5. Fathead Minnow Measurements From the Conasauga River During the April and September Sampling Events.

Date	Site	Sex[1]	Body Weight (g)[2]	Vitellogenin (mg/ml)
4/26/01	Forested	f	1.89	5.915
4/26/01	Forested	f	1.49	3.751
4/26/01	Forested	f	2.28	0.808
4/26/01	Forested	f	2.76	5.963
4/26/01	Forested	f	1.84	12.418
4/26/01	Forested	f	2.56	13.932
4/26/01	Forested	f	3.46	8.352
4/26/01	Forested	f	3.06	<0.0005
4/26/01	Forested	f	2.35	1.472
4/26/01	Forested	f	2.47	6.806
4/26/01	Forested	m	6.52	0.003
4/26/01	Forested	m	3.46	<0.0005
4/26/01	Forested	m	5.92	0.014
4/26/01	Forested	m	3.39	<0.0005
4/26/01	Forested	m	3.46	<0.0005
4/26/01	Forested	m	1.97	<0.0005
4/26/01	Forested	m	4.59	0.022
4/26/01	Forested	m	6.91	<0.0005
4/26/01	Forested	m	3.04	0.024
4/26/01	Forested	m	3.22	<0.0005
4/26/01	Agricultural	f	2.35	19.746
4/26/01	Agricultural	f	2.74	6.506
4/26/01	Agricultural	f	2.18	0.135
4/26/01	Agricultural	f	2.63	5.830
4/26/01	Agricultural	f	2.24	<0.0005
4/26/01	Agricultural	f	2.53	2.534
4/26/01	Agricultural	f	2.26	0.446
4/26/01	Agricultural	f	2.83	8.889
4/26/01	Agricultural	f	3.33	15.425
4/26/01	Agricultural	f	2.01	7.149
4/26/01	Agricultural	m	5.23	0.024
4/26/01	Agricultural	m	4.83	<0.0005
4/26/01	Agricultural	m	4.88	0.086
4/26/01	Agricultural	m	4.32	0.029
4/26/01	Agricultural	m	4.27	<0.0005

Date	Site	Sex[1]	Body Weight (g)[2]	Vitellogenin (mg/ml)
4/26/01	Agricultural	m	5.09	0.009
4/26/01	Agricultural	m	4.74	0.035
4/26/01	Agricultural	m	4.98	<0.0005
4/26/01	Agricultural	m	2.55	<0.0005
4/26/01	Agricultural	m	3.97	0.009
4/26/01	Urban	f	2.37	9.450
4/26/01	Urban	f	3.09	4.085
4/26/01	Urban	f	3.05	19.829
4/26/01	Urban	f	2.19	14.690
4/26/01	Urban	f	2.2	0.308
4/26/01	Urban	f	2.98	12.160
4/26/01	Urban	f	1.83	14.068
4/26/01	Urban	f	2.53	5.177
4/26/01	Urban	m	3.9	<0.0005
4/26/01	Urban	m	4.09	<0.0005
4/26/01	Urban	m	4.13	<0.0005
4/26/01	Urban	m	3.4	<0.0005
4/26/01	Urban	m	3.01	<0.0005
4/26/01	Urban	m	2.44	8.201
4/26/01	Urban	m	4.53	0.053
4/26/01	Urban	m	4.78	0.005
4/26/01	Urban	m	2.71	0.017
4/26/01	Urban	m	4.47	0.013
4/26/01	Urban	m	4.52	<0.0005
9/19/01	Forested	f	2.04	20.464
9/19/01	Forested	f	2.67	18.798
9/19/01	Forested	f	2.02	4.955
9/19/01	Forested	f	2.51	16.671
9/19/01	Forested	f	2.19	31.659
9/19/01	Forested	f	1.61	1.191
9/19/01	Forested	f	2.21	15.596
9/19/01	Forested	f	1.87	4.747
9/19/01	Forested	f	2.11	12.584
9/19/01	Forested	m	1.96	<0.0005
9/19/01	Forested	m	1.98	<0.0005
9/19/01	Forested	m	1.40	0.003
9/19/01	Forested	m	1.50	<0.0005

Date	Site	Sex[1]	Body Weight (g)[2]	Vitellogenin (mg/ml)
9/19/01	Forested	m	1.05	<0.0005
9/19/01	Forested	m	1.67	<0.0005
9/19/01	Forested	m	1.14	<0.0005
9/19/01	Forested	m	1.46	<0.0005
9/19/01	Forested	m	1.35	<0.0005
9/19/01	Forested	m	1.66	<0.0005
9/19/01	Forested	m	1.73	<0.0005
9/19/01	Forested	m	1.55	<0.0005
9/19/01	Forested	m	1.75	<0.0005
9/19/01	Forested	m	1.52	0.004
9/19/01	Forested	u	1.48	<0.0005
9/19/01	Forested	u	1.31	<0.0005
9/19/01	Forested	u	1.17	<0.0005
9/19/01	Agricultural	f	2.37	15.846
9/19/01	Agricultural	f	2.12	19.587
9/19/01	Agricultural	f	2.21	3.606
9/19/01	Agricultural	f	2.08	5.703
9/19/01	Agricultural	f	1.88	14.838
9/19/01	Agricultural	m	1.48	0.004
9/19/01	Agricultural	m	2.28	0.003
9/19/01	Agricultural	m	1.26	<0.0005
9/19/01	Agricultural	m	2.01	0.008
9/19/01	Agricultural	m	1.82	<0.0005
9/19/01	Agricultural	m	1.36	0.002
9/19/01	Agricultural	m	1.48	<0.0005
9/19/01	Agricultural	m	1.84	<0.0005
9/19/01	Agricultural	m	1.43	<0.0005
9/19/01	Agricultural	u	1.63	<0.0005
9/19/01	Urban	f	2.23	8.215
9/19/01	Urban	f	1.92	7.893
9/19/01	Urban	f	2.02	8.395
9/19/01	Urban	f	1.84	0.061
9/19/01	Urban	f	1.76	4.06
9/19/01	Urban	f	1.77	7.868
9/19/01	Urban	f	1.83	0.983
9/19/01	Urban	f	1.37	2.265
9/19/01	Urban	m	2.02	<0.0005

Date	Site	Sex[1]	Body Weight (g)[2]	Vitellogenin (mg/ml)
9/19/01	Urban	m	2.22	<0.0005
9/19/01	Urban	m	1.95	<0.0005
9/19/01	Urban	m	1.48	<0.0005
9/19/01	Urban	m	1.98	<0.0005
9/19/01	Urban	m	1.56	<0.0005
9/19/01	Urban	m	1.34	<0.0005
9/19/01	Urban	m	1.22	<0.0005
9/19/01	Urban	u	1.15	<0.0005
9/19/01	Urban	u	1.24	<0.0005
9/19/01	Urban	u	0.83	0.017
9/19/01	Urban	u	1.25	<0.0005
9/19/01	Urban	u	1.50	0.071
9/19/01	Urban	u	1.40	<0.0005
9/19/01	Urban	u	1.34	<0.0005
9/19/01	Urban	u	1.46	0.012

[1] f = female, m = male, u = undetermined sex
[2] Body weight was not measured during the spring sampling
[3] N = nodule, F = fungus, FE = fin erosion, UL = external ulcer, P = parasites event

APPENDIX C5

BASS

C5.1 Overview of Bass Field Study

The targeted species for the study were largemouth bass. However, species of bass varied at the seven locations. The species of bass included largemouth bass (*Micropterus salmoides*), spotted bass (*Micropterus punctulatus*), and redeye bass (*Micropterus coosae*). All three species prefer slightly different habitats and individually could not be found at all seven sites; however, they are all from the same genus and are closely related.

C5.1.1 Monoclonal Antibody Cross Reactivity

The cross reactivity of redeye and spotted Bass vitellogenin to the anti-largemouth bass vitellogenin monoclonal antibody (Mab-3G2) was validated by direct ELISA. Male redeye and spotted bass were collected from the wild and transported to the University of Florida. Each species was prepped and then implanted with 17β-estradiol (5 mg/kg). About 10 days later the fish were bled again and euthanized. The estrogenized plasma was used to purify vitellogenin for each species using ion exchange chromatography on POROS 20HQ (BioCad Sprint Perfusion System). A direct ELISA was conducted using standards (0, 0.01, 0.1, 0.2, 0.4, 0.6, 0.8, & 1.0 μg/ml) prepared from largemouth, redeye, and the spotted bass in order to compare the curves and Mab cross-reactivity. It was apparent that the curves were very similar to largemouth bass (>50-percent homology). Hence, ELISA quantification of the redeye and spotted bass samples was determined using largemouth bass vitellogenin as the standard to simplify the assay.

C5.1.2 Wild Fish Collection

Wild fish were collected using boat-mounted electroshocking equipment where allowable, and angling methods in river segments containing threatened and endangered species. Two federally endangered species, the Amber darter *(Persina antesella*) and Conasauga logperch (*Percina jenkinsi*), and one federally threatened species, blue shiner (*Cyprinella caerulea*), are found in the middle and upper reaches of the Conasauga, therefore electroshocking equipment cannot be utilized.

A total of 20 fish, cumulatively from all three species, was the target sample size at each site. In some cases, due to the small size of the fish caught, making it unable to sex the specimen, or the inability to catch enough bass within the sample reach, the target number was not achieved.

The weight of each fish was measured to the nearest 0.01 g and a blood sample was collected from the caudal vein. For the large fish, a syringe was used to collect blood. Blood collection from the small fish was identical to that described for the fathead minnow using a razor blade and hematocrit tube. Blood collected using the syringe was injected into a heparinzed vacuutainer, mixed, and stored on ice until ready to be centrifuged. A minimum of 0.2 mL of blood was collected from each specimen. Each sample was centrifuged for 10 minutes at

5000 revolutions per minute (rpm). Plasma was collected and transferred into a labeled microcentrifuge tube for future vitellogenin analysis.

The fish were killed after blood was drawn by a quick puncture into the pith. The weight of the liver and gonad were measured to the nearest 0.01 g in order to determine gonadosomatic (GSI) and hepatosomatic (HSI) indices. Separate subsamples of the liver and gonad (1 cm^3) were preserved in 10-percent buffered formalin for histology analysis. At the University, samples were stored at $-20^{\circ}C$ until processed. The histology samples were stored in the formalin solution for seven days and then exchanged into 70-percent ethanol until processed.

Tissue samples from the tail or filet from the lateral side of the caudal peduncle were collected from 3 fish per station for potential future screening for pesticide residues if necessary.

C5.1.3 Results

All measurements for Bass samples collected in the Chattahoochee and Conasauga Rivers during the April and September sampling events are shown in Table C-6. Box plots of vitellogenin, GSI, and HSI are shown in Figures C-5 through C-7.

Table C-6. Bass Measurements From the Chattahoochee and Conasauga Rivers During the April and September Sampling Events

Date	River Site	River Site	Species[1]	Sex[2]	Total Length (cm)[4]	Body (g)	Liver (g)	Gonad (g)	HSI (%)[5]	GSI (%)[5]	Health[6]	Vtg (mg/ml)[7]	Plasma (uL)	Protein (mg/dL)	Glucose (mg/dL)	Lactate (mg/dL)	E2 (ng/mL)	T (ng/mL)
4/30/01	CO - 1	C	SPB	F1	-	14.5	1.8	-	12.41	-	-	0.049	0	-	-	-	-	-
4/30/01	CO - 1	C	SPB	F1	-	11	-	-	-	-	-	0.002	24	28.9	2.99	0.85	0.691	0.583
4/30/01	CO - 1	C	SPB	F1	-	11.5	-	-	-	-	-	0.002	0	-	-	-	-	-
4/30/01	CO - 1	C	SPB	F1	-	16	-	-	-	-	-	0.020	0	-	-	-	-	-
4/30/01	CO - 1	C	SPB	F3	-	175.3	5.5	6.5	3.14	3.71	-	12.099	105	15.7	2.81	1.12	0.985	0.309
4/30/01	CO - 1	C	SPB	F3	-	35	-	-	-	-	-	6.936	0	-	-	-	-	-
4/30/01	CO - 1	C	SPB	M	-	50	0.5	0.3	1.00	0.60	-	0.022	13	23.4	1.19	1.84	2.325	0.077
4/30/01	CO - 1	C	LMB	M3	-	21.7	0.15	-	0.69	-	-	0.034	13	33.0	1.55	1.34	1.400	2.177
4/30/01	CO - 1	C	SPB	M3	-	53.5	0.55	-	1.03	-	FE	0.098	38	23.3	2.08	0.56	0.791	1.253
4/30/01	CO - 1	C	SPB	M3	-	100	1	0.3	1.00	0.30	-	0.032	228	13.0	1.35	0.38	0.126	0.055
4/30/01	CO - 1	C	SPB	M3	-	186	2.4	1.7	1.29	0.91	-	0.031	75	36.1	2.93	0.63	0.775	0.288
4/30/01	CO - 1	C	SPB	M3	-	108	0.9	0.8	0.83	0.74	-	0.164	46	18.1	2.16	0.71	1.380	0.280
4/30/01	CO - 1	C	SPB	M3	-	112	1.2	0.7	1.07	0.63	-	0.191	68	30.8	3.80	0.65	0.576	0.243
4/30/01	CO - 1	C	SPB	M3	-	89	1.5	0.5	1.69	0.56	-	0.002	0	-	-	-	-	-
4/30/01	CO - 1	C	SPB	M3	-	11.5	-	-	-	-	-	0.048	9	23.5	7.90	2.64	3.191	0.114
4/30/01	CO - 1	C	SPB	M3	-	208	-	0.8	-	0.38	-	0.002	76	40.3	0.89	0.77	0.316	0.365
4/30/01	CO - 1	C	SPB	M3	-	500	3.5	1.5	0.70	0.30	-	0.022	144	23.1	1.98	0.62	0.157	0.170
4/30/01	CO - 1	C	SPB	U	-	13	-	-	-	-	-	0.013	0	-	-	-	-	-
4/30/01	CO - 1	C	SPB	U	-	10.9	-	-	-	-	-	0.002	17	11.2	1.81	0.47	0.678	0.521
4/30/01	CO - 1	C	SPB	U	-	8.6	-	-	-	-	-	0.002	0	-	-	-	-	-
5/1/01	CO - 2	B	SPB	F1	-	86.5	0.9		1.04	-	-	0.009	49	51.7	1.22	1.04	0.527	0.476
5/1/01	CO - 2	B	SPB	F1	-	45	-	-	-	-	-	0.020	22	3.0	0.80	1.69	1.910	1.538
5/1/01	CO - 2	B	SPB	F1	-	53	-	-	-	-	-	0.021	167	17.2	1.07	0.67	0.278	0.282
5/1/01	CO - 2	B	SPB	F3	-	245	1.8	1.2	0.73	0.49	-	0.002	85	24.1	0.73	0.79	0.533	0.464
5/1/01	CO - 2	B	SPB	F3	-	121	1.2	1.1	0.99	0.91	-	8.236	52	67.9	1.90	0.98	1.036	3.850
5/1/01	CO - 2	B	SPB	F3	-	146	1.5	1	1.03	0.68	FE	1.089	141	19.7	0.63	0.65	0.297	0.192
5/1/01	CO - 2	B	SPB	F3	-	139	1.1	1.7	0.79	1.22	-	14.047	136	35.8	2.21	0.89	0.374	0.033
5/1/01	CO - 2	B	SPB	M3	-	85.5	0.9		1.05	-	-	0.002	36	69.0	4.91	1.19	0.756	1.025
5/1/01	CO - 2	B	LMB	M3	-	197	1.3	0.5	0.66	0.25	-	0.033	103	20.6	2.40	0.61	0.224	0.215

Date	River Site	River Site	Species[1]	Sex[2]	Total Length (cm)[4]	Body (g)	Liver (g)	Gonad (g)	HSI (%)[5]	GSI (%)[5]	Health[6]	Vtg (mg/ml)[7]	Plasma (uL)	Protein (mg/dL)	Glucose (mg/dL)	Lactate (mg/dL)	E2 (ng/mL)	T (ng/mL)
5/1/01	CO - 2	B	SPB	M3	-	155	1.8	1	1.16	0.65	-	0.002	92	26.4	3.38	0.79	0.373	0.286
5/1/01	CO - 2	B	SPB	M3	-	274	1.5	2.3	0.55	0.84	-	0.555	103	52.4	0.00	0.64	0.549	1.185
5/1/01	CO - 2	B	SPB	M3	-	157	1.8	0.6	1.15	0.38	-	0.002	192	27.3	3.31	0.78	0.213	0.015
5/1/01	CO - 2	B	SPB	M3	-	162	0.9	0.5	0.56	0.31	-	0.021	26	35.1	0.72	1.26	0.973	0.039
5/2/01	CO - 2	B	SPB	F3	-	506	4.4	23	0.87	4.55	-	18.922	273	27.9	1.14	0.57	0.689	0.112
5/2/01	CO - 2	B	SPB	F3	-	383	3.7	14.3	0.97	3.73	-	18.598	19	102.3	1.94	2.00	5.831	0.677
5/2/01	CO - 2	B	SPB	M3	-	181	1.8	0.6	0.99	0.33	-	0.006	110	0.4	5.49	0.75	0.514	0.554
5/2/01	CO - 2	B	SPB	M3	-	122	0.78	1.06	0.64	0.87	-	0.025	95	64.6	1.58	0.65	0.498	0.291
5/2/01	CO - 2	B	SPB	M3	-	123	1.1	0.47	0.89	0.38	-	0.032	185	32.4	3.15	0.76	0.192	0.108
5/2/01	CO - 2	B	SPB	M3	-	73	0.53	0.09	0.73	0.12	-	0.016	148	18.6	0.33	0.67	0.381	0.190
5/2/01	CO - 2	B	SPB	U	-	28	0.3	-	1.07	-	-	0.044	143	26.6	0.43	0.46	0.319	0.007
5/2/01	CO - 3	A	REB	F	-	110	2.5	8.1	2.27	7.36	-	4.698	73	52.8	1.75	1.01	1.313	0.482
5/2/01	CO - 3	A	REB	F2/3	-	30	-	-	-		-	0.181	65	37.3	1.20	0.61	0.605	0.464
5/2/01	CO - 3	A	REB	F3	-	216	3.3	7.5	1.53	3.47	-	19.927	37	60.2	5.84	1.54	1.717	0.478
5/2/01	CO - 3	A	REB	F3	-	71	1.8	4.3	2.54	6.06	-	12.501	64	49.1	2.37	0.92	0.878	1.666
5/2/01	CO - 3	A	REB	F3	-	73	1.3	3.5	1.78	4.79	LCL	4.159	197	15.8	1.39	0.69	0.212	0.016
5/2/01	CO - 3	A	REB	F3	-	173	5.3	14.5	3.06	8.38	-	7.682	200	61.9	3.10	0.95	0.695	0.005
5/2/01	CO - 3	A	REB	F3	-	192	4.6	22.2	2.40	11.56	-	9.272	192	29.7	2.94	0.76	0.520	0.227
5/2/01	CO - 3	A	REB	F3	-	75	1.4	7.3	1.87	9.73	-	5.102	146	28.2	2.26	0.58	1.679	0.654
5/2/01	CO - 3	A	REB	F3	-	72	1.7	5.2	2.36	7.22	-	4.708	146	28.2	2.56	0.76	0.281	0.201
5/2/01	CO - 3	A	REB	F3	-	48.5	0.8	4	1.65	8.25	-	2.786	41	55.4	0.55	0.65	0.908	0.241
5/2/01	CO - 3	A	REB	F3	-	183	3.9	12.4	2.13	6.78	LCL	5.466	83	36.7	2.70	0.88	0.894	0.085
5/2/01	CO - 3	A	REB	F3	-	61	1.2	3.5	1.97	5.74	-	2.786	8	91.7	18.70	3.22	4.119	0.461
5/2/01	CO - 3	A	REB	F3	-	39	0.9	2.6	2.31	6.67	-	3.489	36	59.2	3.16	1.23	1.755	0.028
5/2/01	CO - 3	A	REB	F3	-	62	1.1	3.1	1.77	5.00	-	5.919	64	38.5	0.44	0.76	1.365	0.339
5/2/01	CO - 3	A	REB	F3	-	52.5	0.8	4.8	1.52	9.14	-	4.980	0	-	-	-	-	-
5/2/01	CO - 3	A	REB	M3	-	87	1.5	1	1.72	1.15	-	0.007	132	29.8	5.21	0.69	0.241	0.166
5/2/01	CO - 3	A	REB	M3	-	164	2.6	1.3	1.59	0.79	-	0.017	145	19.5	3.14	0.69	0.262	0.007
5/2/01	CO - 3	A	REB	M3	-	98	2.1	1.2	2.14	1.22	LCL	0.002	158	24.2	4.03	0.93	0.192	0.168
5/2/01	CO - 3	A	REB	M3	-	151	1	0.8	0.66	0.53	-	0.028	22	66.8	5.77	1.64	1.517	0.233
5/2/01	CO - 3	A	SPB	U	-	103	1	-	0.97	-	SG	0.002	114	41.4	5.54	0.99	0.345	0.111

Date	River Site	River Site	Species[1]	Sex[2]	Total Length (cm)[4]	Body (g)	Liver (g)	Gonad (g)	HSI (%)[5]	GSI (%)[5]	Health[6]	Vtg (mg/ml)[7]	Plasma (uL)	Protein (mg/dL)	Glucose (mg/dL)	Lactate (mg/dL)	E2 (ng/mL)	T (ng/mL)
5/7/01	CH - 1	D1	LMB	F1	-	85	0.57	0.33	0.67	0.39	-	0	234	22.4	1.52	0.73	0.158	0.004
5/7/01	CH - 1	D1	LMB	F2	-	530	7	3	1.32	0.57	LP	0.115	118	0.0	2.82	1.31	0.467	0.380
5/7/01	CH - 1	D1	LMB	F2	-	138	0.76	0.7	0.55	0.51	-	0.002	94	38.1	1.15	0.90	0.330	0.793
5/7/01	CH - 1	D1	LMB	F3	-	1270	14.8	26.5	1.17	2.09	-	33.875	142	44.2	3.16	0.81	0.492	0.234
5/7/01	CH - 1	D1	LMB	F3	-	700	6.4	30	0.91	4.29	-	5.595	201	24.6	1.93	0.69	0.569	0.093
5/7/01	CH - 1	D1	SPB	F3	-	187	1.25	4.32	0.67	2.31	-	11.297	134	42.2	0.24	0.76	0.298	0.274
5/7/01	CH - 1	D1	LMB	F3	-	1310	15.5	58	1.18	4.43	ML	1.519	116	0.0	2.55	1.20	0.340	0.339
5/7/01	CH - 1	D1	LMB	F3	-	239	1.59	2.01	0.67	0.84	-	0.075	139	30.1	0.23	1.11	0.248	0.215
5/7/01	CH - 1	D1	LMB	M	-	107	0.67	0.11	0.63	0.10	-	0.002	146	46.5	1.22	0.88	0.307	0.403
5/7/01	CH - 1	D1	SPB	M3	-	268	2	0.22	0.75	0.08	-	0.014	113	59.2	0.46	0.79	0.353	0.026
5/7/01	CH - 1	D1	LMB	M3	-	229	1	0.19	0.44	0.08	-	0.041	233	25.5	1.00	0.63	0.219	0.004
5/7/01	CH - 1	D1	SPB	M3	-	112	1.14	0.13	1.02	0.12	LA	0.054	86	26.0	4.95	1.14	0.319	0.188
5/7/01	CH - 1	D1	LMB	M3	-	100	0.62	0.14	0.62	0.14	-	0.002	63	33.0	2.38	1.11	0.471	0.369
5/7/01	CH - 1	D1	LMB	M3	-	87	0.6	0.1	0.69	0.11	-	0.002	88	23.8	0.84	0.83	0.259	0.168
5/7/01	CH - 1	D1	LMB	M3	-	117	0.87	0.08	0.74	0.07	-	0.002	140	21.5	3.46	0.72	0.193	0.058
5/7/01	CH - 1	D1	SPB	M3	-	128	0.84	0.08	0.66	0.06	-	0.023	212	0.0	1.12	1.31	0.238	0.253
5/7/01	CH - 1	D1	LMB	M3	-	315	2.08	0.31	0.66	0.10	-	0.017	76	15.9	3.18	0.76	0.570	0.402
5/7/01	CH - 1	D1	LMB	M3	-	272	1.7	0.6	0.63	0.22	ML	0.002	146	35.2	1.85	0.81	0.259	0.378
5/7/01	CH - 1	D1	LMB	M3	-	246	1.08	0.36	0.44	0.15	-	0.034	131	30.5	0.88	1.14	0.416	0.094
5/7/01	CH - 1	D1	SPB	U	-	67	0.39	-	0.58	-	-	0.021	148	34.2	3.90	0.85	0.149	0.007
5/8/01	CH - 2	D2	LMB	F1	-	41.4	0.21	0.17	0.51	0.41	-	0.021	38	12.5	1.02	1.34	0.732	0.026
5/8/01	CH - 2	D2	SPB	F1	-	641	4.3	8.43	0.67	1.32	L	0.345	184	17.5	2.35	1.79	0.206	0.275
5/8/01	CH - 2	D2	SPB	F1	-	136	1.12	0.19	0.82	0.14	-	0.024	205	11.9	5.13	1.80	0.091	0.098
5/8/01	CH - 2	D2	SPB	F1	-	25.9	0.2	0.09	0.77	0.35	-	0.002	25	16.1	4.51	2.03	0.830	0.619
5/8/01	CH - 2	D2	SPB	F2	-	361	2.76	0.81	0.76	0.22	-	0.003	198	27.7	3.66	1.80	0.164	0.021
5/8/01	CH - 2	D2	SPB	F3	-	324	3.63	8.4	1.12	2.59	-	2.194	64	39.3	2.57	1.99	0.656	0.016
5/8/01	CH - 2	D2	LMB	F3	-	1270	11.7	28.2	0.92	2.22	FL	5.562	222	37.3	3.24	1.14	0.528	0.053
5/8/01	CH - 2	D2	SPB	F3	-	630	7.9	22.7	1.25	3.60	FL, LP	21.20	132	47.7	4.39	1.58	0.415	0.008
5/8/01	CH - 2	D2	SPB	F3	-	683	5.54	14.78	0.81	2.16	-	7.879	86	21.7	3.87	1.37	0.460	0.012
5/8/01	CH - 2	D2	LMB	F3	-	1017	11.04	38.2	1.09	3.76	-	7.398	204	22.3	2.05	1.49	0.764	0.035
5/8/01	CH - 2	D2	LMB	F3	-	262	1.98	5.91	0.76	2.26	-	12.60	211	17.0	2.61	1.55	0.192	0.052

Date	River Site	River Site	Species[1]	Sex[2]	Total Length (cm)[4]	Body (g)	Liver (g)	Gonad (g)	HSI (%)[5]	GSI (%)[5]	Health[6]	Vtg (mg/ml)[7]	Plasma (uL)	Protein (mg/dL)	Glucose (mg/dL)	Lactate (mg/dL)	E2 (ng/mL)	T (ng/mL)
5/8/01	CH - 2	D2	SPB	M3	-	58	0.55	0.53	0.95	0.91	-	0.022	14	43.9	1.39	2.39	1.669	0.561
5/8/01	CH - 2	D2	SPB	M3	-	344	3.4	0.49	0.99	0.14	-	0.060	131	30.5	3.10	1.19	0.331	0.081
5/8/01	CH - 2	D2	SPB	M3	-	204	1.54	0.32	0.75	0.16	-	0.002	69	30.9	2.26	1.88	0.594	0.015
5/8/01	CH - 2	D2	LMB	M3	-	1035	7.64	2.14	0.74	0.21	FE, L, ML	0.002	99	33.2	2.98	1.78	0.452	0.010
5/8/01	CH - 2	D2	LMB	M3	-	834	4.9	1.3	0.59	0.16	-	0.002	186	33.3	2.77	1.40	0.252	0.196
5/8/01	CH - 2	D2	LMB	M3	-	417	2.4	0.6	0.58	0.14	-	0.026	169	28.6	0.65	1.52	0.183	0.063
5/9/01	CH - 2	D2	SPB	F3	-	480	3.45	12.31	0.72	2.56	ML	0.568	195	22.5	1.81	0.90	0.185	0.123
5/9/01	CH - 2	D2	SPB	M	-	87	0.8	0.13	0.92	0.15	-	0.002	199	17.2	3.15	0.92	0.320	0.005
5/9/01	CH - 2	D2	SPB	M3	-	257	1.58	0.25	0.61	0.10	-	0.002	139	18.7	0.87	1.03	0.257	0.007
5/9/01	CH - 3	D3	SPB	F1	-	34	0.2	0.16	0.59	0.47	-	0.002	460	15.2	0.49	0.48	0.071	0.002
5/9/01	CH - 3	D3	SPB	F1	-	475	4.6	8.68	0.97	1.83	-	-	296	16.1	2.63	1.20	0.090	0.114
5/9/01	CH - 3	D3	SPB	F2	-	220	2.27	0.65	1.03	0.30	-	0.121	129	19.0	2.84	1.11	0.373	0.008
5/9/01	CH - 3	D3	LMB	F3	-	986.5	8.9	28.8	0.90	2.92	FE	18.604	158	23.6	1.81	1.04	0.596	0.055
5/9/01	CH - 3	D3	SPB	F3	-	295.6	3.34	2.17	1.13	0.73	-	13.592	111	26.0	3.36	1.33	0.145	0.086
5/9/01	CH - 3	D3	SPB	F3	-	361	3.12	7.63	0.86	2.11	-	4.634	163	45.2	3.29	1.08	0.118	0.006
5/9/01	CH - 3	D3	SPB	F3	-	543	4.08	6.45	0.75	1.19	-	19.638	85	35.0	3.46	1.41	0.292	0.012
5/9/01	CH - 3	D3	SPB	F3	-	559	6.08	21.37	1.09	3.82	Lots of eggs	6.954	172	17.7	2.99	1.01	0.505	0.130
5/9/01	CH - 3	D3	LMB	F3	-	1985	17.31	34.3	0.87	1.73	SL	6.136	133	30.4	3.09	1.19	0.619	0.008
5/9/01	CH - 3	D3	LMB	F3	-	1680	15.72	54.61	0.94	3.25	FL	3.187	232	26.8	2.81	0.87	0.726	0.004
5/9/01	CH - 3	D3	SPB	F3	-	868	11.49	6.9	1.32	0.79	-	1.417	268	26.2	3.54	1.11	0.108	0.152
5/9/01	CH - 3	D3	SPB	M3	-	176	0.97	0.15	0.55	0.09	-	0.021	100	26.8	2.06	0.82	0.200	0.049
5/9/01	CH - 3	D3	SPB	M3	-	411	2.53	0.88	0.62	0.21	-	0.029	155	23.7	1.38	0.93	0.126	0.006
5/9/01	CH - 3	D3	LMB	M3	-	86	0.47	0.05	0.55	0.06	-	0.015	255	6.4	1.57	0.99	0.133	0.013
5/9/01	CH - 3	D3	SPB	M3	-	99	0.88	0.08	0.89	0.08	-	0.002	205	11.9	2.30	1.29	0.090	0.005
5/9/01	CH - 3	D3	LMB	M3	-	428	3.03	1.22	0.71	0.29	L	0.002	132	13.2	2.36	1.10	0.299	0.008
5/9/01	CH - 3	D3	LMB	M3	-	475	4.35	0.95	0.92	0.20	-	0.055	233	11.6	5.17	1.40	0.161	0.177
5/9/01	CH - 3	D3	SPB	M3	-	465	3.01	0.69	0.65	0.15	-	0.037	183	17.5	2.01	1.37	0.232	0.017
5/9/01	CH - 3	D3	SPB	U	-	45.6	0.3	-	0.66	-	SG	0.002	57	2.3	1.95	1.35	0.353	0.240
5/10/01	CH - 4	D3	LMB	F	-	116	1.3	7.37	1.12	6.35	LP	16.975	207	6.7	1.13	0.38	0.142	0.020

Date	River Site	River Site	Species[1]	Sex[2]	Total Length (cm)[4]	Body (g)	Liver (g)	Gonad (g)	HSI (%)[5]	GSI (%)[5]	Health[6]	Vtg (mg/ml)[7]	Plasma (uL)	Protein (mg/dL)	Glucose (mg/dL)	Lactate (mg/dL)	E2 (ng/mL)	T (ng/mL)
5/10/01	CH - 4	D3	LMB	F1	-	47.3	0.69	0.08	1.46	0.17	LP	0.002	105	14.1	4.81	1.34	0.551	0.079
5/10/01	CH - 4	D3	LMB	F1	-	87	1.05	0.86	1.21	0.99	LP	0.015	107	7.9	0.64	0.55	0.271	0.009
5/10/01	CH - 4	D3	LMB	F1	-	127	1.75	0.32	1.38	0.25	LP	0.002	104	1.8	4.02	1.38	0.339	0.010
5/10/01	CH - 4	D3	LMB	F1	-	31.3	0.28	0.1	0.89	0.32	-	0.020	648	1.3	0.40	0.82	0.041	0.002
5/10/01	CH - 4	D3	LMB	F3	-	378	6.3	10.8	1.67	2.86	-	3.922	224	16.8	2.58	1.59	0.240	0.004
5/10/01	CH - 4	D3	LMB	F3	-	101	0.88	3.95	0.87	3.91	-	0.002	0	-	-	-	-	-
5/10/01	CH - 4	D3	LMB	F3	-	930	12.49	22.04	1.34	2.37	LP	17.512	169	23.2	2.22	1.22	0.360	0.006
5/10/01	CH - 4	D3	LMB	F3	-	599	9.12	31.74	1.52	5.30	LP	19.295	0	-	-	-	-	-
5/10/01	CH - 4	D3	LMB	M3	-	184	3.56	0.37	1.93	0.20	LP	0.002	165	1.6	3.01	1.45	0.197	0.006
5/10/01	CH - 4	D3	LMB	M3	-	116	1.03	0.07	0.89	0.06	LP	0.002	213	11.8	1.70	0.81	0.199	0.013
5/10/01	CH - 4	D3	LMB	M3	-	159	2.33	0.46	1.47	0.29	LP	0.055	189	6.8	1.57	0.91	0.175	0.005
5/10/01	CH - 4	D3	LMB	M3	-	217	2.77	0.53	1.28	0.24	LP	0.002	67	9.4	3.82	1.40	0.478	0.101
5/10/01	CH - 4	D3	LMB	M3	-	489	3.56	1.54	0.73	0.31	LP	0.018	171	17.8	2.04	0.90	0.205	0.006
5/10/01	CH - 4	D3	LMB	U	-	45	0.38	-	0.84	-	-	0.020	30	14.2	1.16	0.89	0.701	0.033
5/10/01	CH - 4	D3	LMB	U	-	41.2	0.46	-	1.12	-	LP	0.002	46	20.0	2.32	1.84	0.822	0.022
5/10/01	CH - 4	D3	LMB	U	-	31.6	0.34	0.14	1.08	0.44	LP	0.021	72	16.2	0.63	0.95	0.280	0.014
9/17/01	CO - 1	C	SPB	F1	14.1	31	0.38	0.13	1.23	0.42	-	0.002	22	17.4	3.99	0.90	1.587	0.564
9/17/01	CO - 1	C	SPB	F1	22.8	119	0.65	0.49	0.55	0.41	-	0.002	191	33.1	3.64	0.77	0.082	0.024
9/17/01	CO - 1	C	SPB	F1	17.3	52	0.38	0.18	0.73	0.35	-	0.003	122	7.6	2.49	0.79	0.242	0.042
9/17/01	CO - 1	C	SPB	F1	17.8	62	0.42	0.19	0.68	0.31	-	0.002	119	25.4	3.44	0.99	0.234	0.087
9/17/01	CO - 1	C	SPB	F1	16.2	39	0.31	0.09	0.79	0.23	-	0.011	45	11.3	1.49	0.44	0.447	0.164
9/17/01	CO - 1	C	SPB	F1	15.4	35	0.31	0.14	0.89	0.40	-	-	90	21.4	3.56	0.79	0.252	0.047
9/17/01	CO - 1	C	SPB	F2	33.8	497	4.6	2.49	0.93	0.50	-	0.006	270	26.2	2.56	0.71	0.065	0.004
9/17/01	CO - 1	C	SPB	F2	45.3	1233	5.31	7.5	0.43	0.61	-	0.02	259	21.4	2.18	0.75	0.101	0.032
9/17/01	CO - 1	C	SPB	F3	48.6	1287	10.04	14.46	0.78	1.12	-	0.026	236	26.8	2.22	0.66	0.107	0.031
9/17/01	CO - 1	C	SPB	F3	43.4	1113	8.06	6.8	0.72	0.61	-	0.017	167	77.5	2.72	1.54	0.218	0.063
9/17/01	CO - 1	C	SPB	F3	27.3	195	2.22	1.5	1.14	0.77	-	0.019	262	11.4	2.49	0.95	0.126	0.004
9/17/01	CO - 1	C	SPB	M2	21.3	93	0.76	0.05	0.82	0.05	-	0.002	119	7.7	3.72	1.16	0.177	0.033
9/17/01	CO - 1	C	SPB	M2/3	19.7	84	0.5	0.09	0.60	0.11	-	0.002	219	11.7	1.52	0.47	0.128	0.005
9/17/01	CO - 1	C	SPB	M2/	13.9	27	0.27	0.09	1.00	0.33	-	0.009	28	0.0	0.51	0.22	0.819	1.193

Date	River Site	River Site	Species[1]	Sex[2] 3	Total Length (cm)[4]	Body (g)	Liver (g)	Gonad (g)	HSI (%)[5]	GSI (%)[5]	Health[6]	Vtg (mg/ml)[7]	Plasma (uL)	Protein (mg/dL)	Glucose (mg/dL)	Lactate (mg/dL)	E2 (ng/mL)	T (ng/mL)
9/17/01	CO - 1	C	SPB	M3	34.2	428	3.34	0.56	0.78	0.13	-	0.004	176	23.0	1.50	0.55	0.206	0.016
9/17/01	CO - 1	C	SPB	M3	19.2	74	0.43	0.1	0.58	0.14	-	0.002	157	23.6	2.32	0.86	0.091	0.006
9/17/01	CO - 1	C	SPB	M3	14.4	32	0.28	0.1	0.88	0.31	-	0.002	47	2.5	2.83	0.63	0.526	0.021
9/17/01	CO - 1	C	SPB	M3	19.3	90	1	0.11	1.11	0.12	-	0.004	175	17.7	3.87	0.89	0.156	0.101
9/17/01	CO - 1	C	SPB	U	13.8	23	0.17	-	0.74	-	-	0.002	0	-	-	-	-	-
9/17/01	CO - 1	C	SPB	U	14.7	28	0.15	-	0.54	-	-	0.001	0	-	-	-	-	-
9/17/01	CO - 1	C	SPB	U	14.2	25	0.2	0.04	0.80	0.16	-	0.002	11	0.0	2.08	0.86	1.986	0.338
9/18/01	CO - 3	A	SPB	F1	22.6	110	0.8	0.27	0.73	0.25	-	0.002	212	32.4	2.42	0.89	0.052	0.018
9/18/01	CO - 3	A	REB	F2	19.6	96	1.31	0.63	1.36	0.66	-	0.002	0	-	-	-	-	-
9/18/01	CO - 3	A	REB	F2	17.3	61	0.36	0.55	0.59	0.90	-	0.002	112	44.0	2.36	0.56	0.184	0.055
9/18/01	CO - 3	A	REB	F2/3	20.7	88	0.86	0.56	0.98	0.64	-	0.003	122	60.6	4.71	0.96	0.199	0.053
9/18/01	CO - 3	A	REB	F3	22.8	147	1.9	1.31	1.29	0.89	-	0.117	268	26.2	1.55	0.46	0.062	0.119
9/18/01	CO - 3	A	REB	F3	19.6	80	1.05	0.64	1.31	0.80	-	0.056	107	38.5	3.26	0.94	0.371	0.039
9/18/01	CO - 3	A	REB	F3	20	95	0.81	0.87	0.85	0.92	-	0.109	120	49.0	4.64	1.25	0.091	0.059
9/18/01	CO - 3	A	REB	M3	24.8	185	1.82	0.65	0.98	0.35	-	0.002	244	26.6	2.12	0.82	0.061	0.049
9/18/01	CO - 3	A	REB	M3	20.4	94	0.6	0.43	0.64	0.46	-	0.002	202	53.6	3.76	1.23	0.071	0.005
9/18/01	CO - 3	A	REB	M3	19.3	85	0.92	0.3	1.08	0.35	-	0.004	154	40.3	2.51	1.24	0.110	0.006
9/18/01	CO - 3	A	REB	M3	20.1	83	1.21	0.23	1.46	0.28	-	0.002	170	61.0	4.07	1.24	0.164	0.026
9/18/01	CO - 3	A	REB	M3	21.8	127	1.07	0.29	0.84	0.23	-	0.002	151	35.0	4.19	0.92	0.108	0.007
9/19/01	CO - 2	B	SPB	F1	14.7	27	0.21	0.04	0.78	0.15	-	0.007	5	186.0	0.55	3.86	2.290	1.351
9/19/01	CO - 2	B	SPB	M3	16.6	47	0.4	0.04	0.85	0.09	-	0.002	11	98.2	8.23	4.68	1.697	4.001
9/19/01	CO - 2	B	SPB	M3	23.7	132	0.72	0.11	0.55	0.08	-	0.002	55	42.1	1.10	0.99	0.414	0.274
9/19/01	CO - 2	B	SPB	M3	19.6	65	0.39	0.03	0.60	0.05	-	0.023	52	51.4	0.36	1.46	0.401	0.315
9/19/01	CO - 2	B	SPB	M3	19.7	72	0.35	0.11	0.49	0.15	-	0.012	59	56.1	0.10	1.79	0.319	0.195
9/19/01	CO - 2	B	REB	M3	16.3	51	0.38	0.08	0.75	0.16	-	0.002	43	48.0	1.40	1.68	0.456	0.097
9/19/01	CO - 3	A	SPB	F2	18.4	232	1.62	0.74	0.70	0.32	-	0.002	163	23.4	3.59	0.94	0.116	0.054
9/19/01	CO - 3	A	REB	F2/3	20	83	0.6	0.67	0.72	0.81	-	0.009	47	80.4	2.84	1.80	0.633	0.149
9/19/01	CO - 3	A	REB	F2/3	18.3	60	0.34	0.41	0.57	0.68	-	0.002	65	61.0	1.14	1.70	0.326	0.076
9/19/01	CO - 3	A	REB	F3	24.3	167	1.81	1.85	1.08	1.11	-	0.575	224	47.4	3.36	1.21	0.109	0.013
9/19/01	CO - 3	A	REB	M3	21	90	0.59	0.21	0.66	0.23	-	0.002	206	43.0	2.76	0.80	0.100	0.048

Date	River Site	River Site	Species[1]	Sex[2]	Total Length (cm)[4]	Body (g)	Liver (g)	Gonad (g)	HSI (%)[5]	GSI (%)[5]	Health[6]	Vtg (mg/ml)[7]	Plasma (uL)	Protein (mg/dL)	Glucose (mg/dL)	Lactate (mg/dL)	E2 (ng/mL)	T (ng/mL)
9/19/01	CO - 3	A	REB	M3	21.3	103	0.75	0.39	0.73	0.38	-	0.014	161	61.8	1.62	1.05	0.130	0.031
9/19/01	CO - 3	A	REB	M3	17	50	0.43	0.08	0.86	0.16	-	0.002	37	71.1	3.66	2.15	0.497	0.101
9/19/01	CO - 3	A	REB	M3	15.2	34	0.14	0.04	0.41	0.12	-	0.002	0	-	-	-	-	-
9/19/01	CO - 3	A	REB	U	12.7	19	0.32	-	1.68	-	-	0.002	41	21.2	2.59	1.50	0.570	0.358
9/25/01	CH - 4	D4	LMB	F	19.3	91	1.58	0.22	1.74	0.24	-	0.018	270	1.4	1.31	0.39	0.041	0.057
9/25/01	CH - 4	D4	LMB	F1	22.2	123	1.47	0.44	1.20	0.36	-	0.002	279	31.0	3.79	1.13	0.061	0.113
9/25/01	CH - 4	D4	LMB	F1	25.7	208	1.68	0.82	0.81	0.39	-	0.002	219	47.7	1.52	1.11	0.245	0.209
9/25/01	CH - 4	D4	LMB	F1	18.7	78	0.98	0.31	1.26	0.40	-	0.002	22	86.2	10.10	3.24	1.201	0.983
9/25/01	CH - 4	D4	LMB	F2	24.3	148	1.43	0.97	0.97	0.66	-	0.005	226	42.3	1.59	0.21	0.059	0.004
9/25/01	CH - 4	D4	LMB	F2	19.8	97	1	0.28	1.03	0.29	-	0.005	108	38.5	2.49	0.80	0.197	0.163
9/25/01	CH - 4	D4	LMB	F2	29	267	2.54	1.34	0.95	0.50	-	0.002	203	22.3	2.69	0.05	0.082	0.048
9/25/01	CH - 4	D4	LMB	F2	28.1	270	4.96	2.53	1.84	0.94	-	0.022	152	46.0	2.34	0.28	0.105	0.067
9/25/01	CH - 4	D4	LMB	F3	39.3	884	17.08	7.6	1.93	0.86	-	0.655	105	57.2	2.70	1.00	0.267	0.065
9/25/01	CH - 4	D4	LMB	F3	22.8	126	2.05	1.08	1.63	0.86	-	0.002	216	37.5	0.63	0.47	0.082	0.005
9/25/01	CH - 4	D4	LMB	F3	27.2	263	2.19	0.94	0.83	0.36	-	0.002	176	39.1	2.77	0.29	0.128	0.070
9/25/01	CH - 4	D4	LMB	F3	28	252	4.15	1.82	1.65	0.72	-	0.002	299	35.5	8.20	1.04	0.068	0.048
9/25/01	CH - 4	D4	LMB	M	20.7	93	1.04	0.15	1.12	0.16	-	0.002	34	-	-	-	43.390	0.195
9/25/01	CH - 4	D4	LMB	M3	37.4	779	6.61	1.41	0.85	0.18	-	0.002	154	34.8	3.12	1.36	0.151	0.062
9/25/01	CH - 4	D4	LMB	M3	34.6	620	6.63	0.85	1.07	0.14	-	0.002	206	53.4	1.86	0.10	0.070	0.055
9/25/01	CH - 4	D4	LMB	M3	34.8	560	9.22	1.39	1.65	0.25	-	0.029	167	39.5	2.38	0.10	0.108	0.055
9/25/01	CH - 4	D4	LMB	M3	25.4	195	2.43	0.08	1.25	0.04	-	0.004	158	40.0	2.49	0.19	0.106	0.034
9/25/01	CH - 4	D4	LMB	M3	22.9	146	2.45	0.12	1.68	0.08	-	0.012	319	35.2	1.73	0.24	0.065	0.055
9/25/01	CH - 4	D4	LMB	M3	22.8	134	1.43	0.13	1.07	0.10	-	0.012	177	6.9	1.67	0.63	0.146	0.176
9/25/01	CH - 4	D4	LMB	U	20.3	101	1.18	-	1.17	-	-	0.002	148	40.7	2.01	1.12	0.131	0.086
9/25/01	CH - 4	D4	LMB	U	22.1	122	1.27	0.03	1.04	0.02	-	0.002	212	32.5	2.83	0.66	0.168	0.033
9/26/01	CH - 3	D3	LMB	F	10.9	14	0.22	0.04	1.57	0.29	-	0.002	45	0.0	0.41	-0.06	0.459	0.812
9/26/01	CH - 3	D3	LMB	F1	12.9	20	0.25	0.08	1.25	0.40	-	0.002	57	0.0	2.12	0.49	1.841	0.248
9/26/01	CH - 3	D3	LMB	F1	17.2	45	0.36	0.14	0.80	0.31	-	0.002	57	0.0	2.82	0.70	1.076	0.304
9/26/01	CH - 3	D3	LMB	M3	36.7	612	3.23	0.54	0.53	0.09	-	0.043	229	17.5	0.77	0.27	0.111	0.123
9/26/01	CH - 3	D3	SPB	U	17.4	50	0.49	0.11	0.98	0.22	-	0.006	0	-	-	-	-	-
9/26/01	CH - 3	D3	SPB	U	15.3	34	0.34	0.03	1.00	0.09	-	0.01	0	-	-	-	-	-

Date	River Site	River Site	Species[1]	Sex[2]	Total Length (cm)[4]	Body (g)	Liver (g)	Gonad (g)	HSI (%)[5]	GSI (%)[5]	Health[6]	Vtg (mg/ml)[7]	Plasma (uL)	Protein (mg/dL)	Glucose (mg/dL)	Lactate (mg/dL)	E2 (ng/mL)	T (ng/mL)
9/26/01	CH - 3	D3	LMB	U	12.1	18	0.12	0.05	0.67	0.28	-	0.002	0	-	-	-	-	-
9/26/01	CH - 3	D3	LMB	U	26.5	211	1.41	0.13	0.67	0.06	-	0.023	245	12.9	2.03	0.63	0.132	0.353
9/27/01	CH - 2	D2	SPB	F	12.1	19	0.19	0.09	1.00	0.47	-	0.013	268	4.0	0.82	0.28	0.038	0.026
9/27/01	CH - 2	D2	SPB	F	12.4	18	0.2	0.08	1.11	0.44	-	0.009	37	16.5	3.51	1.86	0.553	0.099
9/27/01	CH - 2	D2	SPB	F1	16.8	53	0.6	0.22	1.13	0.42	-	0.002	60	0.0	3.21	1.44	0.457	0.353
9/27/01	CH - 2	D2	SPB	F1	15.1	31	0.41	0.08	1.32	0.26	-	0.006	76	5.6	5.60	1.16	0.370	0.213
9/27/01	CH - 2	D2	LMB	F2	52.5	2196	21.06	14.4	0.96	0.66	-	0.123	113	47.4	2.70	0.83	1.395	0.222
9/27/01	CH - 2	D2	LMB	F2	39.6	826	5.29	4.78	0.64	0.58	-	0.004	112	36.8	3.25	1.29	0.156	0.131
9/27/01	CH - 2	D2	SPB	F2	35.8	554	3.64	2.51	0.66	0.45	-	0.014	127	35.5	4.00	1.01	0.336	0.140
9/27/01	CH - 2	D2	SPB	F2	34.8	499	3.86	2.7	0.77	0.54	-	0.002	268	17.1	4.00	1.19	0.153	0.284
9/27/01	CH - 2	D2	SPB	M3	34.4	564	5.3	1.03	0.94	0.18	-	0.004	118	31.0	4.35	1.21	1.502	0.293
9/27/01	CH - 2	D2	SPB	M3	37.1	690	6.35	1.27	0.92	0.18	-	0.011	72	30.6	5.27	1.21	0.388	0.307
9/27/01	CH - 2	D2	SPB	M3	25.4	180	1.33	0.3	0.74	0.17	-	0.002	166	23.5	5.11	0.73	0.244	0.308
9/27/01	CH - 2	D2	LMB	M3	31.8	383	1.92	0.26	0.50	0.07	-	0.012	200	22.6	1.33	0.99	0.125	0.085
9/27/01	CH - 2	D2	LMB	M3	24.7	182	1.68	0.19	0.92	0.10	-	0.004	218	26.8	4.60	1.57	0.132	0.068
9/27/01	CH - 2	D2	SPB	M3	21.2	104	1.1	0.2	1.06	0.19	-	0.002	97	10.7	5.02	0.92	0.198	0.526
9/27/01	CH - 2	D2	SPB	M3	17.7	71	0.84	0.12	1.18	0.17	-	0.002	150	14.3	4.49	1.50	0.258	0.076
9/27/01	CH - 2	D2	SPB	M3	16.7	53	0.66	0.12	1.25	0.23	-	0.002	79	0.0	4.49	1.01	0.513	0.479
9/27/01	CH - 2	D2	SPB	U	13.4	22	0.21	0.03	0.95	0.14	-	0.005	0	-	-	-	-	-
9/27/01	CH - 2	D2	SPB	U	12.3	13	-	-	-	-	-	0.007	8	446.7	14.28	2.62	2.778	0.780
9/27/01	CH - 3	D3	SPB	F	18.7	73	0.64	0.12	0.88	0.16	-	0.035	28	112.9	6.66	1.62	0.788	0.736
9/27/01	CH - 3	D3	SPB	F1	13.6	23	0.21	0.12	0.91	0.52	-	0.002	26	20.5	2.02	0.12	0.957	0.131
9/27/01	CH - 3	D3	LMB	F2	39.2	787	6.94	5.6	0.88	0.71	-	0.018	106	97.1	1.57	0.71	0.405	0.282
9/27/01	CH - 3	D3	SPB	F2	35.4	539	4.06	2.99	0.75	0.55	-	0.005	57	124.2	5.40	0.98	0.527	0.277
9/27/01	CH - 3	D3	LMB	F2	30.3	353	3.2	1.03	0.91	0.29	-	0.002	163	71.8	7.49	1.15	0.135	0.549
9/27/01	CH - 3	D3	LMB	F2	27.2	253	1.23	0.98	0.49	0.39	-	0.002	266	82.7	2.30	0.99	0.087	0.173
9/27/01	CH - 3	D3	LMB	M3	40.4	914	5.83	1.83	0.64	0.20	-	0.008	226	85.1	3.58	0.75	0.074	0.037
9/27/01	CH - 3	D3	SPB	U	13.8	21	0.17	-	0.81	-	-	0.002	13	161.1	7.67	0.61	1.328	0.555
9/27/01	CH - 3	D3	SPB	U	11.7	18	0.18	-	1.00	-	-	0.002	10	62.1	8.69	1.83	3.002	0.444
9/28/01	CH - 1	D1	?	F1	18.6	65	0.68	0.11	1.05	0.17	-	0.002	111	26.2	3.80	0.90	0.463	0.080
9/28/01	CH - 1	D1	SPB	F1	18.2	55	0.32	0.31	0.58	0.56	-	0.002	0	-	-	-	-	-

Date	River Site	River Site	Species[1]	Sex[2]	Total Length (cm)[4]	Body (g)	Liver (g)	Gonad (g)	HSI (%)[5]	GSI (%)[5]	Health[6]	Vtg (mg/ml)[7]	Plasma (uL)	Protein (mg/dL)	Glucose (mg/dL)	Lactate (mg/dL)	E2 (ng/mL)	T (ng/mL)
9/28/01	CH - 1	D1	SPB	F1	17.8	59	0.57	0.25	0.97	0.42	-	0.002	96	33.1	4.65	1.15	0.190	0.160
9/28/01	CH - 1	D1	SPB	F1	16.8	46	5.46	0.11	11.87	0.24	-	0.002	0	-	-	-	-	-
9/28/01	CH - 1	D1	LMB	F2	39.9	686	4.01	4.51	0.58	0.66	-	0.045	164	47.6	2.04	0.86	0.189	0.077
9/28/01	CH - 1	D1	LMB	F2	31.2	411	4.13	2.37	1.00	0.58	-	0.002	245	44.0	1.81	0.84	0.127	0.162
9/28/01	CH - 1	D1	SPB	F2	22.9	166	1.18	0.71	0.71	0.43	-	0.002	112	26.1	3.79	0.73	0.412	0.026
9/28/01	CH - 1	D1	LMB	F2	40	787	6.77	4.18	0.86	0.53	-	0.034	114	47.2	3.50	1.19	0.340	0.060
9/28/01	CH - 1	D1	LMB	F2	50.8	1994	21.12	16.47	1.06	0.83	-	0.009	287	47.2	1.54	0.63	0.129	0.010
9/28/01	CH - 1	D1	?	M3	41.1	907	7.84	0.7	0.86	0.08	-	0.005	16	135.5	9.79	2.32	2.196	0.336
9/28/01	CH - 1	D1	SPB	M3	30.8	350	1.69	0.26	0.48	0.07	-	0.002	161	43.0	0.90	0.72	0.245	0.054
9/28/01	CH - 1	D1	LMB	M3	28.7	298	2.15	0.31	0.72	0.10	-	0.004	179	23.1	3.03	1.09	0.189	0.081
9/28/01	CH - 1	D1	SPB	M3	33.7	551	4.95	1.45	0.90	0.26	-	0.002	95	50.1	3.90	1.03	0.317	0.045
9/28/01	CH - 1	D1	SPB	M3	26.6	240	2.68	0.33	1.12	0.14	-	0.021	111	36.9	4.34	1.04	0.391	0.070
9/28/01	CH - 1	D1	SPB	M3	24.3	166	1.4	0.32	0.84	0.19	-	0.003	82	46.9	3.62	1.10	0.326	0.116
9/28/01	CH - 1	D1	LMB	M3	25.7	199	1.27	0.134	0.64	0.07	-	0.002	173	47.0	2.49	0.74	0.134	0.030
9/28/01	CH - 1	D1	LMB	M3	35.2	580	4.78	0.57	0.82	0.10	-	0.004	141	49.4	2.54	0.89	0.206	0.007
9/28/01	CH - 1	D1	SPB	M3	35.4	613	5.09	0.96	0.83	0.16	-	0.012	68	25.0	3.25	0.62	0.538	0.184
9/28/01	CH - 1	D1	LMB	M3	29.5	309	2.56	0.16	0.83	0.05	-	0.002	281	30.0	3.23	0.84	0.070	0.048
9/28/01	CH - 1	D1	SPB	M3	22.4	123	1.16	0.15	0.94	0.12	-	0.002	97	49.7	4.06	0.89	0.273	0.082
9/28/01	CH - 1	D1	SPB	U	35.1	535	3.46	0.09	0.65	0.02	-	0.019	244	35.1	4.06	1.07	0.147	0.117
9/28/01	CH - 1	D1	SPB	U	15.8	40	0.4	0.04	1.00	0.10	-	0.002	67	44.3	5.00	1.22	0.393	0.144
9/28/01	CO - 2	B	SPB	F1	19.8	80	0.49	0.22	0.61	0.28	-	0.002	162	47.8	7.66	0.40	0.210	0.327
9/28/01	CO - 2	B	SPB	F2	23.7	114	0.69	0.67	0.61	0.59	-	0.036	88	68.8	4.59	0.47	0.326	0.665
9/28/01	CO - 2	B	SPB	M3	33.3	441	3.02	0.93	0.68	0.21	-	0.006	389	53.7	2.11	0.37	0.116	0.053
9/28/01	CO - 2	B	SPB	M3	31.2	305	1.42	0.64	0.47	0.21	-	0.002	187	41.6	3.71	0.21	0.148	0.095
9/28/01	CO - 2	B	SPB	M3	27.6	190	1.22	0.44	0.64	0.23	-	0.019	293	42.7	1.32	0.08	0.142	0.122
9/28/01	CO - 2	B	REB	M3	21.8	120	1.11	0.93	0.93	0.78	-	0.026	84	69.9	3.08	0.68	0.822	0.306

[1] LMB = Largemouth bass, SPB = Spotted bass, REB = Redeye bass

[2] M1 = spermatogonia, M2 = spermatocytes, M3 = sperm, F1 = previtellogenic, F2 = corticol alveoli, F3 = vitellogenesis, M = male (stage unidentified), F = female (stage unidentified), U = unidentified. Bold = sex verified or determined from histopathology

[3] L-HYP = Cell hypertrophy, L-SPC = Space between the cells, L? = unusual looking liver

[4] Total length was not measured during the spring sampling event.

[5] HSI = liver weight/total weight*100. GSI = gonad weight/total weight*100.

[6] L = lesion, N = nodule, P = parasite, FE = fin erosion, LCL = light colored liver, LP = liver parasites, SG = small gonad, LN = liver nodules, ML = mouth lesions, LA = liver abnormal, FL = fin lesion, SL = spots on liver

[7] Vtg = vitellogenin detection limit = 0.002 mg/mL.

NOTES:

An "-" in a space indicates that either the analysis/measurement was not taken or a sample was not collected

U = Not Detected

River Site Designations, as used on figures:

C	=	CO - 1 = Conasauga River Location, Urban Site
B	=	CO - 2 = Conasauga River Location, Agricultural Site
A	=	CO - 3 = Conasauga River Location, Forested Site
D1	=	CH - 1 = Chattahoochee River Location, Station #1
D2	=	CH - 2 = Chattahoochee River Location, Station #2
D3	=	CH - 3 = Chattahoochee River Location, Station #3
D4	=	CH - 4 = Chattahoochee River Location, Station #4

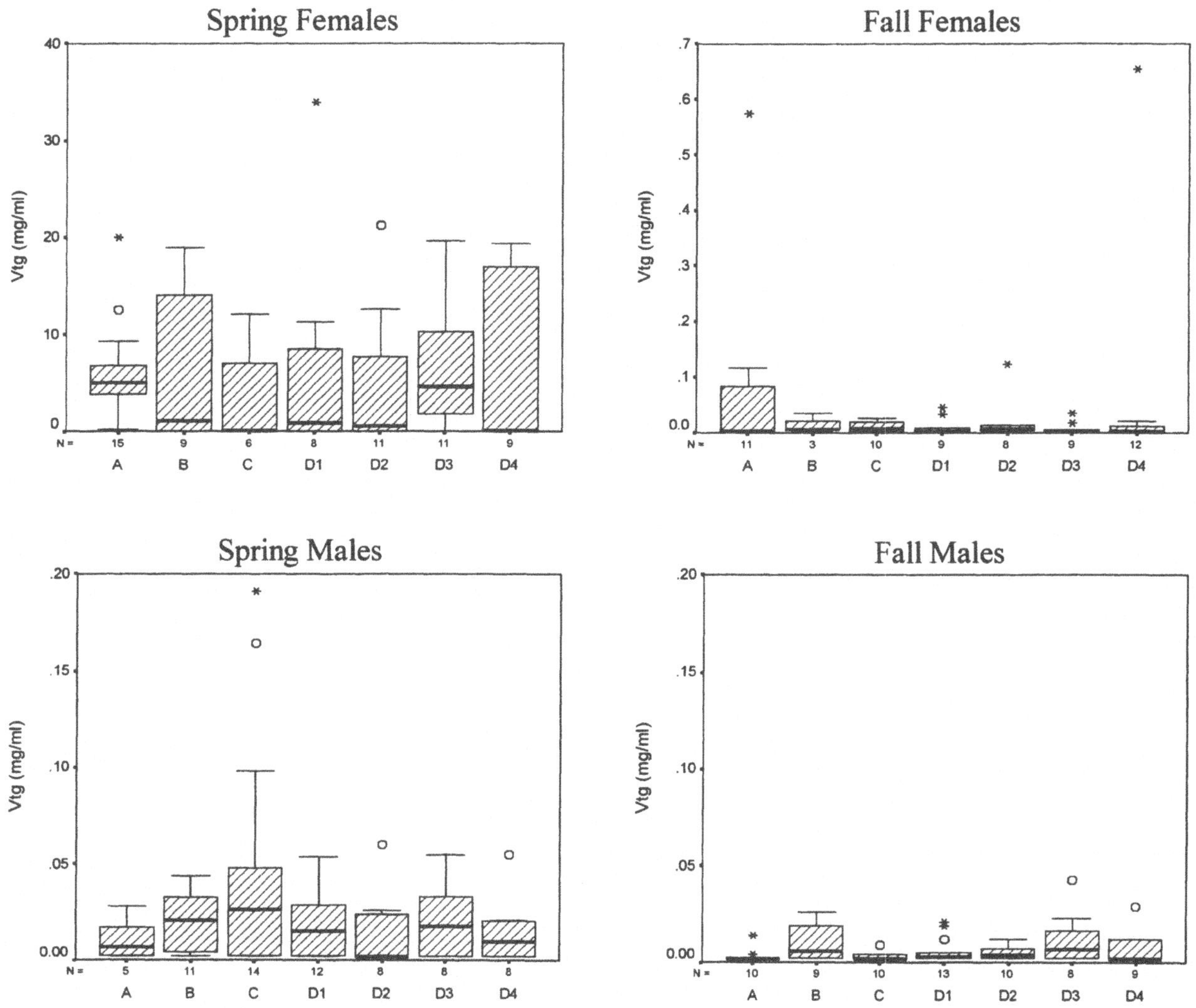

Figure C-5. Comparison of Plasma Vitellogenin Levels For Bass Collected at Sampling Locations on the Conasauga and Chattahoochee Rivers (Spring and Fall 2001). Note: A = Forested; B = Agricultural; C = Urban. (Conasauga Stations shown in Figure C-3.) D1 = Station #1; D2 = Station #2; D3 = Station #3; D4 = Station #4. (Chattahoochee Stations shown in C-2.) Station A predominantly redeye.

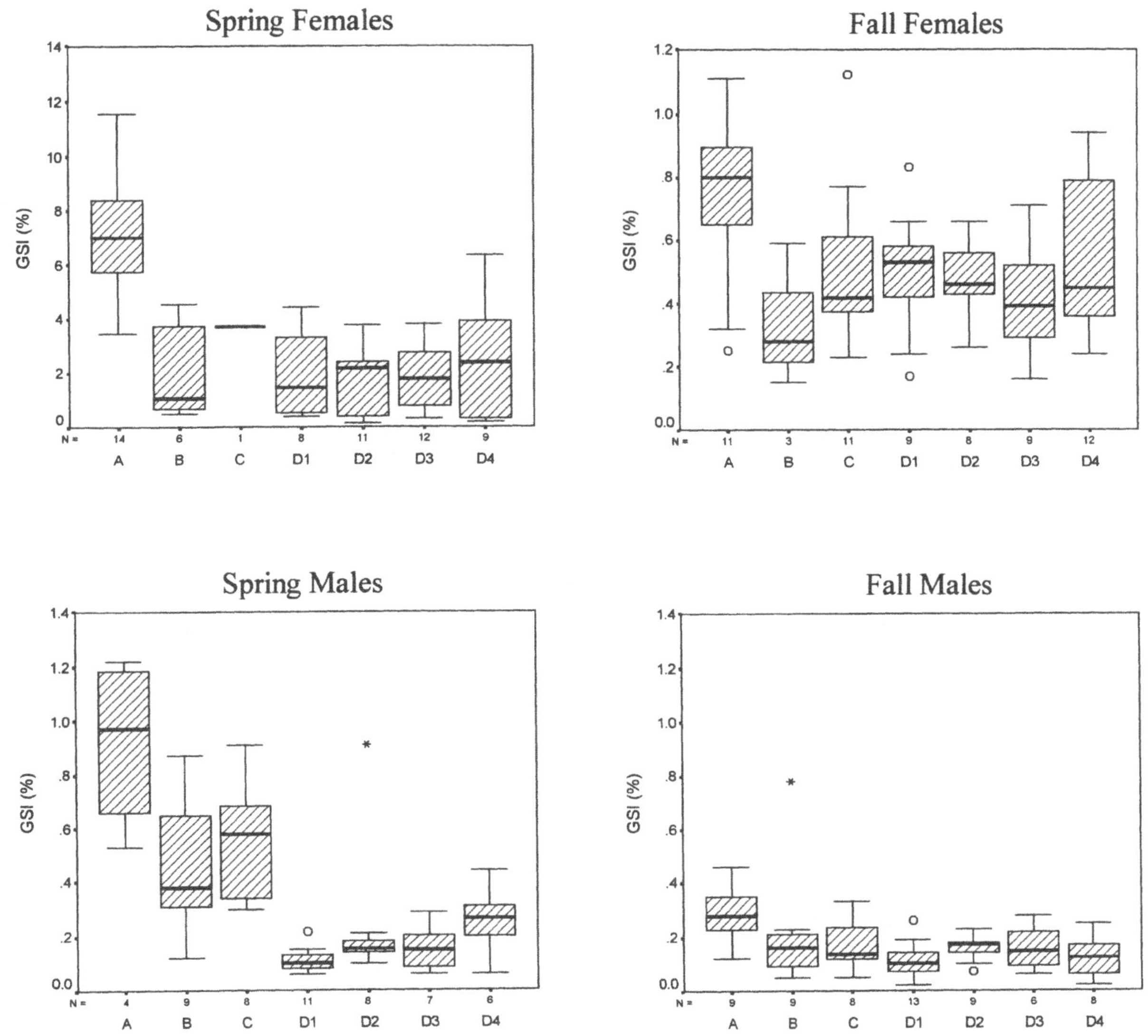

Figure C-6. Comparison of Gonadal Somatic Index (GSI) For Bass Collected Sampling Locations on the Conasauga and Chattahoochee Rivers (Spring And Fall 2001). Note: A = Forested; B = Agricultural; C = Urban. (Conasauga Stations shown in Figure C-3.) D1 = Station #1; D2 = Station #2; D3 = Station #3; D4 = Station #4. (Chattahoochee Stations shown in C-2.)

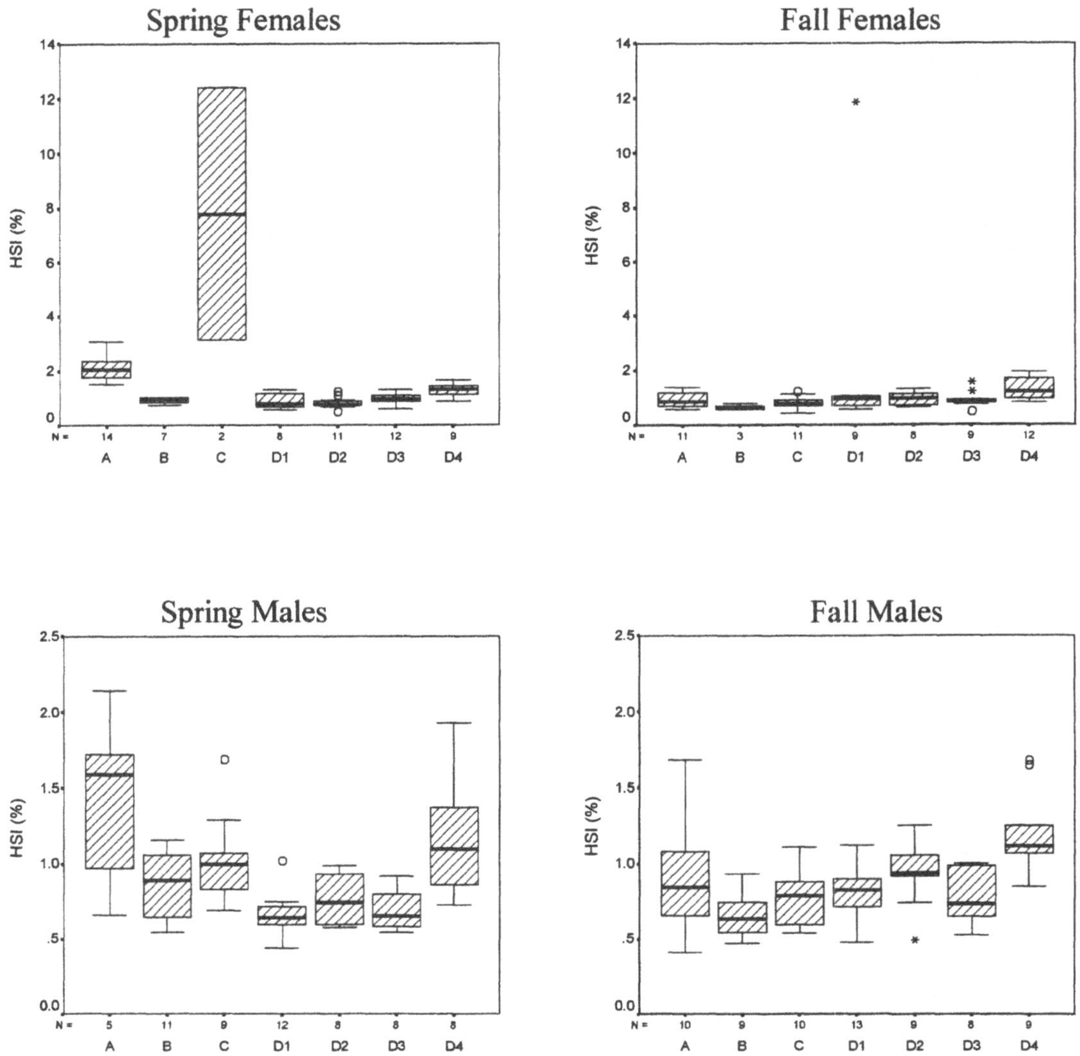

Figure C-7. Comparison of Hepatosomatic Index (HIS) For Bass Collected Sampling Locations of the Conasauga and Chattahoochee Rivers (Spring and Fall 2001). Note: A = Forested; B = Agricultural; C = Urban. (Conasauga Stations shown in Figure C-3.) D1 = Station #1; D2 = Station #2; D3 = Station #3; D4 = Station #4. (Chattahoochee Stations shown in C-2.)

APPENDIX C6

WATER SAMPLES

C6.1 Water Sample Quality Measurements

Water samples were collected an analyzed for general water quality parameters at each of the seven fish sampling locations in both rivers. These *in situ* water quality measurements are summarized in Table C-7.

Table C-7. Summary of In-Situ Water Chemistry Measurements at Each of the Seven Fish Stations

ID	Date	Turbidity	Specific Conductivity	Dissolved Oxygen	Temperature	pH
C	4/12/01	25.00	156	7.03	20.48	7.42
	4/26/01	27.40	165	8.16	16.56	7.58
	9/5/01	66	149	5.38	24.2	7.17
	9/17/01	M	229	8.36	21.6	7.79
B	4/12/01	29.50	98	7.81	20.63	7.55
	4/26/01	18.50	101.7	8.58	15.66	7.58
	9/5/01	11	78	6.52	22	6.99
	9/19/01	9.5	113	8.37	20.7	7.93
A	4/12/01	8.40	14	8.67	17.5	7.02
	4/26/01	8.80	13.7	8.36	16.44	6.98
	9/5/01	2	29	8.32	19.6	6.98
	9/18/01	0.3	20.1	8.61	18.7	7.22

	Date	Turbidity	Specific Conductivity	Dissolved Oxygen	Temperature	pH
D1	5/7/01	13.30	203	8.65	28.02	7.2
	9/28/01	9.6	182.1	7.75	25.4	7.7
D2	5/8/2001 *	18.30	237.4	7.43	24.31	6.97
	9/16/2001 **	M	185	8.1	22.74	7.65
D3	9/16/2001 **	M	185	8.1	22.74	7.65
D4	5/10/01	4.40	102.7	10.1	15.8	6.76
	9/25/01	36.8	69.5	8.31	18.06	7.48

In addition, grab samples of water were collected to evaluate response using the YES assay. A total of 19 samples (including one 20 ng/L spike and one blank) and a total of 24 samples (two duplicates, three spiked samples, and one blank) were processed from three stations on the Conasauga River and four stations on the Chattahoochee River during the April and September sampling, respectively. These locations are shown in Figures C-2 and C-3. In the fall, chemical analyses were preformed on selected samples.

A processing station was set up near each sampling site. Samples were collected from the river's surface from the shore or from a bridge. Sample collection methods were altered after the spring survey for the fall survey. In the spring, grab samples collected in a 5-gallon cubitainer were filtered through a 47-mm glass microfibre filter, then 2-L samples were peristaltically pumped through (20cc/5g) C18 columns, then cooled (4°C) until eluted with methanol in the laboratory. In the fall, there was an opportunity to split samples to compare YES assay results with chemical analyses performed by Ed Kolodziej, working with Dr. Sedlak at University of California, Berkley on chemicals analysis of EDCs. Consistent with their procedures, three-gallon fluorinated jerricans were used for sample collection, then following filtration through the glass fiber filters, the two liters of each sample were pressure filtered through conditioned C18 disks. As with the columns, the C-18 disks were refrigerated at 4°C for transportation back to the laboratory.

Figure C-8 shows the results of the YES assay spikes into milli-Q water and river samples. This suggests that a possible inhibition of the response of the E_2 in this matrix. This has consistent with findings of (Garcia-Reyero et al, 2001) where it was determined that inhibition in river samples or effluents may be hormone specific, or reflect more general response in the yeast systems.

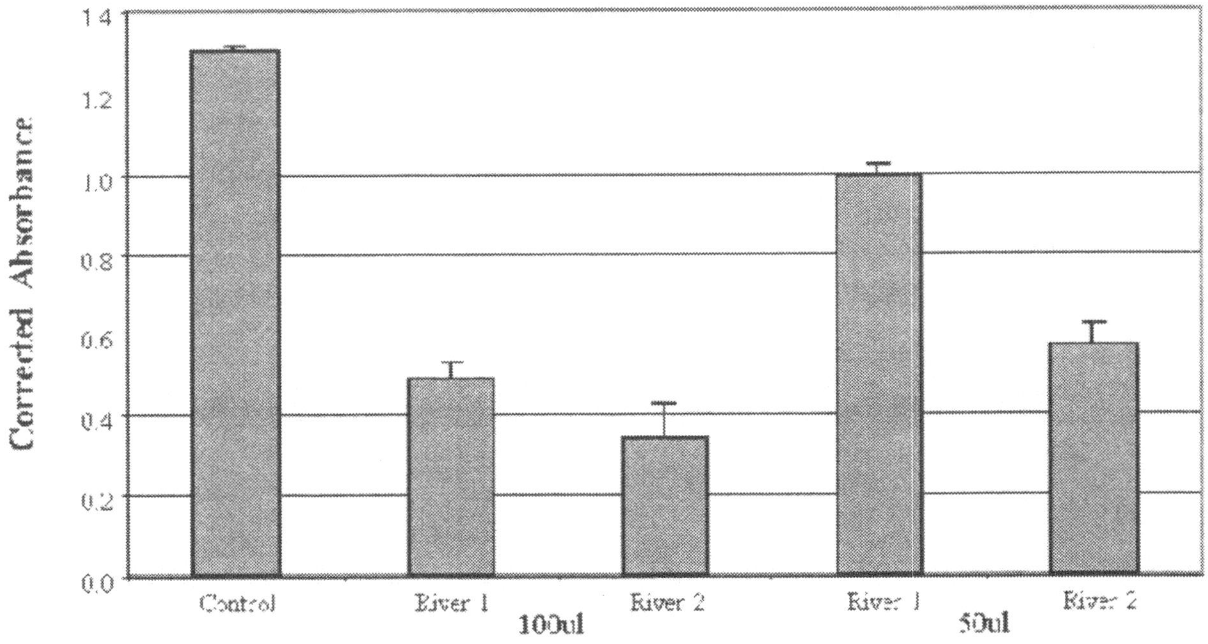

Figure C-8. In Vitro Screening: Inhibition of Response Present in Spiked River Water Samples. Control and river samples spike at 20 ng/L in the field. After concentration on C-18 and elution with menthanol, 100 µl methanol extract was added to wells. Spikes into river samples showed significantly lower response than control water. Increased response with reduced volume (50µl) suggests reduced interference.

The responses in the YES assay for the various stations in the spring and fall are shown in Figure C-9.

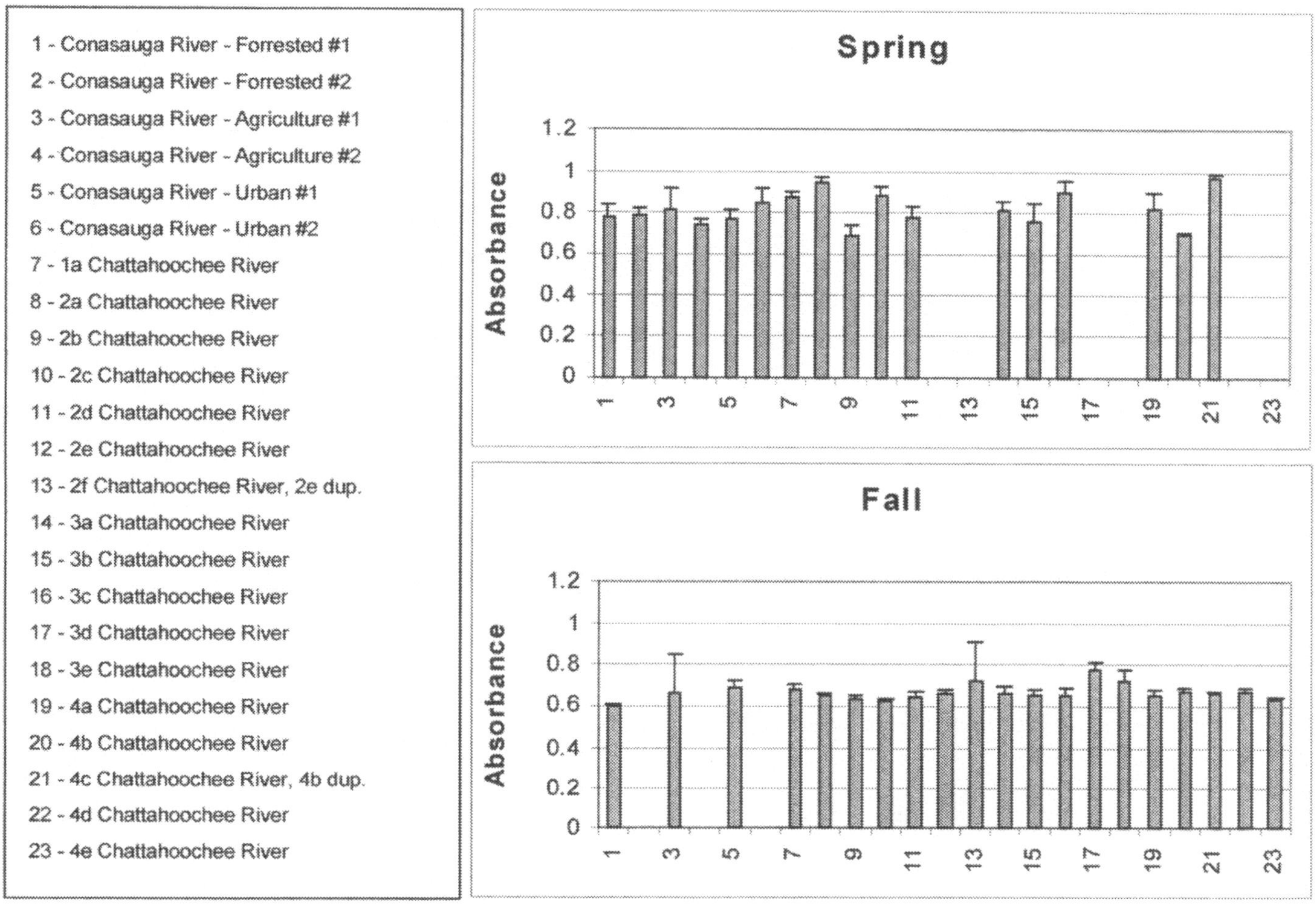

Figure C-9. In Vitro Screening: Inhibition of Response Present in Spiked River Water Samples. To see site location, refer to Figures C-2 and C-3,

In the spring all river stations showed response above the field blank. In the fall, the highest response in the YES Assay corresponded to Station B - (Conasauga agricultural station) where a slight elevation of vitellogenin levels were shown in the fall samples (see Figure C-9).

APPENDIX C7

CHEMICAL ANALYSES FOR EDCs

C7.1 Chemical Analyses

The results of the chemical analyses done on the fall samples are shown in Table C-8. Estrone was the only chemical detected at a concentration of 8.5ng/L at Station 3E. Estrone is a degraded product of E2. Recoveries were 17 to 38 percent for estradiol.

C7.1.1 Sediment Samples

Five sediment samples were collected from three stations on the Conasauga River and two stations on the Chattahoochee River in September 2001. These samples were collected from just off shore at all locations. The sediments were refrigerated on wet ice, transported back to the laboratory and stored in a cold room at 4°C. 15-mL of each sample was extracted with 15-mL of hexane. The YES assay was performed on 100-μL aliquots of the hexane extracts, evaporated to dryness prior to addition of the yeast media.

The results of the sediment screening are shown in Figure C-10. Previous studies of various sediment samples screened with fathead minnows suggested some potential for estrogenic responses.

Table C-8. Data Summary for Atlanta Area Sampling Trip September, 2001

Sample	Sample Site Code	Estradiol Conc. [ng/L]	Estrone Conc. [ng/L]
Conasauga Creek (Forested Site) 9-24-01	A	N.D.	N.D.
Conasauga Creek (Urban Site) 9-24-01	C	N.D.	N.D.
Chattahoochee R. @ Georgia Power 9-28-01	D-1a	N.D.	N.D.
Chattahoochee R. @ Bankhead Drive 9-28-01	D-2a	N.D.	N.D.
Chattahoochee R. @ Camp Creek Pkwy 9-26-01	D-2c	N.D.	N.D.
Sweetwater Creek 9-26-01	D-2d	N.D.	N.D.
Utoy WWTP 9-27-01	D-2e	N.D.	**B.Q.**
Chattahoochee R. @ Rt. 166 9-26-01	D-3a	N.D.	N.D.
Chattahoochee R. @ Rt. 92 9-26-01	D-3c	N.D.	N.D.
Camp Creek WWTP 9-27-01	D-3d	N.D.	**B.Q.**
Sweetwater WWTP 9-27-01	D-3e	**B.Q.**	**8.5**
Big Creek 9-25-01	D-4a	N.D.	N.D.
Chattahoochee R. @ Willeo Creek 9-25-01	D-4b	N.D.	N.D.
Chattahoochee R. @ Erica's 9-25-01	D-4d	N.D.	N.D.
Limits of Detection:	Estrone:	**0.5ng/L**	
	17β-Estradiol	**0.4ng/L**	
Limits of Quantification:	Estrone:	**1.5ng/L**	
	17β-Estradiol	**1.2ng/L**	
Spike Recoveries (measured by 17β-Estradiol):			
Utoy WWTP, 2e (9-27-01)			**38%**
Camp Creek Pkwy, 3b (9-27-01)			**17%**

N.D. - Sample concentration is below limit of detection. B.Q, Sample concentration is below limit of quantification.
See Section A6-1 for Methods

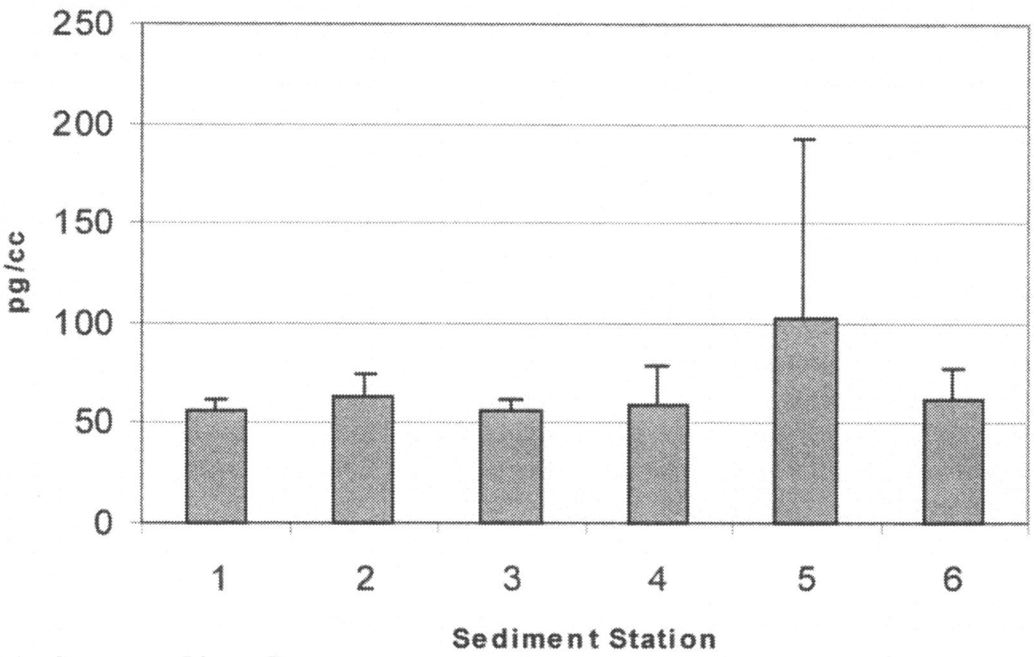

1 - Conasauga River - Forest
2 - Conasauga River - Agriculture
3 - Conasauga River - Urban
4 - Chattahoochee River - #1
5 - Chattahoochee River - #3
6 - Hexane Blank

Figure C-10. In Vitro Screening: One of The Five Stations Showed Positive Response. To see site locations, refer to Figures C-2 and C-3.

APPENDIX C8

RNA STUDIES

C8.1 Introduction

Fathead minnows were placed in cages in the spring and in the fall at three sampling stations on the Conasauga River as described previously. Three locations are included: forest (reference area), mainly agricultural use (nominally of intermediate contaminant level) and urban (nominally of high contaminant level). The fathead minnows were placed at the location in cages for 2 weeks. For RNA measurements the entire body cavity (internal organs including liver and gonad) was removed from the fathead minnows and placed in RNAlater fixative. The samples were processed for RNA as described in Appendix A.

C8.1.1 Males

In male fish, vitellogenin mRNA is an immediate biomarker for estrogenic substances since it is induced within 12 hours of exposure and reaches its maximum level still increasing at 48 hours. With no further exposure, no further synthesis of mRNA is seen and what is present is quickly degraded. This is different from plasma vitellogenin protein levels that reach their maximum concentration in males after 3 days from a single exposure and then remain in the blood for up to a month, since plasma vitellogenin in males is not quickly turned over. Thus the measured levels of mRNA found in the deployed fathead minnows would have resulted from exposures sensed during the second week of deployment. The mRNA levels are compared to plasma vitellogenin protein at each location and for each season. We also measure ERα mRNA in all of the fish. ERα is present normally in both males and females. Its levels are also increased by exposure to E2, but the range of response (3-8 fold) is much less than that seen for vitellogenin mRNA that can vary by over a thousand-fold. These results are illustrated in Figure C-11 for males, C-12 for females, and data presented in Table C-9.

Table C-9. Data Summary for mRNA Measurements

Golder ID	Sample ID	VTG Protein Conc.	ERa mRNA (copies/µg)	Vtg mRNA (copies/µg)
Spring Urban Female				
101	Spring-1	9.450	149707.3	76400651
102	Spring-2	4.085	1060.8	1623255
105	Spring-5	19.829	35140.2	16895400
114	Spring-14	12.160	62005.1	27902156
116	Spring-16	14.068		
119	Spring-19	5.177	11996.3	396596
Average		10.795	51981.9	24643611
Standard Error		2.398	26576.5	13905572

Golder ID	Sample ID	VTG Protein Conc.	ERa mRNA (copies/µg)	Vtg mRNA (copies/µg)
Spring Urban Male				
103	Spring-3	0.0005	13371.1	74064
104	Spring-4	0.0005	2002.4	14631
106	Spring-6	0.0005	11542.4	290109
107	Spring-7	0.0005	12424.3	18640
110	Spring-10	0.0005	5247.5	10724
112	Spring-12	0.053	1695.9	1262
113	Spring-13	0.005	2279.8	5942
115	Spring-15	0.017	13081.8	344156
117	Spring-17	0.013	9718.4	175929
118	Spring-18	0.0005	3965.6	75539
removed sample 111 (Spring-11) because M and vtg 8.2mg probably F				
Average		0.009	7532.9	101099
Standard Error		0.005	1563.0	39896
Spring Agricultural Female				
122	Spring-22	19.746	88060.0	42544487
124	Spring-24	6.506	14540.6	4165920
126	Spring-26	0.135	6039.1	6841
128	Spring-28	5.83	4215.2	12090
129	Spring-29	0.0005	25341.8	2648020
132	Spring-32	2.534	5524.6	2250983
135	Spring-35	0.446	2383.1	511
136	Spring-36	8.889	12975.9	6377927
137	Spring-37	15.425	3200.5	2498074
139	Spring-39	7.149	108675.8	10911197
Average		6.666	27095.7	7141605
Standard Error		2.099	12177.4	4076238
Spring Agricultural Male				
138	Spring-38	0.009	33982.5	17340
120	Spring-20	0.024	12143.5	18898
121	Spring-21	0.0005	27209.4	58910
125	Spring-25	0.029	17282.2	83023
127	Spring-27	0.0005	8490.2	5195
130	Spring-30	0.009	3086.8	23723
131	Spring-31	0.035	5684.9	2207

Golder ID	Sample ID	VTG Protein Conc.	ERa mRNA (copies/µg)	Vtg mRNA (copies/µg)
133	Spring-33	0.0005	8016.9	4295
134	Spring-34	0.0005	6952.4	497
Average		0.012	13649.9	23788
Standard Error		0.004578027	3517.7	9532
Spring Forest Female				
140	Spring-40	5.915	88167.3	12952511
145	Spring-45	3.751	10814.0	606379
146	Spring-46	0.808	4201.2	6345
147	Spring-47	5.963	4004.7	128606
148	Spring-48	12.418	2096.6	9585
149	Spring-49	13.932	38317.8	27779204
152	Spring-52	8.352	13589.5	4292301
154	Spring-54	0.0005	2477.5	2847
155	Spring-55	1.472	24234.3	792699
158	Spring-58	6.806	3800.2	32096
Average		5.94	19170.3	4660257
Standard Error		1.49	8507.7	2870752
Spring Forest Male				
141	Spring-41	0.003	2255.8	14190
142	Spring-42	0.0005	2902.0	11528
143	Spring-43	0.014	23963.5	7430
144	Spring-44	0.0005	4167.6	2480
150	Spring-50	0.0005	16481.8	2681
151	Spring-51	0.0005	2047.9	1127
153	Spring-53	0.022	7815.0	34345
156	Spring-56	0.0005	43897.3	4887
157	Spring-57	0.024	11145.9	1205
159	Spring-59	0.0005	1147.0	700
Average		0.007	11582.4	8057
Standard Error		0.003	4285.7	3266
Fall Forest Female				
162	Fall-3	20.464	18636.1	295785
164	Fall-5	18.798	1242.9	16600
165	Fall-6	4.955	8234.3	19148

Golder ID	Sample ID	VTG Protein Conc.	ERa mRNA (copies/µg)	Vtg mRNA (copies/µg)
166	Fall-7	16.671	16948.1	318482
167	Fall-8	31.659	30006.8	158150
179	Fall-20	15.596	45285.4	228840
184	Fall-25	4.747	41218.3	70059
185	Fall-26	12.584	27594.3	108856
Average		15.684	23645.8	151990
Standard Error		3.085	5415.7	42023
Fall Forest Male				
160	Fall-1	0.0005	21921.9	4153
161	Fall-2	0.0005	39306.4	8090
170	Fall-11	0.003	18408.3	16870
171	Fall-12	0.0005	13563.3	20916
172	Fall-13	0.0005	31080.1	4145
173	Fall-14	0.0005	57889.2	63525
177	Fall-18	0.0005	24487.3	3235
182	Fall-23	0.0005	32054.8	63718
183	Fall-24	0.004	89162.3	41018
Average		0.0012	36430.4	25074
Standard Error		0.0004	7903.9	8293
Fall Agricultural Female				
186	Fall-27	15.846	3021.1	15576
188	Fall-29	19.587	35464.7	3010980
190	Fall-31	3.606	10694.3	924068
195	Fall-36	5.703	2771.9	5191
199	Fall-40	14.838	4203.6	48704
Average		11.916	11231.1	800904
Standard Error		3.086	6228.6	579453
Fall Agricultural Male				
187	Fall-28	0.004	67258.2	2059
191	Fall-32	0.0005	15481.3	305
192	Fall-33	0.008		
193	Fall-34	0.0005	2204.4	234
194	Fall-35	0.002	13429.0	2019
Average		0.003	24593.2	1154

Golder ID	Sample ID	VTG Protein Conc.	ERa mRNA (copies/µg)	Vtg mRNA (copies/µg)
Standard Error		0.001	14517.9	511
Fall Urban Female				
202	Fall-43	8.215	57793.6	7031
203	Fall-44	7.893	11742.4	1822
206	Fall-47	8.395	4685.5	227
209	Fall-50	0.061		
210	Fall-51	4.06	1672.2	465
213	Fall-54	7.868	5992.8	1387
214	Fall-55	0.983	98905.9	156525
223	Fall-64	2.265		
Average		4.968	30132.1	27910
Standard Error		1.249	16200.2	25743
Fall Urban Male				
201	Fall-42	0.0005		
205	Fall-46	0.0005	2372.95	134.61
215	Fall-56	0.0005		
219	Fall-60	0.0005		
220	Fall-61	0.0005		
Average		0.0005		
Standard Error		0		

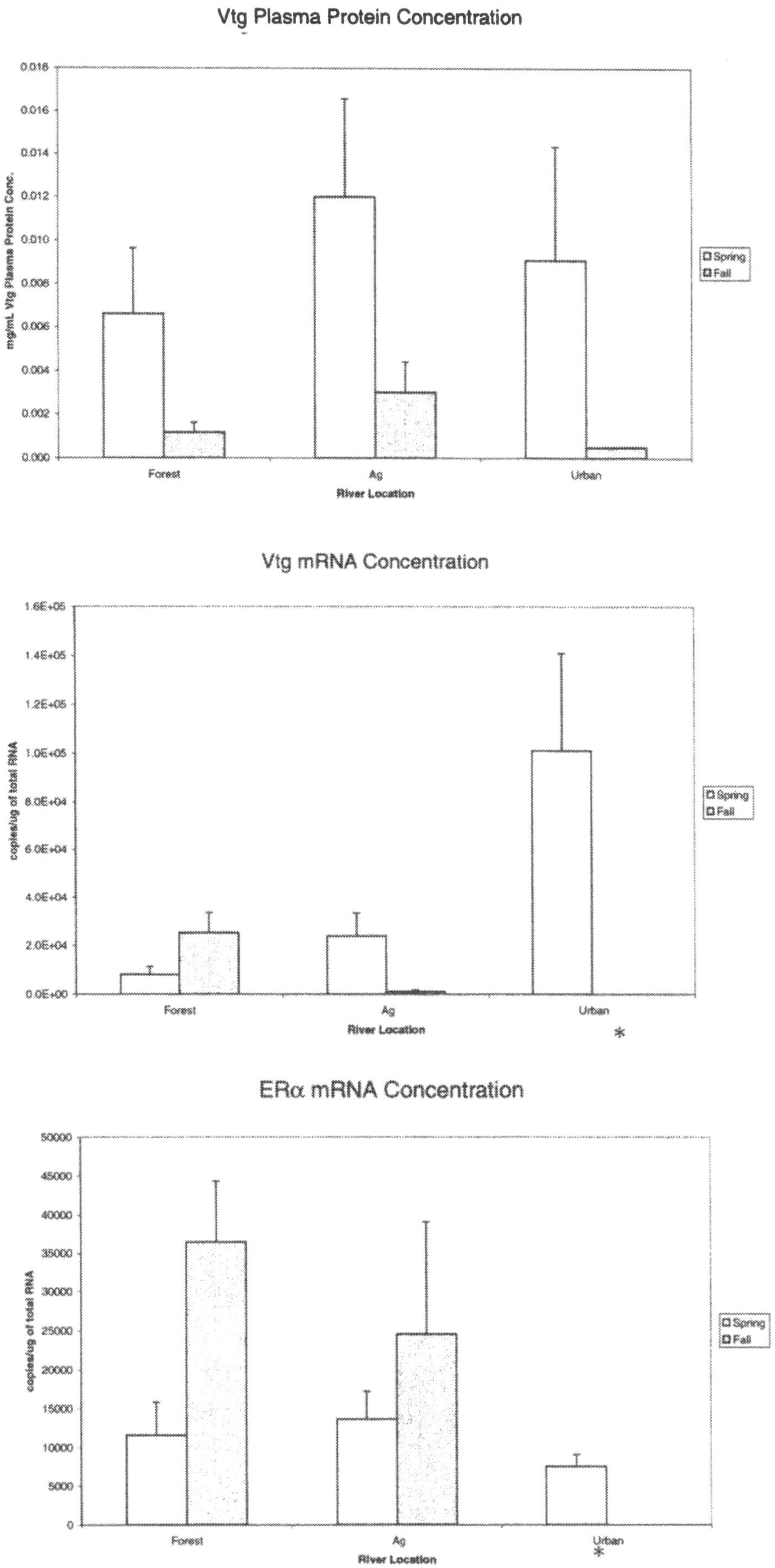

Figure C-11. Male Fathead Minnows: Vitellogenin Protein, Vitellogenin mRNA and Estrogen Receptor α Measurements in Fish Exposed to Test Sites on the Conasauga River.
*denotes samples that were lost.

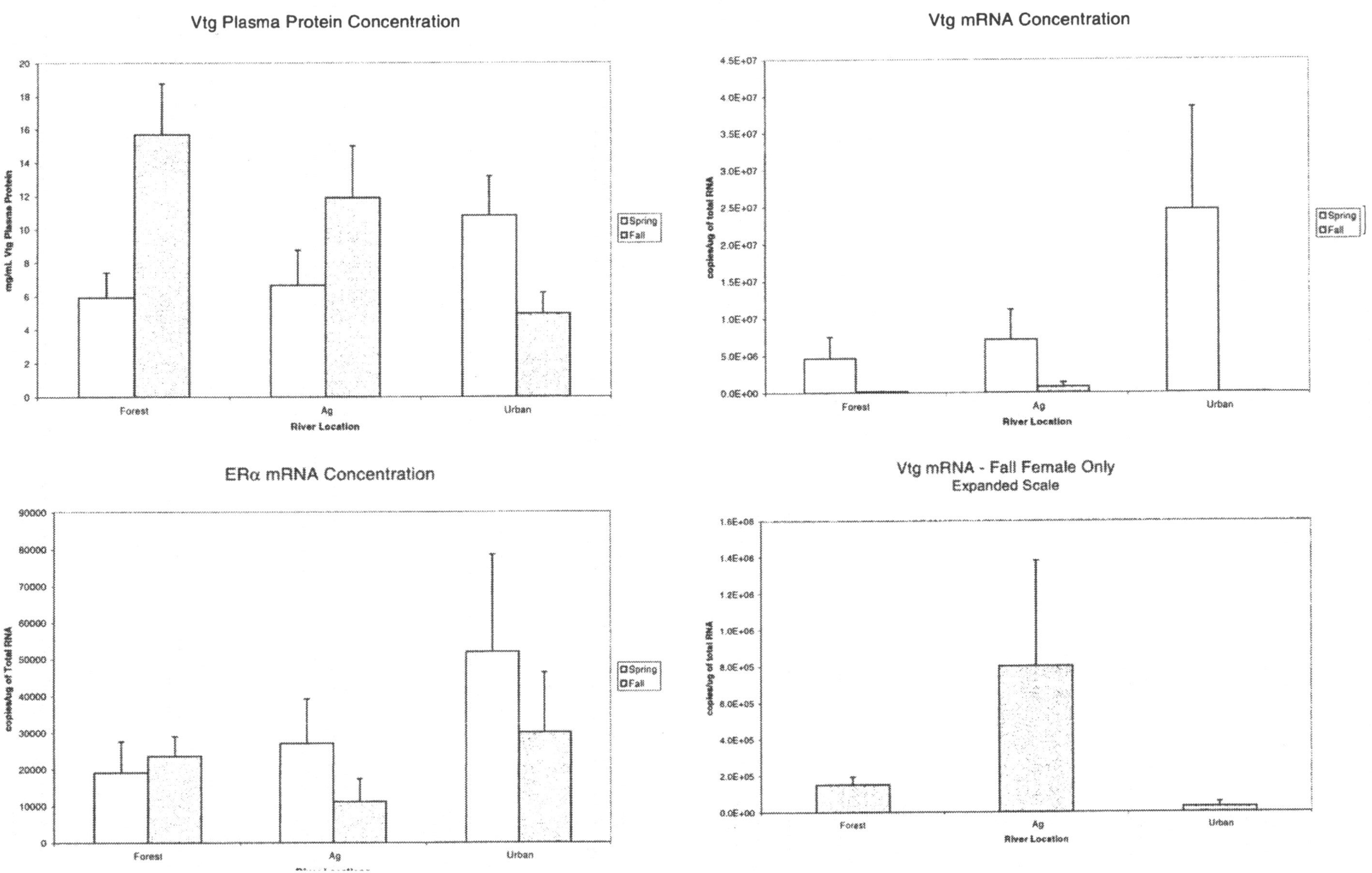

Figure C-12. Female Fathead Minnows: Vitellogenin Protein, Vitellogenin mRNA and Estrogen Receptor α Measurements in Fish Exposed to Test Sites on the Conasauga River.

In the spring we saw high mortality of fish deployed at the urban site, possibly due to toxic levels of exposure, whereas fish deployed at the other two sites did not experience high mortality. Male fish contained plasma vitellogenin at all sites and they showed the highest average value at the agricultural site. The vitellogenin mRNA data is similar to the protein data. Fathead minnows exposed to the forest location had the lowest amount of vitellogenin mRNA (roughly 10,000 copies/μg total RNA), this increased in the Agricultural site (25,000 copies/μg total RNA) and was highest in fish exposed to the urban site (150,000 copies/μg total RNA). All male fish had ERα mRNA, as expected and the copy number did not increase in parallel with vitellogenin mRNA at each of the exposure sites.

In the fall, high mortality of deployed fish was seen only at the agricultural site. The levels of plasma vitellogenin protein vary among the sites but are much lower than what was seen in the spring. The variation among the sites does not appear to be significant. Vitellogenin mRNA levels are also much lower than in the spring. In this case, fathead minnows exposed to the forest site appear to have the highest vitellogenin mRNA levels (25,000 copies/μg total RNA). Unfortunately due to sampling error, the RNA samples for the males from the urban site were degraded when they were prepared and we don't have values for this season. ERα mRNAs are higher in the fall than they were in the spring (average range from 23,000 – 37,000 copies/μg total RNA), for both the forest and the agricultural locations. No values are reported for the urban site since we did not obtain quality RNA for this analysis. No data exist in the literature to help us evaluate the significance of changes in ERα mRNAs as a consequence of exposure or season.

In summary, there is a fairly good correspondence between plasma vitellogenin levels and vitellogenin mRNA levels in these fish. The vitellogenin mRNA levels suggest that male fish were exposed to estrogenic substances at both the agricultural and urban sites when compared to the forest site in the spring. In the fall, the response levels are much lower, suggesting that either the fish are less sensitive during this period of time or that exposures are much lower.

C8.1.2 Females

For females, the presence of plasma vitellogenin protein and hepatic vitellogenin mRNA is normal and because the fish are asynchronous in their cycles, these values will fluctuate somewhat over their 3-4 day cycle (Ankley et al, 2001). However, when a large number are grouped together this variation is averaged and results in a wider spread for controls. Plasma vitellogenin is constantly recruited into the next batch of developing eggs, so it stays at a relatively stable level. Loss of vitellogenin protein and vitellogenin mRNA when compared to controls can point to anti-estrogens or to contaminants that lower the endogenous levels of E2.

In the spring in exposed females, we see approximately the same amount of plasma vitellogenin in fish deployed at the forest and agricultural sites and a slight increase in fish deployed at the urban site. Their vitellogenin mRNA levels and their ERα mRNA levels parallel this. Again the urban site was the one where many of the fish died, suggesting that contaminant levels might have been in the toxic range.

In the fall, plasma vitellogenin levels were higher in fish exposed to water at the forest and agricultural locations and lower in fish exposed to the urban site. Vitellogenin mRNAs were much lower than in the spring but the location that produced the highest values was the agricultural location (Fig. C-12). ERα mRNAs appear to be similar at the forest and agricultural location and slightly higher at the urban location.

APPENDIX D

Overview of Target Species' Lifecycles

D.1 Fathead Minnows

Fathead minnows (*Pimephales promelas*) were used for the caged fish portion of the study. Fathead minnows were chosen as the species for laboratory tests and the caged fish portion of the field study. Fathead minnows are the species identified for evaluating impact of EDCs on fish (Ankley et al, 2001). Fathead minnows were selected for the following criteria:

- Represent one of the largest and most ecologically diverse families (Cyprinidae);
- Are the most extensively tested fish species in the world;
- Are routinely used in regulatory programs for whole effluent testing, pesticide registration, water quality criteria derivation;
- Due to the extensive testing, there is extensive toxicological database for the species;
- Many methods have been developed for culturing and handling the species;
- Have a rapid life cycle by reaching reproductive maturity in four to five months after hatching;
- Are sexually dimorphic (see discussion below); and
- Reproduction can be controlled in the laboratory through temperature and photoperiod year round.

Fathead minnow sex is simple to determine by external observation once they have reached sexual maturity. Males have tubercles on their snout and have an elongated, fleshy or spongy pad that extends from the nape to the dorsal fin, and females have a fleshy ovipositer (Ankley et al, 2001). Because of the ease to distinguish the sex of adult specimens, it is important to use adults in any endocrine disruption studies. It is difficult to find vendors that have adults for sale and that are disease-free individuals. Therefore, planning ahead is important for these projects, so there is plenty of time to grow juveniles or "cleanse" adults purchased from a vendor.

Fathead minnows lay approximately 50 to 150 eggs per spawn, and under laboratory conditions fathead minnows will typically spawn every three to four days (Ankley et al, 2001). Scientists have extensive knowledge about the cycles of fathead minnows but still know little about the basic reproductive physiology and endocrinology.

D.2 Bass Species

The targeted wild specimens were freshwater bass of the family Centrarchidae, genus *Micropterus*, the largemouth (*M. salmoides*), spotted (*M. punctalatus*), and redeye bass (*M. coosae*). Freshwater bass are targeted because they are known to be sensitive, have a defined life cycle, are of recreational importance, and have been targeted in previous field studies.

A single species of bass could not be targeted, due to the variation in physical and chemical characteristics of the Conasauga River between the headwaters and lower reaches. The three species of bass have different habitat preferences based on water temperature, flow velocity, and substrate. Generally, the redeye bass occurs in the cooler, swifter headwaters, largemouth bass prefer warmer, more sluggish waters, and the spotted bass is intermediate of the two. Some overlapping does occur, and areas containing either spotted and redeye bass or spotted and largemouth bass are common. Spotted, redeye, and largemouth bass are closely related species and use overlapping food resources. All three species primarily feed on insects, crayfishes and fishes. In general, young of all three species consume small prey items, including microcrustaceans, small insects, small fishes and small crayfishes. Adults utilize larger prey items including large insects, larger fishes and larger crayfishes. Adult largemouth bass tend to be more piscivorous than adult spotted bass or redeye bass. Spotted, redeye, and largemouth bass are higher order predators in the food chain, and as such, they may bioaccumulate hydrophobic chemicals.

A substantial database of information on species of bass has been developed, including hormonal and vitellogenic cycles in streams of the southeastern United States. Because bass reproduce only once a year, in a fairly synchronous manner, each part of their cycle can be easily described and isolated. Therefore, the temporal measurements of biomarkers would be influenced both by this natural cycle as well as external environmental hormones.

D.2.1 Largemouth Bass

The native range of the largemouth bass (*Micropterus salmoides*) is from the lower Great Lakes to the Mississippi River drainages, the Gulf Coast, Florida and part of the Rio Grande, and along the Atlantic slope as far north as Virginia (Scott and Crossman, 1973). The largemouth bass has been stocked throughout North America and other warm regions of the world as a recreational sportfish. Largemouth bass in the southern United States will usually spawn after one year, at a length of at least 10 inches. Males select a nest site in the spring once water temperatures are above 60°F, typically between April and June, peaking in early May. Normal spawning water temperature ranges from 65°F to 75°F in water depths of 1 to 4 feet near the shoreline. Hard substrates such as sand and gravel in the vicinity of cover or structure are preferred spawning areas. Spawning with multiple females is common. Females typically release ½ of their eggs during the first spawn and half of the remaining eggs during the second spawn. A third spawn occurring up to 1 month later is common. Females average about 4,000 eggs per pound of body weight. Usually at spawning the weight of eggs in females comprises 10 percent or more of her body weight. Males do not eat during the spawning season, and attempt to guard the nest and fan silt away from the eggs until they hatch. Eggs hatch in 2 to 4 days, depending on water temperature. The male will continue to guard the fry until they disperse, which may be a period of 2 weeks or more depending on temperature. Nesting success is reduced in the wild due to predation by sunfish and when temperature drops below 60°F. (Davis and Lock, 1997).

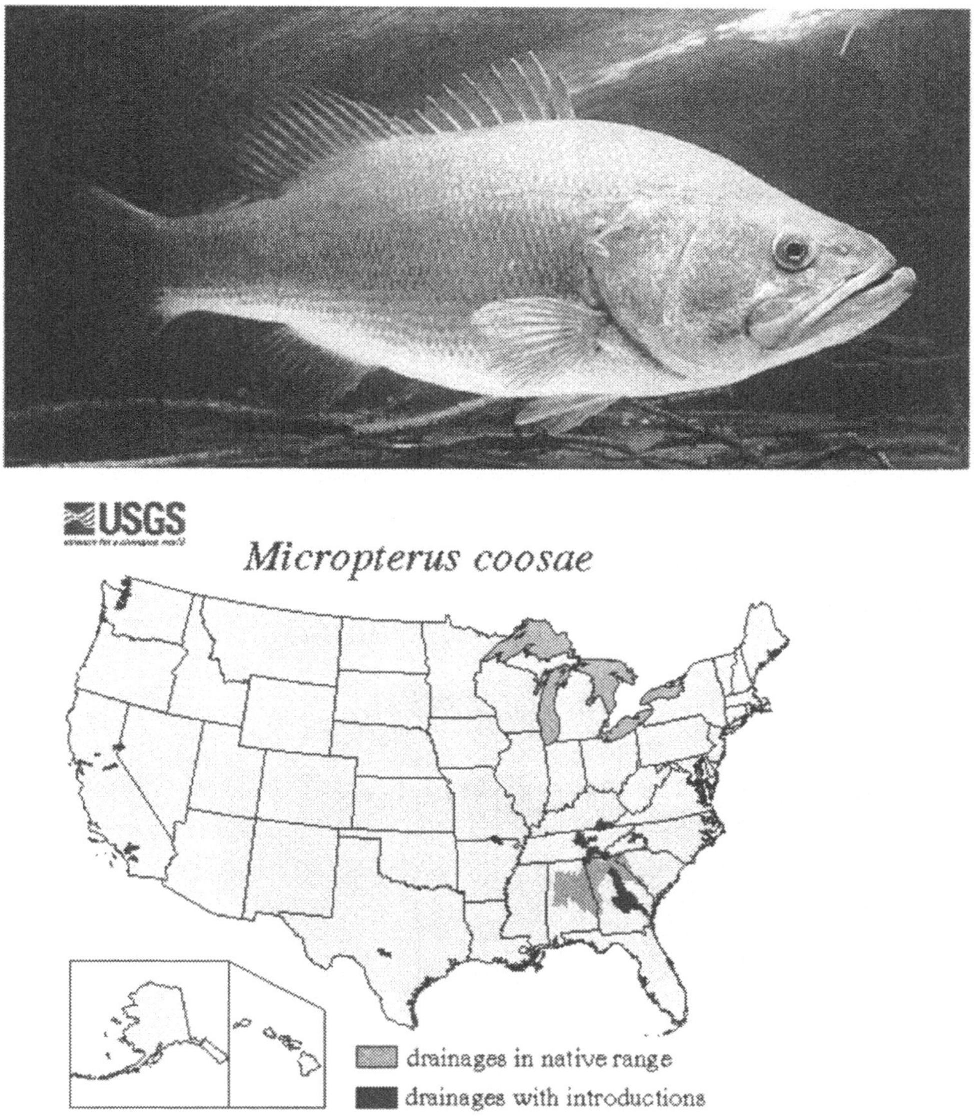

Figure D-1. Largemouth Bass (Photograph copyright F. Eugene Hester, 1996. Reprinted with permission of Nature's Images, 2001) and Distribution (Courtesy of Pam Fuller, USGS, Nonindigenous Aquatic Species Program, Gainesville, Fla., 2001).

D.2.2 Spotted Bass

The native range of the spotted bass (*Micropterus punctulatus*) includes the Mississippi River basin from southern Ohio and West Virginia to southeastern Kansas, and south to the Gulf of Mexico; Gulf slope drainage from the Chattahoochee River, Georgia (where possibly introduced), to the Guadalupe River, Texas (Page and Burr, 1991). Spotted bass are considered intermediate between the largemouth and smallmouth bass, and are segregated from the former by habitat preference. Spotted bass tend to be found in areas with more current than is preferred by the largemouth bass, and in areas that are too warm, turbid, and sluggish for smallmouth bass. Spawning occurs at temperatures from 54 to 74°F in rock or gravel substrate. Females lay between 1,150 and 47,000 eggs, which are guarded by the male. Spawning typically takes place in spring, from April to June, and peaks in late April to early May (Texas Parks and Wildlife Division, 2001).

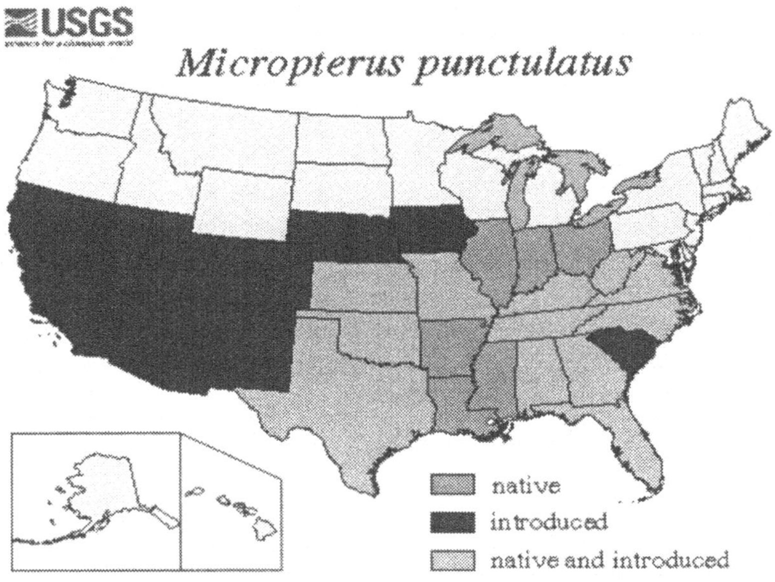

Figure D-2. Spotted Bass Photograph (Reprinted with permission of Noel Burkhead, USGS, 2001) and Distribution (Courtesy of Pam Fuller, USGS, Nonindigenous Aquatic Species Program, Gainesville, Fla., 2001).

D.2.3 Redeye Bass

The native range of the redeye bass (*Micropterus coosae*) includes the area above the Fall Line in the Savannah, Chattahoochee, and Mobile Bay basins, of North Carolina, South Carolina, Georgia, Tennessee, and Alabama (Page and Burr, 1991). Successful introductions have been accomplished in Kentucky, Tennessee, and California streams. They prefer clear, fast-flowing mountain streams with rocky runs, moderately deep pools, and substrate of sand, gravel, and cobbles. They commonly inhabit waters that are too cool for other species of bass, yet to warm for trout. The growth rate of redeye bass is slow compared to other species of bass. Redeyes grow quickly during their first year, after which growth is approximately one inch per year.

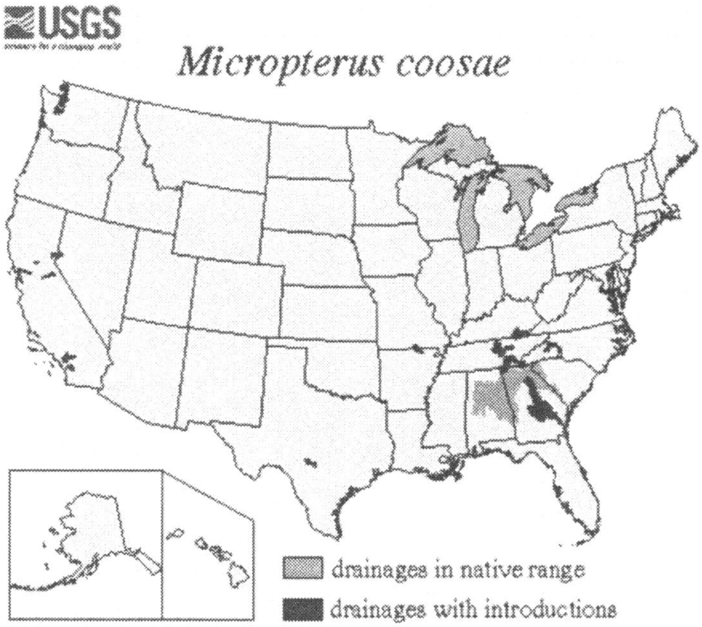

Figure D-3. Redeye Bass Photograph (Reprinted with permission of Noel Burkhead, USGS, 2001) and Distribution (Courtesy of Pam Fuller, USGS, Nonindigenous Aquatic Species Program, Gainesville, Fla., 2001).

Sexual maturity is reached after approximately four years. Redeye bass spawn in the spring when water temperatures are between 62 and 68° F. A typical 5-inch female contains about 2,300 eggs. Nests are constructed in coarse gravel in the upstream portions of pools, and are guarded by males, who remain for approximately 7 days following hatching while the young disperse (Springer, 2001).

APPENDIX E

BIBLIOGRAPHY

Acharia, K., Lal, B., and Singh, T.P. (2000) Modulatory effect of temperature on the influence of dietary linolenic (18 : 3 n-3) and linoleic (18:2 n-6) acids on the gonadal recrudescence in (*Clarias batrachus L.*). *J. Fish Biol.* 57:4:968.

Alhadeff, S.J., Musser, J.W., Sandercock, A.C. and Dyar, T.R. (2000) *Digital Environmental Atlas of Georgia* (a CD-ROM). Atlanta, Georgia.

Allen, Y., Matthiessen, P., Haworth, S., and Thain, J.E. (1999) Survey of estrogenic activity in United Kingdom estuarine and coastal waters and its effects on gonadal development of the flounder *Platichthys flesus*. *ET&C* 18:1791.

Allen, Y., Scott, A.P., Matthiessen, P., Haworth, S., Thain, J.E., and Feist, S. (1999) Survey of estrogenic activity in United Kingdom estuarine and coastal waters and its effects on gonadal development of the flounder (*Platichthys flesus*). *Environmental Toxicology and Chemistry* 18:8:1791.

Ankley, G., Mihaich, E., Stahl, R., Jr., Tillitt, D., Colborn, T., McMaster, S., Miller, R., Bantle, J., Campbell, P., Denslow, N., Dickerson, R., Folmar, L., Fry, M., Giesey, J., Grey, L.E., Guiney, P., Hutchinson, T., Kennedy, S., Peterson, R., Purdy, R., Ringer, R., Thomas, P., Touart, L., Van der Kraak, G., Zacharewski, T. (1998) Overview of a workshop on screening methods for detecting potential (anti-) estrogenic/androgenic chemicals in wildlife. *Environmental Toxicology and Chemistry* 17:68-87.

Ankley, G.T., Jensen, K.M., Kahl, M.D., Korte, J.J., and Makynen, E.A. (2001) Description and evaluation of a short-term reproduction test with the fathead minnow (*Pimephales promelas*). *Environ. Toxicol. Chem.* 20:6:1276.

Arcand-Hoy, L.D. and Benson, W.H. (1998) Fish reproduction: an ecologically relevant indicator of endocrine disruption. *Environmental Toxicology and Chemistry.* 17:49-57.

Arukwe, A., Forlin, L., and Goksøyr, A. (1997) Xenobiotic and steroid biotransformation enzymes in Atlantic salmon (*Salmo salar*) liver treated with an estrogenic compound 4-nonylphenol. *Environmental Toxicology and Chemistry* 16:12:2576.

Arukwe, A., Goksøyr, A. R., Thibaut, R., and Cravedi, J.P., (2000) Metabolism and organ distribution of nonylphenol in Atlantic salmon (*Salmo salar*), *Marine Environmental Research* 50:1-5:141.

Arukwe, A., Kullman, S.W., and Hinton, D.E. (2001) Differential biomarker gene and protein expressions in nonylphenol and estradiol-17 beta treated juvenile rainbow trout (*Oncorhynchus mykiss*). *Comp. Biochem. Physiol. C-Toxicol. Pharmacol.* 129:1:1.

Ashfield, L. A., Pottinger, T.G., and Sumpter, J.P. (1998) Exposure of female juvenile rainbow trout to alkylphenolic compounds results in modifications to growth and ovosomatic index. *Environmental Toxicology and Chemistry* 17:3:679.

Atlanta Regional Commission (1998) *Water Resources of the Atlanta Region* (a map). Atlanta, Georgia

Atlanta Regional Commission (2000). Big Creek Watershed Study. Master Plan. Prepared for The Cities of Alpharetta, Cumming and Roswell as well as Cherokee, Forsyth, and Fulton Counties. Prepared by Camp Dresser & McKee.

Ausubel., F.M., Brent, R., Kingston, R.E., Moore, DD, Seidman, J.G., Smith J.A. and Struhi, K (eds) (1987) *Current Protocols in Molecular Biology*, New York, New York: GreenPub. Assoc. and Wiley Interscience.

Avall Lundqvist, E., Nordstroem, L., Sjoevall, K., and Eneroth, P. (1989) Evaluation of seven different tumor markers for the establishment of tumor marker panels in gynecologic malignancies. *European Journal Gynaecol. Oncol.* 10:395.

Baker, M.A., Cerniglia, G.J., and Zaman, A. (1990) Microtiter plate assay for the measurement of glutathione and glutathione disulfide in large numbers of biological samples. *Anal. Biochem.* 190:360.

Baronti, C., Curini, R., D'Ascenzo, G., Di Corcia, A., Gentili, A., and Samperi, R. (2000) Monitoring Natural and Synthetic Estrogens at Activated Sludge Sewage Treatment Plants and in a Receiving River Water. *Environmental Science & Technology* 34:24:5059.

Batty, J.L.R. (1999) Morphological and reproductive characteristics of male mosquitofish (*Gambusia affinis* holbrook) inhabiting sewage-contaminated waters in New South Wales, Australia. *Arch. Environ. Contam. Tox.* 36:301.

Baylis, J.R., Wiegmann, D.D., and Hoff, M.H. (1993) Alternating life-histories of smallmouth bass. *Trans. Am. Fish. Soc.* 122:3:500.

Beitinger, T.L. (1990) Behavioral reactions for the assessment of stress in fishes. *J. Gt. Lakes Res.* 16:4:495.

Belfroid, A.C., Van der Horst, A., Vethhak, A.D., Schäfer, A.J., Rijs, G.B.J., Wegener, J., Cofino, W.P. (1999) Analysis and occurrence of estrogenic hormones and their glucuronides in surface water and wastewater in The Netherlands. *Sci Total Environ* 225:101-108.

Bell, A.M., Effects of an endocrine disrupter on courtship and aggressive behaviour of male three-spined stickleback, (*Gasterosteus aculeatus*). *Anim. Behav.* 62:4:775.

Bennetau-Pelissero, C., Breton, B., Bennetau, B., Corraze, G., Le Menn, F., Davail-Cuisset, B., Helou, C., and Kaushik, S.J. (2001) Effect of genistein-enriched diets on the endocrine process of gametogenesis and on reproduction efficiency of the rainbow trout (*Oncorhynchus mykiss*). *Gen. Comp. Endocrinol.* 121:2:173.

Bergeron, J.M., Crews, D., and McLachlan, J.A. (1994) PCBs as environmental estrogens: Turtle sex determination as a biomarker of environmental contamination. *Environmental Health Perspectives* 102:780.

Berlinsky, D.L., Jackson, L.F., Smith, T.I.J., and Sullivan, C.V. (1995) The annual reproductive cycle of the white bass (*Morone chrysops*). *J. of the World Aquaculture Society* 26:3:252.

Bevans, H.E. Goodbred, S.L., Miesner, J.F. Watkins, S.A., Gross, T.S, Denslow, N.D., and Bowman, C.J., Kroll, K.J., Hemmer, M.J., Flomar, L.C., Denslow, N.D. (2000). Estrogen-induced vitellogenin mRNA and protein in sheepshead minnow (*Cyprindon variegates*). *Gen. Comp. Endocrinol.* 120:300.

Bevans, H.E., Goodbred, S.L., Miesner, J.F., Watkins, S.A., Denslow, N.D., and Schoeb, T. (1996) *Synthetic organic compounds and carp endocrinology and histology in Las Vegas Wash and Las Vegas and Callville Bays of Lake Mead, Nevada, 1992 and 1995.* US. Department of the Interior, U.S. Geological Survey, Report 96-4266.

Bieniarz, K., and Epler, P. (1992) Advances in reproductive endocrinology of fish. *J. Physiol. Pharmacol.* 43:4 Supl 1:215.

Bjerselius, R., Lundstedt-Enkel, K., Olsen, H., Mayer, I., and Dimberg, K. (2001) Male goldfish reproductive behaviour and physiology are severely affected by exogenous exposure to 17β-estradiol. *Aquatic Toxicology* 53:139. © 2001 Elsevier Science B.V. All Rights Reserved.

Black, R.W., Haggland, A.L., and Voss, F.D. (2000) Predicting the probability of detecting organochlorine pesticides and poly chlorinated biphenyls in stream systems on the basis of land use in the Pacific Northwest, USA. *Environmental Toxicology and Chemistry* (ET&C) 19:1044.

Blair, J.B., Miller, M.R., Pack D., Barnes, R., Ten, S.J., and Hinton, D.E. (1990) Isolated trout liver cells: establishing short-term primary cultures exhibiting cell-to-cell interactions. *In Vitro Cell. Dev. Biol.* 26:237.

Bowman, C.J., and Denslow, N.D. (1999) Development and validation of a species- and gene-specific molecular biomarker: Vitellogenin mRNA in largemouth bass (*Micropterus salmoides*). *Ecotoxicology* 8:5:399.

Bowman, C.J., Kroll, K.J., Hemmer, M.J., Folmar, L.C., and Denslow, N.D. (2000) Estrogen-induced vitellogenin mRNA and protein in sheepshead minnow (*Cyprinodon variegatus*). *Gen. Comp. Endocrinol.* 120:3:300.

Brock, M.I. and Shapiro, D.J. (1983) Estrogen stabilizes vitellogenin mRNA against cytoplasmic degradation. *Cell* 34:207.

Brown and Caldwell, Golder Associates Inc., and Jordan, Jones and Goulding (2000) *Source water assessment for Dalton Utilities.* Atlanta, Georgia.

Brown, G.E., LeBlanc, V.J., and Porter, L.E. (2001) Ontogenetic changes in the response of largemouth bass (*Micropterus salmoides, centrarchidae, perciformes*) to heterospecific alarm pheromones. *Ethology* 107:5:401.

Brüschweiler, B.J., Würgler, F.E., and Fent, K. (1996) An elisa assay for cytochrome P4501A in fish liver cells. *Environmental Toxicology and Chemistry* 15:4:592.

Bulger, W.H., Feil, V.J. and Kupfer, D. (1985) Role of hepatic monooxygenases in generating estrogenic metabolites from methoxychlor and from its identified contaminants. *Mol Pharmacol.* 27:115.

Bulger, W.H., Muccitelli, R.M., and Kupfer, D. (1978) Studies on the *in vivo* and in vitro estrogenic activities of methoxychlor and its metabolism. Role of hepatic mono-oxygenase in methoxychlor activation. *Biochem. Pharmacol.* 27:2417.

Burke, R. (2000) *Presentation handout to the Atlanta Regional Commission from presentation given August 14, 2000 by the Georgia Environmental Protection Division.* Atlanta, Georgia.

Burse, J.R., Cheek, A.O. and Bart, H.L. Jr. (2000). Effects of exposure to paper mill effluents in *Lepomis macrochirus* from Elevenmile Creek, Florida. Poster #12 Abstract presented at *e .hormone 2000 Conference*, October 15-18, 2000, New Orleans, Louisiana.

Cardwell, J.R., and Liley, N.R. (1990) Current status of behavioral endocrinology in tropical marine teleosts. *Prog Clin Biol Res.* 342:673.

Cardwell, J.R., and Liley, N.R. (1991) Androgen control of social status in males of a wild population of stoplight parrotfish, (*Sparisoma varide*) (Scaridae). *Horm Behav.* 25:1:1.

Cardwell, J.R., Sorensen, P.W., Van der Kraak, G.J., and Liley, N.R. (1996) Effect of dominance status on sex hormone levels in laboratory and wild-spawning male trout. *Gen Comp Endocrinol.* 101:3:333.

Carnevali, O. and Belveldere, P. (1991) Comparative studies of fish, amphibian and reptilian vitellogenins. *J. Exp. Zoology.*

Celius, T., Matthews, J.B., Giesy, J.P., and Zacharewski, T.R. (2000) Quantification of rainbow trout (*Oncorhynchus mykiss*) zona radiata and vitellogenin mRNA levels using real-time PCR after *in vivo* treatment with estradiol-17 beta or alpha-zearalenol. *J. Steroid Biochem. Mol. Biol.* 75:2-3:109.

Chang, C.F. and Chen, M.R. (1990) Fluctuation in sex steroids and sex steroid-binding protein during the development and annual cycle of the male common carp, (*Cyprinus carpio*). *Comparative Biochemistry and Physiology.* 97A-4:565-568.

Chang, X.T., Kobayashi, T., Todo, T., Yoshiura, Y., Ikeuchi, T., Kajiura, H., Nakamura, M., and Nagahama, Y. (1999) cDNA sequence of tilapia type beta estrogen receptor, *NCBI.*

Cheek, A.O., Brouwer, T.H., Carroll, S., Manning, S., McLachlan, J.A., and Brouwer, M. (2001) Experimental evaluation of vitellogenin as a predictive biomarker for reproductive disruption. *Environ. Health Perspect.* 109:7:681.

Cheek, A.O., Brouwer, T.H., Carroll, S., Manning, S., McLachlan, J.A., and Brouwer, M. (2000) Experimental evaluation of vitellogenin as a predictive biomarker for reproductive disruption. *Am. Zool.* 40:6:972.

Cheskis, B.J., Karathanasis, S., and Lyttle, C.R. (1997) Estrogen receptor ligands modulate its interaction with DNA. *J. Biol. Chem.* 272:11384.

Chomczynski, P., Sacchi, N., (1987) Single step method of RNA isolation by acid guanidinium thiocyanate phenol chloroform extraction. *Analytical Biochem.* 162: 156.

Christiansen, T., Korsgaard, B., and Jespersen, A. (1998) Induction of vitellogenin synthesis by nonylphenol and 17β-estradiol and effects on the testicular structure in the eelpout (*Zoarces viviparus*). *Marine Environ. Research* 46:141.

City of Atlanta (1998). Watershed Management Guidance Document. Metro Atlanta Urban Watersheds Initiative. Prepared for the City of Atlanta Department of Public Works Division of Wastewater Services. Prepared by CH2MHill in association with W.L. Jorden and R&D Environmental, Inc.

City of Atlanta (1998). West Watershed Impacts Assessment Report. Prepared for the City of Atlanta Department of Public Works Division of Wastewater Services. Prepared by CH2MHill.

Clark, M.E., Rose, K.A., Chandler, J.A., Richter, T.J., Orth, D.J., and Van Winkle, W. (1998) Simulating smallmouth bass reproductive success in reservoirs subject to water level fluctuations. *Environ. Biol. Fishes* 51:2:161.

Clem, L.W., Bly, J.E., Ellsaesser, C.F., Lobb, C.J., and Miller, N.W. (1990) Channel catfish as an unconventional model for immunological studies. *J. Exp Zool Suppl.* 4:123.

Clemons, J.H., Allan, L.M., Marvin, C.H., Wu, Z., McCarry, B.E., Zacharewski, T.R. (1998) Evidence of estrogen- and TCDD-like activities in crude and fractionated extracts of PM_{10} air particulate material using in vitro gene expression assays. *Environ. Sci. Technol.* 32:1853.

Colborn, T. and C. Clement (1992) Chemically-induced alterations in sexual and functional development: the wildlife/Human Connection, *Advances in Modern Environmental Toxicology Vol. 21*. Princeton Scientific, Princeton, NJ.

Colucci, M.S., Bork, H. and Topp, E. (2001) Persistence of estrogenic hormones in agricultural soils: I. 17β-estradiol and estrone. *J. Environ. Qual.* Vol. 30.

Colucci, M.S. and Topp, E. (2001) Persistence of estrogenic hormones in agricultural soils: II 17α-ethynylestradiol. *J. Environ. Qual.* Vol 30.

Comas, D., Piulachs, M.D., and Belles, X. (2001) Induction of vitellogenin gene transcription in vitro by juvenile hormone in *Blattella germanica*. *Mol. Cell. Endocrinol.* 183:1-2:93.

Copeland, P.A. and Thomas, P. (1988) The measurement of plasma vitellogenin levels in a marine teleost, the spotted seatrout (*Cynoscion nebulosus*) by homologous radioimmunoassay. *Comparative and Biochem. Physiol.* 91B:17.

Copeland, P.A., Sumpter, J.P., Walker, T.K., and Croft, M. (1986) Vitellogenin levels in male and female rainbow trout (*Salmo gairdneri* Richardson) at various stages of the reproductive cycle. *Comp Biochem. Physiol.* 83B:487.

Couillard, C.M. and Nellis, P. (1999) Organochlorine contaminants in Mummichog (*Fundulus heteroclitus*) living downstream from a bleached-kraft pulp mill in the Miramichi Estuary, New Brunswick, Canada. *ET&C* 18:2545.

Crowther, J.R. (1995) *Methods in Molecular Biology, Volume 42, ELISA Theory and Practice*, Humana Press, Towowa, New Jersey.

Cummings, A.M. (1997) Methoxychlor as a model for environmental estrogens. *Critical Reviews in Toxicol.* 27:367.

Cyr, D.G., and Eales, J.G. (1996) Interrelationships between thyroidal and reproductive endocrine systems in fish. *Review in Fish Biology and Fisheries* 6:165.

Danylchuk, A.J., and Tonn, W.M. (2001) Effects of social structure on reproductive activity in male fathead minnows (*Pimephales promelas*). *Behav. Ecol.* 12:4:482.

Davis, J.T., and Lock, J.T. (1997) Culture of largemouth bass fingerlings. *Southern Regional Aquaculture Center* Pub. 201.

Davis, J.T., and Lock, J.T. (1997) Largemouth bass: Biology and life history. *Southern Regional Aquaculture Center.*

Dean, D.M. and Sanders, M.M. (1996) Ten years after: reclassification of steroid-responsive genes. *Mol. Endocrinol* 10:1489.

Debouck, C. and Goodfellow, P.N. (1999) DNA microarrays in drug discovery and development. *Nature Genetics, (Suppl)* 21:48.

Dehal, S.S. and Kupfer, D. (1994) Metabolism of the proestrogenic pesticide methoxychlor by hepatic P450 monooxygenases in rats and humans. Dual pathways involving novel ortho ring-hydroxylation by CYP2B. *Drug Metab. Dispos.* 22:937.

Delcuve, G.P., Sun, J. M., and Davie, J.R. (1992) Expression of rainbow trout apolipoprotein A-I genes in liver and hepatocellular carcinoma. *J. Lipid Res.* 33:251.

Dellovade, T., Schwanzel-Fukuda, M., Gordan, M., Pfaff, D. Aspects of GnRH neurobiology conserved across vertebrate forms. *Gen Comp Endocrinol.* 112(3):276-82.

Denslow, N.D., Bowman, C.J., Ferguson, R.J., Lee, H.S., Hemmer, M.J., and Folmar, L.C. (2001) Induction of gene expression in sheepshead minnows (*Cyprinodon variegatus*) treated with 17 beta-estradiol, diethylstilbestrol, or ethinylestradiol: The use of mRNA fingerprints as an indicator of gene regulation. *Gen. Comp. Endocrinol.* 121:3:250.

Denslow, N.D., Bowman, C.J., Lee, H.S., Ferguson, R.J., Hemmer, C.J., and Folmar, L.C. (1999) "Biomarkers of endocrine disruption at the mRNA level," In *Environmental Toxicology and Risk Assessment*: 8th Volume. ASTM STP 1364, (D. Henshel, Ed.), American Society for Testing and Materials, (Accepted).

Denslow, N.D., Chow, M., Chow, M.M., Bonomelli, S., Folmar, L.C., Heppell, S.A., and Sullivan, C.V. (1997) "Development of biomarkers for environmental contaminants affecting fish," *Chemically Induced Alterations in Functional Development and Reproduction of Fishes.* (R. Rolland, M. Gilbertson and R.E. Peterson, eds.) SETAC Press, pp. 73-83.

Denslow, N.D., Chow, M., Kroll, K., Reiskind, M., Johnson, B., Wieser, C., Wiebe, J., and Gross, T.S. (1997) *Determination of baseline seasonal information on vitellogenin production in female and male largemouth bass collected from lakes in central Florida.* Abstract to SETAC 18th Annual Meeting, San Francisco, CA.

Denslow, N.D., Chow, M.M., Folmar, L.C., Bonomelli, S., Heppell, S.A. and Sullivan, C.V. (1996) Development of antibodies to teleost vitellogenins. Potential biomarkers for environmental estrogens, *Environmental Toxicology and Risk Assessment* 5:STP 1306, D.A. Bengston and D.S., Henshel, Eds., American Society for Testing and Materials, Philadelphia, PA, pp. 23-36.

Denslow, N.D., Folmar, L.C., Gross, T.S., Goodbred, S.L., and Gilliom, R.J. (1999) Exposure of fish to estrogenic xenobiotics in rivers and lakes in the United States. (accepted for publication, *International Symposium on Environmental Endocrine Disruptors '98, Dec 11-13, 1998*, Kyoto, Japan).

Denslow, N.D., Kroll, K., Schoeb, T., Johnson, B., Wiebe, J., Wieser, C., and Gross, T.S. (1998) Documentation of endocrine-disruption and altered seasonal reproductive parameters for brown bullhead catfish from several lakes in central Florida. *Abstract to SETAC 19th Annual Meeting*, Charlotte, NC.

Denslow, N.D., Lee, H.S., Bowman, C.J., Hemmer, M.J., and Folmar, L.C. (2001) Multiple responses in gene expression in fish treated with estrogen. *Comp. Biochem. Physiol. B-Biochem. Mol. Biol.* 1209:2-3:277.

Desbrow, C., Routledge, E., Sheehan, D., Waldlock, M., and Sumpter, J. (1996) *The identification and assessment of oestrogenic substances in sewage treatment works effluents, Report P2-490/7,* United Kingdom Environment Agency, MAFF Fisheries, Almondsbury. Bristol, UK, pp. 1-74.

Desbrow, C., Routledge, E.J., Brighty, G.C., Sumpter, J.P., and Waldock, M. (1998) Identification of estrogenic chemicals in STW effluent: 1. Chemical fractionation and in vitro biological screening. *Environ. Sci. Technol.* 32:11:1549.

Desbrow, C., Routledge, E.J., Brighty, G.C., Sumpter, J.P., and Waldock, M. (1999) Identification of estrogenic chemicals in STW effluent: Chemical fractionation and in vitro biological screening. *Environ. Sci. Technol.* 32:11:1549.

Deutsch, W.G., Webber, E.C., Bayne, D.R., and Reed, C.W. (1992) Effects of largemouth bass stocking rate on fish populations in aquatic mesocosms used for pesticide research. *Environ. Toxicol. Chem.* 11:1:5.

DeWoody, J.A., Fletcher, D.E., Wilkins, S.D., Nelson, W.S., and Avise, J.C. (2000) Genetic monogamy and biparental care in an externally fertilizing fish, the largemouth bass (Micropterus salmoides). *Proc. R. Soc. Lond. Ser. B-Biol. Sci.* 267:1460:2431.

Dierckx, P. J. (1985) Hepatic glutathione *S*-transferases in rainbow trout and their interaction with 2,4-dichlorophenoxyacetic acid and 1,4-benzoquinone. *Comp. Biochem. Physiol.* 82C:495.

Ding, J.L., Hee, P.L., and Lam, T.J. (1989) Two forms of vitellogenin in the plasma and gonads of male *Oreochromis aureus*. *Comp Biochem Physiol* 93B:363.

Donohoe, R.M. and Curtis, L.R. (1996) Estrogenic activity of chlordecone, o,p'-DDT, and o,p'-DDE in juvenile rainbow trout: Induction of vitellogenesis and interaction with hepatic estrogen binding sites. *Aquatic Toxicology* 36:31.

Douglas, D.R., Johnson, W.E., Denslow, N.D., Wieser, C., Wiebe, J., and Gross, T.S (1998) Endocrine disruption in largemouth bass in a re-flooded agricultural area. *Abstract to SETAC 19th Annual Meeting*, Charlotte, NC.

Douglasville-Douglas County Water and Sewer Authority (1999). Gothard's Creek Watershed Assessment Study: Volume I – Watershed Characterization and Volume II – Watershed Management Plan. Prepared for Douglasville-Douglas County Water and Sewer Authority and R.J. Wood and Company. Prepared by Jordan, Jones & Goulding, Inc.

Douglasville-Douglas County Water and Sewer Authority (2000). Final Report on the Anneewakee Creek Watershed Assessment Study. Prepared for Douglasville-Douglas County Water and Sewer Authority and R.J. Wood and Company. Prepared by CH2MHill.

Down, N.E., Peter, R.E., and Leatherland, J.F. (1990) Seasonal changes in serum gonadotropin, testosterone, 11-ketotestosterone, and estradiol-17β levels and their relation to tumor burden in gonadal tumor-bearing carp x goldfish hybrids in the Great Lakes. *General and Comparative Endocrinology* 77:192.

Duan, R., Porter, W., and Safe, S. (1998) Estrogen-induced c-fos protooncogene expression in MCF-7 human breast cancer cells: role of estrogen receptor SP1 complex formation. *Endocrinol.* 139:1981.

Duggan, A., Paolucci, M., Tercyak, A., Gigliotti, M., Small, D., and Callard, I. (2001) Seasonal variation in plasma lipids, lipoproteins,apolipoprotein A-I and vitellogenin in the freshwater turtle (*Chrysemys picta*). *Comp. Biochem. Physiol. A-Mol. Integr. Physiol.* 130:2:253.

Duggan, D.J., Bittner, M., Chen, Y., Meltzer, P., and Trent, M.J. (1999) Expression profiling using cDNA microarrays. *Nature Genetics (Suppl)* 21:10.

ECME: Environmental Estrogens (The Endocrine System). 1998. Website: tmc.tulane.edu/ECME. Tulane/Xavier Center for Bioenvironmental Research.

Egaas, E., Falls, J.G., and Dauterman, W.C. (1995) A study of gender, strain and age differences in mouse liver glutathione-*S*-transferase. *Comp Biochem Physiol & Pharmacol Toxicol Endocrinol.* 110:35.

Englebienne, P. (2000) Immune and Receptor Assay in Theory and Practice. CRC Press LLC, Boca Raton , Florida.

Evans, D.H. (1993) *The Physiology of Fishes, Chapter in Reproductive Physiology*. CRC Press. 503.

Fairchild, W.L., Swansburg, E.O., Arsenault, J.T., and Brown, S.B. (1999) Does an association between pesticide use and subsequent decline in catch of Atlantic salmon (*Salmo salar*) represent a case of endocrine disruption? *Environmental Health Perspectives* 107:349.

Feldman, D. and Krishnan, A. (1995) Estrogens in unexpected places: Possible implications for researchers and consumers. *Environmental Health Perspectives* 103:129.

Felip, A., Piferrer, F., Carrillo, M., and Zanuy, S. (2001) Comparison of the gonadal development and plasma levels of sex steroid hormones in diploid and triploid sea bass, (*Dicentrarchus labrax* L.). *J. Exp. Zool.* 290:4:384.

Fenske, M., van Aerle, R., Brack, S., Tyler, C.R., and Segner, H. (2001) Development and validation of a homologous zebrafish (*Danio rerio Hamilton-Buchanan*) vitellogenin enzyme-linked immunosorbent assay (ELISA) and its application for studies on estrogenic chemicals. *Comp. Biochem. Physiol. C-Toxicol. Pharmacol.* 129:3:217.

Flammarion, P., Brion, F., Babut, M., Garric, J., Migeon, B., Noury, P., Thybaud, E., Tyler, C.R., Palazzi, X., and Tyler, C.R. (2000) Induction of fish vitellogenin and alterations in testicular structure: Preliminary results of estrogenic effects in chub (*Leuciscus cephalus*). *Ecotoxicology* 9:1-2:127.

Flammarion, P., Devaux, A., and Garric, J. (2001) Biochemical markers of environmental contamination in continental aquatic ecosystems. Examples of use and prospects for management. *Bull. Fr. Pech. Piscic.* 357-60:209.

Flouriot, G., Pakdel, F., and Valotaire, Y. (1996) Transcriptional and post-transcriptional regulation of rainbow trout estrogen receptor and vitellogenin gene expression. *Mol. Cell Endocrinol.* 124:173.

Flouriot, G., Vaillant, C., Salbert, G., Pelissero, C., Guiraud, S.M., and Valotaire, Y. (1993) Monolayer and aggregate cultures of rainbow trout hepatocytes: long term and viable liver-specific expression in aggregates. *J. Cell Sci.* 105:407.

Folmar, L.C., Denslow, N.D., Kroll, K., Orlando, E.F., Enblom, J., Marcino, J., Metcalfe, C., and Guillette, L.J. (2001) Altered serum sex steroids and vitellogenin induction in walleye (*Stizostedion vitreum*) collected near a metropolitan sewage treatment plant. *Arch. Environ. Contam. Toxicol.* 40:3:392.

Folmar, L.C., Gardner, G.R., Schreibman, M.P., Magliulo-Cepriano, L., Mills, L.J., Zaroogian, G., Gutjahr-Gobell, R., Haebler, R., Horowitz, D.B. and Denslow, N.D. (2001) Vitellogenin-induced pathology in male summer flounder (*Paralichthys dentatus*). *Aquat. Toxicol.* 51:4:431.

Folmar, L.C., Hemmer, M., Webb, R., Bowman, C., Kroll, K., and Denslow, N.D. (1999) Relative estrogenicity of estradiol, ethynyl estradiol and diethylstilbestrol in an *in vivo*, male sheepshead minnow (*Cyprinodon variegatus*) vitellogenin bioassay, submitted.

Folmar, L., Hemmer, M., Webb, R., Robinson, G., Chow, M., Kroll, K., and Denslow, N. (1997) Development of an *in vivo* estrogen responsive bioassay in the sheepshead minnow. *Abstract to SETAC 18th Annual Meeting.* San Francisco, CA.

Folmar, L.C., Denslow, N.D., Rao, V., Chow, M., Crain, A.D., Enblom, J., Marcino, J., and Guillette Jr., L.J. (1996) Vitellogenin induction and reduced serum testosterone concentrations in feral male carp (*Cyprinus carpio*) captured near a major metropolitan sewage treatment plant. *Environ. Health Perspect.* 104:1096.

Folmar, L.C., Denslow, N.D., Wallace, R.A., LaFleur, G., Gross, T.S., Bonomelli, S., and Sullivan, C.V. (1995) A highly conserved N-terminal sequence for teleost vitellogenin with potential value to the biochemistry, molecular biology and pathology of vitellogenesis. *J. Fish Biol.* 46:255.

Folmar, L.D., Denslow, N.D., Wallace, R.A., LaFleur, G., Gross, T.S., Bonomelli, S., and Sullivan, C.V. (1995) A highly conserved N-terminal sequence for teleost vitellogenin with potential value to the biochemistry, molecular biology and pathology of vitellogenesis. *J. Fish Biol.* 46:255.

Folmar, L.C. (1992) *Effects of chemical contaminants on blood chemistry of teleost fish: A bibliography and synopsis of selected effects.* U.S. Environmental Protection Agency, Environmental Research Laboratory, Sabine Island, Gulf Breeze, Florida 32561.

Fontainhas-Fernandes, A., Monteiro, M., Figueiredo, A., Gomes, E., Coimbra, J., and Reis-Henriques, M.A. (2000) Partial or total replacement of fish meal by plant protein affects gonadal development and plasma 17 beta-estradiol levels in female Nile tilapia. *Aquac. Int.* 8:4:299.

Foran, C.M., Bennett, E.R., and Benson, W.H. (2000) Exposure to environmentally relevant concentrations of different nonylphenol formulations in Japanese medaka. *Marine Environmental Research* 50:1-5:135.

Foster, E.P., Fitzpatrick, M.S., Feist, G.W., Schreck, C.B., Yates, J., Spitsbergen, J.M., and Heidel, J.R. (2001) Plasma androgen correlation, EROD induction, reduced, condition factor, and the occurrence of organochlorine pollutants in reproductively immature white sturgeon (*Acipenser transmontanus*) from the Columbia River, USA. *Arch. Environ. Contam. Toxicol.* 41:2:182.

Freeman, B.J. 25 January 2001. Personal Communication.

Freidig, A.P. and Hermens, J.L.M. (2001) Narcosis and chemical reactivity QSARs for acute fish toxicity. *Quant. Struct.-Act. Relat.* 19:6:547.

Frick, B.A. 23 April 2001. Personal Communication.

Frick, B.A., Hippe, D.J., Buell, G.R., Couch, C.A., Hopkins, E.H., Wangsness, D.J., and Garrett, J.W. (1995) *Water Quality in the Apalachicola-Chattahoochee-Flint River Basin, Georgia, Alabama, and Florida, 1992-1995*. U.S. Geological Survey Circular 1164.

Fry, D.M. (1995) Reproductive effects in birds exposed to pesticides and industrial chemicals. *Environmental Health Perspectives*. 103:165.

Fukada, H., Haga, A., Fujita, T., Hiramatsu, N., Sullivan, C.V., and Hara, A. (2001) Development and validation of chemiluminescent immunoassay for vitellogenin in five salmonid species. *Comp. Biochem. Physiol. A-Mol. Integr. Physiol.* 130:1:163.

Fulton County (2000). Fulton County – Big Creek Water Resources Management Plan. Watershed Characterization Findings to Date. Fact Sheet. Retrieved on December 19, 2001 from http://www.fultonco-wrmp.com/big_creek.htm

Fulton County (2000). Fulton County Sandy Springs Watershed Assessment. Summary of Watershed Characterization Activities. Retrieved on December 19, 2001 from http://www.fultonco-wrmp.com/sandy_springs.htm

Fulton County (2001). Characterization Summary for the Johns Creek Study Area. Retrieved on December 19, 2001 from http://www.fultonco-wrmp.com/johns_creek.htm

Funkenstein, B. (2001) Developmental expression, tissue distribution and hormonal regulation of fish (*Sparus aurata*) serum retinol-binding protein. *Comp. Biochem. Physiol. B-Biochem. Mol. Biol.* 1029:2-3:613.PY 2001

Funkenstein, B., Bowman, C.J., Denslow, N.D., Cardinali, M., and Carnevali, O. (2000) Contrasting effects of estrogen on transthyretin and vitellogenin expression in males of the marine fish, Sparus aurata. *Mol. Cell. Endocrinol.* 167:1-2:33.

Gagne, F., and Blaise, C. (2000) Evaluation of environmental estrogens with a fish cell line. *Bull. Environ. Contam. Toxicol.* 65:4:494.

Gagne, F., Marcogliese, D.J., Blaise, C., and Gendron, A.D. (2001) Occurrence of compounds estrogenic to freshwater mussels in surface waters in an Urban area. *Environ. Toxicol.* 16:3:260.

Gaido, K.W., Leonard, L.S., Lovell, S., Gould, J.C., Babai, D., Portier, C.J., and McDonnell, D.P. (1997) Evaluation of chemicals with endocrine modulating activity in a yeast-based steroid hormone receptor gene transcription assay. *Toxicology and Applied Pharmacology*. 143:205.

Gaido, K.W., Maness, S.C., McDonnell, D.P., Dehal, S.S., Kupfer, D., and Safe, S. (2000) Interaction of methoxychlor and related compounds with estrogen receptor alpha and beta, and androgen receptor: structure-activity studies. *Mol. Pharmacol.* 58:852.

Gallagher, E., Sheehy, K., Janssen, P., Eaton, D.L., and Collier, T.K. (1999) Isolation and cloning of homologous glutathione *S*-transferase cDNAs from English sole and starry flounder liver. *Aquat. Toxicol.* in press.

Gallagher, E.P. and Sheehy, K.M. (1999) Decreased glutathione *S*-transferase expression in brown bullheads from a contaminated field site. *Mar. Environ. Res.,* in press.

Gallagher, E.P. and Sheehy, K.M. (1999) Effects of phenytoin on oxidative stress biomarkers and tissue viability in cultured fetal liver slices. *American Association for Cancer Research 1999 Meeting. abstr.*

Gallagher, E.P., Buetler, T.M., Wang, C., Stapleton, P.L., Stahl, D.L., and Eaton, D.L. (1995) The effects of diquat and ciprofibrate on hepatic cytochrome P450, glutathione *S*-transferase, UDP-glucuronosyltransferase and antioxidant enzyme messenger RNA expression and catalytic activities in rat liver. *Toxicol. Appl. Pharmacol.* 134:81.

Gallagher, E.P., Canada, A.T., and DiGiulio, R.T. (1992) The protective role of glutathione in chorothalonil-induced toxicity to channel catfish. *Aquat. Toxicol.* 23:155.

Gallagher, E.P., Kedderis, G., and DiGiulio, R.T. (1991) Glutathione *S*-transferase mediated chlorothalonil metabolism in liver and gill subcellular fractions of channel catfish. *Biochem. Pharmacol.* 42:139.

Gallagher, E.P., Sheehy, K.M., Lame, M.W., and Segall, H.J. (1999b) *In vitro* kinetics of hepatic glutathione S-transferase conjugation in largemouth bass and brown bullheads. *Environ. Toxicol. and Chem.* in press.

Gallagher, E.P., Stapleton, P.L., Slone, D.H., Schlenk, D., and Eaton, D.L. (1996) Channel catfish glutathione *S*-transferase isoenzyme activity toward (+)-anti-benzo[a]pyrene-trans-7,8-diol-9,10-epoxide. *Aquat. Toxicol.* 34:135.

Gamble, A.V., Bennie, D., Brown, S., Hodson, P.V., Parrott, J., Servos, M., Solomon, K.R., Sherry, J.P. (1998) "Use of rainbow trout (*Onchorhynchus mykiss*) and brown trout (*Salmo trutta*) to screen Hamilton Harbour water for estrogenic effects." In *Aquatic Toxicity Workshop.* Quebec City, Quebec, Canada. Canadian Technical Report of Fisheries and Aquatic Sciences No. 2260, p. 29.

Garcia-Reyero, N. Grau E., Castillo, M., de Alda, M.J.L., Barcelo, D., Pina, B., (2001) Monitoring of endocrine disruptors in surface waters by the yeast recominant assay. *Environmental Toxicology and Chemistry* 20:6:1152

Georgia Department of Natural Resources Environmental Protection Division (2000) *Water Quality in Georgia. 1998-1999.* Atlanta, Georgia.

Giesy, J.P., Pierens, S.L., Snyder, E.M., Miles-Richardson, S., Kramer, V.J., Snyder, S.A., Nichols, K.M., and Villeneuve, D.A. (2000) Effects of 4-nonylphenol on fecundity and biomarkers of estrogenicity in fathead minnows (*Pimephales promelas*). *Environmental Toxicology and Chemistry*. 1 9(5):1368. © SETAC.

Giesy, J.P., Tillittt, D.E., Verbrugge, D.A., Sanderson, J.T., Jude, D.J., Gale, R.W., Meadows, J.C., Zajieck, J.L., Peterman, P.H., Schwartz, T.R., and Tuchman. M.L. (1997) Polychlorinated dibenzo-*p*-dioxins, dibenzofurans, biphenyls and 2,3,7,8-tetrachlorodibenzo-*p*-dioxin equivalents in fishes from Saginaw Bay, Michigan. *ET&C* 16:713.

Gillesby, B.E., Zacharewski, T.R. (1997) Exoextrogens: Mechanisms of Action and Strategies for Identification and Assessment. *Environmental Toxicology and Chemistry* 17:3.

Gimeno, S., Gerritsen, A., Bowmer, T., and Komen, H. (1994) Feminization of male carp. *Nature*. 384:221.

Gimeno, S., Komen, H., Venderbosch, P.W.M., and Bowmer, T. (1997) Disruption of sexual differentiation in genetic male common carp (*Cyprinus carpio*) exposed to an alkylphenol during different life stages. *Environmental Science and Technology*. 31:2884.

Gimeno, S., Komen, H., Venerbosch, P.W.M., and Bowmer, T. (1997) Disruption of sexual differentiation in genetic male common carp (*Cyprinus carpio*) exposed to an alkylphenol during different life stages. *Environmental Science and Technology*. 31:2884.

Gleason, T.R., and Nacci, D.E. (2001) Risks of endocrine-disrupting compounds to wildlife: Extrapolating from effects on individuals to population response. *Hum. Ecol. Risk Assess.* 7:5:1027.

Godwin, J., Crews, D., and Warner, R.R. (1996) Behavioral sex change in the absence of gonads in a coral reef fish. Proc R. Soc. Lond B Biol Sci. 263(1377):1683-1688.

Goodbred, S.L., Gilliom, R.J., Gross, T.S., Denslow, N.D., Bryant, W.B., and Schoeb, T.R. (1997) *Reconnaissance of 17β-estradiol, 11-ketotestosterone, vitellogenin and gonad histopathology in common carp of united states streams: potential for contaminant-induced endocrine disruption*. US Geological Survey, Open-File Report 96-627.

Goodbred, S.L., Gilliom, R.J., Gross, T.S., Denslow, N.P. ,Bryant, W.L. and Schoeb, T.R (1996). *Reconnaissance of 17β-estradiol, 11-ketotestosterone, vitellogenin, and gonad histopathology in common carp of United States streams: Potential for contaminant-induced endocrine disruption*. U.S. Geological Survey Open-File Report 96-627. National Water Quality Assessment, Pesticide National Synthesis Project, USGS.

Goodbred, S.L., Gillion, R.J., Gross, T.S., Denslow, N.P., Bryant, W.B., and Schoeb, T.R (1996) *Reconnaissance of 17β-estradiol, 11-ketotestosterone, vitellogenin, and gonad histopathiology in common carp of United States Streams: Potential for contaminant-induced endocrine disruption*. US. Department of the Interior, U.S. Geological Survey, Report 96-627.

Goodwin, A.E., Grizzle, J.M., Bradley, J.T., and Estridge, B.H. (1992) Monoclonal antibody-based immunoassay of vitellogenin in the blood of male channel catfish (*Ictalurus punctatus*). *Comparative Biochem. and Phys*. 101B:441-446.

Gorski, K., Carneiro, M., and Schibler, U. (1986) Time-specific *in vitro* transcription from the mouse albumin promoter. *Cell*. 47:767-776.

Gray, L.E., Jr. (1998) Xenoendocrine disrupters: laboratory studies on male reproductive effects. *Toxicol Lett*. 102-103:331.

Gray, M.A., Niimi, A.J., and Metcalfe, C.D. (1999) Factors affecting the development of testis-ova in medaka, (*Oryzias latipes*), exposed to octylphenol. *Env. Tox. and Chem.* 18(8):1835.

Gray, M.A., and Metcalf, C.D. (1997) Induction of the testis-ova in Japanese medaka (*Oryzias latipes*) exposed to *p*-nonylphenol. *Env. Tox. and Chem.* 18(5):1082-1086.

Greenberg, M. E. and Bender, T. P. (1997) *Identification of newly transcribed RNA*. Current protocols in molecular biology. (V. B. Chanda, Ed.), pp. 4.10.1. John Wiley and Sons.

Gronen, S., Denslow, N., Manning, S., Barnes, S., Barnes, D., and Brouwer, M. (1999) Serum vitellogenin levels and reproductive impairment of male Japanese medaka (*Oryzias latipes*) exposed to 4-*tert*-octylphenol. *Environmental and Health Perspectives*, accepted.

Gross, M.L., and Kapuscinski, A.R. (1997) Reproductive success of smallmouth bass estimated and evaluated from family-specific DNA fingerprints. *Ecology* 78:5:1424.

Gross, T.S., Sepulveda, M.S., Wieser, C.M., Wiebe, J.J., Schoeb, T.R., Denslow, N.D., Johnson, W.E. (2001) "Seasonal reproductive cycles of Florida largemouth bass (*Micropterus salmoides floridanus*)." Submitted to the *Proceedings of the 2000 Black Bass Symposium*, American Fisheries Society.

Gross, T.S., Wiebe, J., Wieser, C., Shrestha, S., and Denslow, N. (1996) Documentation of endocrine-disruption and altered seasonal reproductive parameters for largemouth bass from several lakes in central Florida. *SETAC 17th Annual Meeting*, Washington, D.C.

Guevel, R.L., Petit, F.G., Goff, P.L., Metivier, R., Valotaire, Y., and Pakdel, F. (2000) Inhibition of rainbow trout (*Orcorhynchus mykiss*) estrogen receptor activity by cadmium. *Biol Reprod.* 63:259.

Guillette, Jr. L.J., Gross, T.S., Masson, G.R., Matte J.M., Percival, H.F., and Woodward, A.R. (1994) Developmental abnormalities of the gonad and abnormal sex hormone concentrations in juvenile alligators living in a contaminated environment. *Gen. Comp Endocrinol.* 101:32-42.

Guillette, Jr., L.J., Gross, T.S., Masson, G.R., Matter, J.M., Percival, H.F., and Woodward, A.R. (1994) Developmental abnormalities of the gonad and abnormal sex hormone concentrations in juvenile alligators from contaminated and control lakes in Florida. *Environ. Health Perspect.* 102:680.

Guillette, L.J., Jr. (1996) Vitellogenin induction and reduced serum testosterone concentrations in feral male carp (*Cyprinus carpio*) captured near a major metropolitan sewage treatment plant. *Environ. Health Perspect.* 104:1096.

Gwinnett County (2000). Gwinnett County Watershed Protection Plan. Prepared for Gwinnett County, Georgia. Prepared by CH2MHill in association with JJ&G, Inc., Tetra Tech, Inc., and R&D Environmental, Inc.

Haddy, J.A., and Pankhurst, N.W. (1999) Stress-induced changes in concentrations of plasma sex steroids in black bream. *J. Fish Biol.* The Fisheries Society of the British Isles. (C) 55:6:1304.

Halm, S., Rand-Weaver, M., Sumpter, J.P., and Tyler, C.R. (2001) Cloning and molecular characterization of an ovarian-derived (brain-like) P450 aromatase cDNA and development of a competitive RT-PCR assay to quantify its expression in the fathead minnow (*Pimephales promelas*). *Fish Physiol. Biochem.* 24:1:49.

Handy, R.D. (1994) Intermittent exposure to aquatic pollutants - assessment, toxicity and sublethal responses in fish and invertebrates. *Comp. Biochem. Physiol. C-Pharmacol. Toxicol. Endocrinol.* 107:2:171.

Harries, J.E., Janbakhsh, A., Jobling, S., Matthiessen, P., Sumpter, J.P., and Tyler, C.R. (1999) Estrogenic potency of effluent from two sewage treatment works in the United Kingdom. *J. Env Tox and Chem.* 18(5):932-937.

Harries, J.E., Runnalls, T., Hill, E., Harris, C.A., Maddix, S., Sumpter, J.P., and Tyler, C.R. (2000) Development of a reproductive performance test for endocrine disrupting chemicals using pair-breeding fathead minnows (*Pimephales promelas*). *Environ. Sci. Technol.* 34:3003.

Harries, J.E., Runnalls, T., Hill, E., Harris, C.A., Maddix, S., Sumpter, J.P., and Tyler, C.R. (2000) Development of a reproductive performance test for endocrine disrupting chemicals using pair-breeding fathead minnows (*Pimephales promelas*). *Environ. Sci. Technol.* 34:14:9.

Harries, J.E., Sheahan, D.A., Jobling, S., Matthiessen, P., Neall, P., Routledge, E.J., Rycroft, R., Sumpter, J.P., and Taylor, T. (1996) A survey of estrogenic activity in United Kingdom inland waters. *Environmental Toxicology and Chemistry* 15:11:1993.

Harries, J.E., Sheahan, D.A., Jobling, S., Matthiessen, P., Neall, P., Sumpter, J.P., Tylor, T., and Zaman, N. (1997) Estrogenic activity in five United Kingdom rivers detected by measurement of vitellogenesis in caged male trout. *Env. Tox. and Chem.* 16(3):534-542.

Harris, C.A., Henttu, P., Parker, M.G., and Sumpter, J.P. (1997) The estrogenic activity of phthalate esters *in vitro. Environmental Health Perspectives*. 105:802.

Harris, C.A., Santos, E.M., Janbakhsh, A., Pottinger, T.G., Tyler, C.R., and Sumpter, J.P. (2001) Nonylphenol affects gonadotropin levels in the pituitary gland and plasma of female rainbow trout. *Environ. Sci. Technol.* 35:14:2909.

Hashimoto, S., Bessho, H., Hara, A., Nakamura, M., Iguchi, T., and Fujita, K. (2000). Elevated serum vitellogenin levels and gonadal abnormalities in wild male flounder (*Pleuronectes yokohamae*) from Tokyo Bay, Japan. *Mar. Environ. Res.* 49:37-53.

Hemmer, M.J., Hemmer, B.L., Bowman, C.J., Kroll, K.J., Folmar, L.C., Marcovich, D., Hoglund, M.D., and Denslow, N.D. (2001) Effects of p-nonylphenol, methoxychlor, and endosulfan on vitellogenin induction and expression in sheepshead minnow (*Cyprinodon variegatus*). *Environ. Toxicol. Chem.* 20:2:336.

Hemming, J.M., Waller, W.T., Chow, M.C., Denslow, N.D. and Venables, B. (2001) Assessment of the estrogenicity and toxicity of a domestic wastewater effluent flowing through a constructed wetland system using biomarkers in male fathead minnows (*Pimephales promelas* rafinesque, 1820). *Environmental Toxicology and Chemistry*. 20(10):2268. © SETAC.

Henson, K. and Gallagher, E. (1999) Conservation of a glutathione *S*-transferase among marine and freshwater fish. Mar. Envr. Res. in review. Heppell, S. A., N.D. Denslow, L.C. Folmar, and C. V. Sullivan. 1995 Universal assay of vitellogenin as a biomarker for environmental estrogens. *Environ. Health Perspectives* 103:9.

Heppell, S.A., Denslow, N.D., Folmar, L.C., and Sullivan, C.V. (1996) Universal assay of vitellogenin as a biomarker for environmental estrogens. *Environmental Health Perspective.* 103(S7):9-15.

Heppell, S.A., Jackson, L.F., Weber, G.M., and Sullivan, C.V. (1999) Enzyme-linked immunosorbent assay (ELISA) of vitellogenin in temperate basses (Genus Morone): Plasma and in vitro analyses. *Trans. Am. Fish. Soc.* 128:3:532.

Herrin, D.L., and Schmidt, G.W. (1988) Rapid, reversible staining of northern blots prior to hybridization. *Biotechniques* 6:196.

Hewitt, M., and Servos, M. (2001) An overview of substances present in Canadian aquatic environments associated with endocrine disruption. *Water Qual. Res. J. Canada* 36:2:191.

Hobby, A.C., Geraghty, D.P., and Pankhurst, N.W. (2000) Differences in binding characteristics of sex steroid binding protein in reproductive and nonreproductive female rainbow trout (*Oncorhynchus mykiss*), black bream (*Acanthopagrus butcheri*), and greenback flounder (*Rhombosolea tapirina*). *Gen. Comp. Endocrinol.* 120:3:249.

Holbech, H., Andersen, L., Petersen, G.I., Korsgaard, B., Pedersen, K.L., and Bjerregaard, P. (2001) Development of an ELISA for vitellogenin in whole body homogenate of zebrafish (*Danio rerio*). *Comp. Biochem. Physiol. C-Toxicol. Pharmacol.* 130:1:119.

Hornung, M.W., Korte, J.J., Tapper, M.A., Lambright, C.R., Kelce, W.R., Schmeider, P.K., Gray, L.E., and Ankley, G.T. (1999) Use of rainbow trout hepatoma cell line (rth-149) for screening estrogenic chemicals: comparison of transfection methods and optimization of reporter activation. *Abstract to SOT*, New Orleans.

Hose, J.E., Cross, J.N., Smith, S.G., and Diehl, D. (1989) Reproductive impairment in a fish inhabiting a contaminated coastal environment off of Southern California. *Environ. Poll.* 57:139.

Howell, A., Osborne, C.K., Morris, C. and Wakeling, A.E. (2000). ICI 182,780 (Faslodex™) Development of a Novel "Pure" Antiestrogen. *Cancer* 89(4):817.

Hughes, M.R., Bailer, A.J., Denton, D.L. (2001) Toxicant- and response-specific comparisons of statistical methods for estimating effective concentrations. *Environ. Toxicol. Chem.* 20:6:1374.

Hwang, U.G., Kagawa, N., and Mugiya, Y. (2000) Aluminum and cadmium inhibit vitellogenin and its mRNA induction by estradiol-17 beta in the primary culture of hepatocytes in the rainbow trout (*Oncorhynchus mykiss*). *Gen. Comp Endocrinol.* 119:69.

Ishibashi, H., Tachibana, K., Tsuchimoto, M., Soyano, K., Ishibashi, Y., Nagae, M., Kohra, S., Takao, Y., Tominaga, N., and Arizono, K. (2001) In vivo testing system for determining the estrogenic activity of endocrine-disrupting chemicals (EDCs) in goldfish (*Carassius auratus*). *J. Health Sci.* 47:2:213.

Jalabert, B., Baroiller, J.F., Breton, B., Fostier, A., Le Gax, F., Guiguen, Y., and Monod, G. (2000) Main neuro-endocrine, endocrine and paracrine regulations of fish reproduction, and vulnerability to xenobiotics. *Ecotoxicology* 9:1-2:25.

Jardine, J.J., Van der Kraak, G.J., and Munkittrick, K.R. (1996) Capture and confinement stress in white sucker exposed to bleached kraft pulp mill effluent. *Ecotoxicol Environ Saf.* 33:3:287.

Jensen, K.M., Korte, J.J., Kahl, M.D., Pasha, M.S., and Ankley, G.T. (2001) Aspects of basic reproductive biology and endocrinology in the fathead minnow (*Pimephales promelas*). *Comp. Biochem. Physiol. C-Pharmacol. Toxicol. Endocrinol.* 128:1:127. Published by Elsevier Science Inc.

Jobling, S. and Sumpter, J.P. (1993) Detergent components in sewage effluent are weakly estrogenic to fish: an *in vitro* study using rainbow trout hepatocytes. *Aquatic Toxicol.* 27:361.

Jobling, S., Nolan, M., Tyler, C.R., Brighty, G., and Sumpter, J.P. (1998) Widespread sexual disruption in wildfish. *Environ. Sci. Technol.* 32:2498.

Jobling, S., Reynolds, T., White, R., Parker, M.G., and Sumpter, J.P. (1995) A variety of environmentally persistent chemicals, including some phthalate plasticizers are weakly estrogenic. *Environmental Health Perspectives.* 103:582.

Jobling, S., Sheahan, D., Osborne., J.A., Matthiessen, P., and Supmter, J.P. (1996) Inhibition of testicular growth in rainbow trout (*Oncorhynchus mykiss*) exposed to estrogenic alkylphenolic chemicals. *Environmental Toxicology and Chemistry.* 15:194.

Johnson, A.C., and Sumpter, J.P. (2001) Removal of endocrine-disrupting chemicals in activated sludge treatment works. *Centre for Ecology and Hydrology,* Wallingford, Oxfordshire OX10 8BB, U.K., and Dept. of Biology and Biochem., Brunel Univ., Uxbridge, Middlesex UB8 3PH, U.K.

Johnson, W.E. and Vogel, D.S. (1985) *DDT residues in Lake Apopka fish: association with use of dicofol in Florida citrus.* Investigative Report, Florida Game and Freshwater Fish Commission.

Jones, P.D., De Coen, W.M., Tremblay, L., and Giesy, J.P. (2001) Vitellogenin as a biomarker for environmental estrogens. *Water Sci. Technol.* 42:7-8:

Kahl, M.D., Jensen, J.J., Korte, G.T., and Ankley, G.T. (2001) Effects of handling on endocrinology and reproductive performance of the fathead minnow. *J. Fish Biol* 59:3:515.

Kane, A.S. and Thohan, S. (1996) *Dynamic culture of fish hepatic tissue slices to assess phase I and phase II biotransformation.* Techniques in Aquatic Toxicology (G.K. Ostrander, Ed.), pp. 371-394. New York, CRC Press.

Karels, A., Markkula, E., and Oikari, A. (2001) Reproductive, biochemical, physiological, and population responses in perch (*Perca fluviatilis L.*) and roach (*Rutilus rutilus L.*) downstream of two elemental chlorine-free pulp and paper mills. *Environ. Toxicol. Chem.* 20:7:1517.

Katritzky, A.R., Tatham, D.B., and Maran, U. (2001) Theoretical descriptors for the correlation of aquatic toxicity of environmental pollutants by quantitative structure-toxicity relationships. *J. Chem. Inf. Comput. Sci.* 41:5:1162.

Kayaba, T., Takeda, N., Adachi, S., and Yamauchi, K. (2001) Ultrastructure of the oocytes of the Japanese eel (*Anguilla japonica)* during artificially induced sexual maturation. *Fish. Sci.* 67:5:870.

Kelce, W.R., Monosson, E., Gamasik, M.P., Laws, S.C., and Earl, Jr., G.L. (1994) Environmental Hormone Disruptors: Evidence That Vinclozolin Developmental Toxicity Is Mediated by Antiandrogenic Metabolites. *Toxicol. Appl. Pharmacol.* 126:276-285.

Kelce, W.R., Stone, C.R., Laws, S.C., Gray, L.E., Kemppainen, J.A., and Wilson, E.M. (1995) Persistent DDT metabolite p,p'-DDE is a potent androgen receptor antagonist. *Nature.* 375:581.

Khan J., Saal L.H., Bittner M.L., Chen Y., (1999) Expression profiling in cancer using cDNA microarrays. *Electrophoresis.* 20:223.

Kime, D.E., Nash, J.P., and Scott, A.P. (1999) Vitellogenesis as a biomarker of reproductive disruption by xenobiotics. *So Aquaculture* (C) 1999 Elsevier Science B.V. All rights reserved. 177:1-4.

Kinnberg, K., Korsgaard, B., and Bjerregaard, P. (2000) Concentration-dependent effects of nonylphenol on testis structure in adult platyfish (*Xiphophorus maculatus*). *Marine Environmental Research* 50:1-5:169.

Kishida, M. and Specker, J. (1993) Vitellogenin in tilapia (*Oreochromis mossambicus*): induction of two forms by estradiol, quantification in plasma and characterization in oocyte extract. *Fish Physiol Biochem.* 12:171.

Kitano, T., Takamune, K., Kobayashi, T., Nagahama, Y., and Abe, S.I. (1999) Suppression of P450 aromatase gene expression in sex-reversed males produced by rearing genetically female larvae at a high water temperature during a period of sex differentiation in the Japanese flounder (*Paralichthys olivaceus*). *J. Mol. Endocrinol* 23:167.

Kloas, W., Schrag, B., Ehnes, C., and Segner, H. (2000) Binding of xenobiotics to hepatic estrogen receptor and plasma sex steroid binding protein in the teleost fish, the common carp (*Cyprinus carpio*). *Gen. Comp. Endocrinol.* 119:3:287.

Klotz, D.M., Beckman, B.S., Hill, S.M., McLachlan, J.A., Walters, M.R. and Arnold, S.F. (1996) Identification of environmental chemicals with estrogenic activity using a combination of *in vitro* assays. *Environmental Health Perspectives*. 104:1084.

Kobayaski, M., Aida, K., and Hanyu, I. (1986) Annual changes in plasma levels of gonadotropin and steroid hormones in goldfish. *Bulletin of the Japanese Society of Scientific Fisheries.* 52(7):1153.

Kokkidis, M.J., Goubier, V., Martin, M., Hardy, I., and Exbrayat, J.M. (2000) Haematological changes in the blood of cultured black-bass (*Micropterus salmoides*) during an annual sexual reproductive cycle. *Cybium* 24:3:113.

Kolpin, D.W., Furlong, E.T., Meyer, M.T., Thurman, E.M., Zaugg, S.D., Barber, L.B., Buxton, H.T. (2002) Pharmaceuticals, Hormones, and Other Organic Wastewater Contaminants in U.S. Streams, 1999-2000: A National Reconnaissance. *Environ. Sci. Technol*. 36-1202.

Koob, T.J., and Callard, I.P. (1999) Reproductive endocrinology of female elasmobranchs: lessons from the little skate (*Raja erinacea*) and spiny dogfish (*Squalus acanthias*). *J. Exp. Zool.* 284(5):557.

Korach, K.S., Couse, J.F., Curtis, S.W., Wasburn, T.F., Lindzey, J., Kimbro, K.S., Eddy, E.M., Migliaccio, S., Snedeker, S.M., and Lubahn, D.B. (1996) Estrogen receptor gene disruption: molecular characterization and experimental and clinical phenotypes. *Recent Prog. Horm. Res.* 51:159.

Kramer, V.J., Miles-Richardson, S., Pierens, S.L., Giesy, J.P. (1998) Reproductive impairment and induction of alkaline-labile phosphate, a biomarker of estrogen exposure, in fathead minnows (*Pimephales promelas*) exposed to waterborne 17 beta-estradiol. *Aquatic Toxicology* 40:4. (C) 1998 Elsevier Science B.V. All rights reserved.

Krege, J.H., Hodgin, J.B., Couse, J.F., Enmark, E., Warner, M., Mahler, J.F., Sar, M., Korach, K.S., Gustafsson, J.A., and Smithies, O. (1998) Generation and reproductive phenotypes of mice lacking estrogen receptor. *Proc. Nat. Acad. Sci. USA.* 95:15677.

Kronovetr, J., Barthova, J., Barth, T., and Ryslava, H. (2000) Purification and characterisation of a vitellogenin derived aminopeptidase from rainbow trout eggs. *Collect. Czech. Chem. Commun.* 65:7:1198.

Kuiper, G.G.J.M., Enmark, E., Pelto-Huikko, M., Nilsson, S., and Gustafsson, J.A. (1996) Cloning of a novel receptor expressed in rat prostate and ovary. *Proc. Natl. Acad. Sci. USA.* 93:5925.

Kumar, V. and Chambon, P. (1988) The estrogen receptor binds tightly to its responsive element as a ligand-induced homodimer. *Cell.* 55:145.

Kwak, H.I., Bae, M.O., Lee, M.H., Lee, Y.S., Lee, B.J., Kang, K.S., Chae, C.H., Sung, H.J., Shin, J.S., Kim, J.H., Mar, W.C., Sheen, Y.Y., and Cho, M.H., Effects of nonylphenol, bisphenol A, and their mixture on the viviparous swordtail fish (*Xiphophorus helleri*) *Environ. Toxicol. Chem.* 20:4:787.

Lange, R., Hutchinson, T.H., Croudace, C.P., and Siegmund, F. (2001) Effects of the synthetic estrogen 17 alpha-ethinylestradiol on the life-cycle of the fathead minnow (*Pimephales promelas*). *Environ. Toxicol. Chem.* 20:6:1216.

Larsson, D. G. J., Hällman, H., and Förlin, L., (2000); Skewed embryonic sex ratios in a viviparous fish: a result of endocrine disruption? *Marine Environmental Research* 50:1-5:191.

Larsson, D. G. J., Hyllner, S.J., and Haux, C. (1994) Induction of vitellin envelope proteins by estradiol-17 in 10 teleost species. *Gen. Comp. Endocrinol.* 96:445.

Larsson, D.G.J., Adolfsson-Erici, M., Parkkonen, J., Pettersson, M., Berg, A.H., Olsson, P.-E., and Förlin, L. (1999). Ethinyloestradiol – an undesired fish contraceptive? *Aquatic Toxicology* 45:91.

Latonnelle, K., Le Menn, F., and Bennetau-Pelissero, C. (2000) In vitro estrogenic effects of phytoestrogens in rainbow trout and Siberian sturgeon. *Ecotoxicology* 9:1-2:115.

Lattier, D.L., Gordon, D.A., Burks, D.J., and Toth, G.P. (2001) Vitellogenin gene transcription: A relative quantitative exposure indicator of environmental estrogens. *Environ. Toxicol. Chem.* 20:9:1979.

Laws, S.C., Carey, S.A., Ferrell, J.M., Bodman, G.J. and Cooper, R.L. (2000) Estrogenic Activity of Octylphenol, Nonylphenol, Bisphenol A, and Methoxychlor in Rats. *Toxicological Sciences* 54:154.

Lazier, C.B., Langley, S., Ramsey, N.B., and Wright, J.M. (1996) Androgen inhibition of vitellogenin gene expression in tilapia (*Oreochromis niloticus*). *Gen. Comp. Endocrinol* 104:321.

Le Gac, F., Thomas, J.L., Mourot, B., and Loir, M. (2001) *In vivo* and in vitro effects of prochloraz and nonylphenol ethoxylates on trout spermatogenesis. *Aquat. Toxicol.* 53:3-4:187.

Le Guevel, R., Petit, F.G., Le Goff, P., Metivier, R., Valotaire, Y., and Pakdel, F. (2000) Inhibition of rainbow trout (*Oncorhynchus mykiss*) estrogen receptor activity by cadmium. *Biol. Reprod.* 63:1:259.

Lee, Y.H., Du, J.L., Yueh, W.S., Lin, B.Y., Huang, J.D., Lee, C.Y., Lee, M.F., Lau, E.L., Lee, F.Y., Morrey, C., Nagahama, Y., and Chang, C.F. (2001) Sex change in the protandrous black porgy, (*Acanthopagrus schlegeli*): A review in gonadal development, estradiol, estrogen receptor, aromatase activity and gonadotropin. *J. Exp. Zool.* 290:7:715.

Lee, Y.H., Lee, F.Y., Yueh, W.S., Tacon, P., Du, J.L., Chang, C.N., Jeng, S.R., Tanaka, H. and Chang, C.F. (2000) Profiles of gonadal development, sex steroids, aromatase activity, and gonadotropin II in the controlled sex change of protandrous black porgy, (*Acanthopagrus schlegeli*) Bleeker. *Gen. Comp. Endocrinol.* 119:1:111.

Legler, J., van den Brink, C.E., Brouwer, A., Murk, A.J., van der Saag, P.T., Vethaak, A.D. and van der Burg, B. (1999) Development of a stably transfected estrogen receptor-mediated luciferase reporter gene assay in the human t47d breast cancer cell line. *Toxicological Sciences* 48:55.

LeGuevel, R., and Pakdel, F. (2001) Assessment of oestrogenic potency of chemicals used as growth promoter by in-vitro methods. *APMIS* 109:103:S473.

Liang, P. and Pardee, A.B. (1992) Differential display of eucaryotic messenger RNA by means of the polymerase chain reaction. *Science.* 257:967.

Liber, K., Gangl, J.G., Corry, T.D., Heinis, L.J., and Stay, F.S. (1999) Lethality and bioaccumulation of 4-nonylphenol in bluegill sunfish in littoral enclosure. *ET&C* 18:394.

Lim, E.H., Teo, B.Y., Lam, T.J., and Ding, J.L. (2001) Sequence analysis of a fish vitellogenin cDNA with a large phosvitin domain. *Gene* 277:1-2:175.

Lindholst, C., Pedersen, S.N., and Bjerregaard, P. (2001) Uptake, metabolism and excretion of bisphenol A in the rainbow trout (*Oncorhynchus mykiss*). *Aquat. Toxicol.* 55:1-2:75.

Loven, M.A., Wood, J.R., and Nardulli, A.M. (2001) Interaction of estrogen receptors alpha and beta with estrogen response elements. *Mol. Cell. Endocrinol.* 181:1-2:151.

Lubahn, D.B., Moyer, J.S., Golding, T.S., Couse, J.F., Korach, K.S., and Smithies, O. (1993) Alteration of reproductive function but not prenatal sexual development after insertional disruption of the mouse estrogen receptor gene. *Proc. Nat. Acad. Sci. USA.* 90:11162.

Lye, C.M., Frid, L.J., Gill, M.E., Cooper, D.W., and Jones, D.M. (1999) Estrogenic alkylphenols in fish tissues, sediments, and waters from the U.K. Tyne and Tees estuaries. *Environ. Sci. Technol.* 33:1009.

Machala, M., Petrivalsky, M., Nezveda, K., Ulrich, R., Dusek, L., Piacka, V., and Svobodova, Z. (1997) Responses of carp hepatopancreatic 7-ethoxyresorufin-*O*-deethylase and glutathione-dependent enzymes to organic pollutants-a field study. *Environ. Toxicol. Chem.* 16:1410.

Mannervik, B. and Danielson, U.H. (1988) Glutathione transferases - Structure and catalytic activity. *CRC Crit. Rev. Biochem.* 23:283.

Marx, A., Sherry, J., Hansen, P.D., and Hock, B. (2001) A new monoclonal antibody against vitellogenin from rainbow trout (*Oncorhynchus mykiss*). *Chemosphere* 44:3:393.

Marzluff, W.F. (1978) Transcription of RNA in isolated nuclei. meth. *Cell Biol.* 19:317.

Matthiessen, P. and Gibbs, P.E. (1998) Critical appraisal of the evidence for tributyltin-mediated endocrine disruption in mollusks. *J. Env. Tox. and Chem.* 17:37.

McArdle, M., Elskus, A., McElroy, A., Larsen, B., Benseon, W. and Schlenk, D. (2000) Estrogenic and CYP1A response of mummichogs and sunshine bass to sewage effluent. *Marine Environmental Research* 50:1-5:175.

McClave, J.T. and Sincich, T. (2000) *Statistics 8th Edition*, Prentice Hall, N.J.

McKinley, R.S., Griffiths, J.S., Kowalyk, H.E., McKenna, G.R., and Cooke, S.J., (2000) Reproductive activity and summer residency patterns of smallmouth bass, (*Micropterus dolomieu*), in a thermal discharge canal on Lake Erie. *J. Freshw. Ecol.* 15:3:307.

McNatt, H.B., Angus, R.A., and Howell, W.M. (2000) Effects of paper mill effluent on a population of eastern mosquitofish, *Gambusia holbrooki*. Poster #13 Abstract presented at *e.hormone 2000 Conference*, October 15-18, 2000, New Orleans, Louisiana.

McTavish, K., Stech, H., and Stay, F. (1998) A modeling framework for exploring the population-level effects of endocrine disruptors. *Environmental Toxicology and Chemistry*. 17:58.

Meek, M.D., Clemons, J.H., Wu, Z.F., and Zacharewski, T.R. (1997) Assessment of the alleged estrogen-receptor mediated activity of eight commercial phthalate esters. *Fundamental Applications in Toxicology*. 36:295.

Menkenyan, O., Ivanov, J., Karabunarliev, S., Bradbury, S., Ankley, G.T., and Karcher, W. (1997) A computationally-based hazard identification algorithm that incorporates ligand flexibility: Identification of potential androgen receptor ligands. *Env. Sci. and Tech.* 31(12):3702.

Mercure, F., Holloway, A.C., Tocher, D.R., Sheridan, M.A., Van der Fraak, G., and Leatherland, J.F. (2001) Influence of plasma lipid changes in response to 17 beta-oestradiol stimulation on plasma growth hormone, somatostatin, and thyroid hormone levels in immature rainbow trout. *J. Fish Biol.* 59:3:605

Merson, R.R., Casey, C.S., Martinez, C., Soffientino, B., Chandlee, M., and Specker, J.L. (2000) Oocyte development in summer flounder: seasonal changes and steroid correlates. *J. Fish Biol.* 57:1:182.

Metcalfe, C.D., Metcalfe, T.L., Kiparissis, Y., Koenig, B.G., Khan, C., Hughes, R.J., Croley, T.R. March, R.E., and Potter, T. (2001) Estrogenic potency of chemicals detected in sewage treatment plant effluents as determined by *in vivo* assays with Japanese medaka (*Oryzias latipes*). *Environ. Toxicol. Chem.* 20:2:297.

Miles-Richardson, S.R., Kramer, V.J., Fitzgerald, S.D., Render, J.A., Render, J.A., Yamini, B., Barbee, S.J., and Giesy, J.P. (1999) Effects of waterborne exposure of 17 beta-estradiol on secondary sex characteristics and gonads of fathead minnows (*Pimephales promelas*). *Aquatic Toxicology* (C) 1999 Elsevier Science B.V. All rights reserved. 1999 NOV.

Milligan, S.R., Balasubramanian, A.V., and Kalita, J.C. (1998) Relative potency of xenobiotic estrogens in an acute in vivo mammalian assay. *Environ. Health Perspect.* 106:23.

Mills, L.J., Gutjahr-Gobell, R.E., Haebler, R.A., Horowitz, D.J.B., Jayaraman, S., Pruell, R.J., McKinney, R.A., Gardner, G.R., and Zaroogian, G.E. (2001) Effects of estrogenic (o,p '-DDT; octylphenol) and anti-androgenic (p,p '-DDE) chemicals on indicators of endocrine status in juvenile male summer flounder (*Paralichthys dentatus*). *Aquat. Toxicol.* 52:2:157.

Mitchell, D.F. (1982) Effects of water level fluctuation on reproduction of largemouth bass, (*Micropterus-salmoides*), at Millerton Lake, California, in 1973. *Calif. Fish Game* 68:2:68.

Moberg, G.P., Doroshov, S.I., Chapman, F.A., Kroll, K.J., Van Eenennaam, J., and Watson, J.C. (1991) Effects of various hormone implants on vitellogenin synthesis and ovarian development in cultured white sturgeon (*Acipenser transmontanus*). In *Acipenser actes due premier colloque international sur l'esturgeon* (Ed. P. Williot) Bordeaux, CEMAGREF, pp 389-399.

Moll, D.M., Frick, E.A., Henderson, A.K., Furlong, E.T. and Meyer, M.T. (2001) "Presence of Pharmaceuticals in Treated Wastewater Effluent and Surface Water Supply Systems, Metropolitan Atlanta, Georgia, July-September 1999." In *Proc. 2001 Georgia Water Resources Conference*. University of Georgia. Athens, Georgia.

Mommsen, T.P., Walsh, P.J. (1998). Vitellogenesis and oocyte assembly. In: Hoar, W.S., Randall, D.J., (Eds.), *Fish Physiology, XIA*. Academic Press, Inc., San Diego, CA, 347-406.

Monteverdi, G.H., and Di Giulio, R.T. (2000) Vitellogenin-associated maternal transfer of exogenous and endogenous ligands in the estuarine fish, *Fundulus heteroclitus. Marine Environmental Research* 50:1-5:191.

Moore, R.E., Davies, M.S., O' Connell, K.M., Harding, E.I., Wiegand, R.C., and Tiemeier, D.C. (1986) Cloning and expression of a cDNA encoding a maize glutathione *S*-transferase in E. Coli. *Nucleic Acids Res.* 14:7227.

Morehead, D.T., (1998) Effect of capture, confinement and repeated sampling on plasma steroid concentrations and oocyte size in female striped trumpeter (*Latris lineata*) (Latrididae). *Mar. Freshw. Res.* 49:5:373.

Mosselman, S., Polman, J., and Dijkema, R. (1996) ER beta: identification and characterization of a novel human estrogen receptor. *FEBS Lett.* 392:49.

Munkittrick, D. R. and Van Der Kraak, G. (1999) "Appropriate Uses of Physiological Techniques for Endocrine Studies" Environmental Toxicology and Risk Assessment: Standardization of Biomarkers for Endocrine Disruption and Environmental Assessment: Eight Volume, ASTM STP 1364, D. S. Henshel, Ed. M. C. Black, M. C. Harrass, American Society of Testing and Materials, West Conshohocken, PA, 1999.

Munkittrick, K.R., Servos, M.R., Carey, J.H., and Van Der Kraak, G.J. (1997) Environmental impacts of pulp and paper wastewater: Evidence for a reduction in environmental effects at North American pulp mills since 1992. *Water Sci. Technol.* 35:329.

Munkittrick, K.R., Van Der Kraak, G.J., McMaster, M.E., Van Den Heuvel, M.R., and Servos, M.R. (1994) Survey of receiving-water environmental impacts associated with discharges from pulp mills. 2. Gonad size, liver size, hepatic EROD activity and plasma sex steroid levels in white sucker. *ET&C* 13:1089.

Munkittrick, K.R., Van Der Kraak, G.J., McMaster, M.E., and Portt, C.B. (1992) Response of hepatic MFO activity and plasma sex steroids to secondary treatment of bleached kraft pulp mill effluent and mill shutdown. *ET&C* 11:1427.

Munkittrick, K.R., Portt, C.B., Van Der Kraak, G.J., Smith, I.R., and Rokosh, D.A. (1991) Impact of bleached kraft mill effluent on population characteristics, liver MFP activity, and serum steroid levels of a Lake Superior white sucker (*Catostomus commersoni*) population. *Can.J. Fish. Aquat. Sci.* 48:1371.

Murata, K., Sugiyama, H., Yasumasu, S., Iuchi, I., Yasumasu, I., and Yamagami, K. (1997) Cloning of cDNA and estrogen-induced hepatic gene expression for choriogenin H, a precursor protein of the fish egg envelope (chorion). *Proceedings of the National Academy of Science, USA.* 94:2050.

Nagler, J.J., Bourma, J., Thorgaard, G.H., and Dauble, D.D. (2001) High incidence of a male-specific genetic marker in phenolic female chinook salmon from the Columbia River. *Environmental Health Perspectives* 109:67.

Nagler, J.J., Krisfalusi, M., and Cyr, D.G. (2001) Quantification of rainbow trout (*Oncorhynchus mykiss*) estrogen receptor-alpha messenger RNA and its expression in the ovary during the reproductive cycle. *J. Mol. Endocrinol.* 20:6:1276.

Nath, P., and Maitra, S., (2001) Role of two plasma vitellogenins from Indian major carp (*Cirrhinus mrigala*) in catfish (*Clarias batrachus*) vitellogenesis. *Gen. Comp. Endocrinol.* 124:1:30.

Navas, J.M. and Segner, H. (2000) Modulation of trout 7-ethoxyresorufin-O-deethylase (EROD) activity by estradiol and octylphenol. *Marine Environmental Research* 50:1-5:157.

Nelson, M. (2001). *Annual reproductive cycle of the largemouth bass* (*Micropterus salmoides*). University of Florida-Gainesville Master's thesis.

Nibel, P.M., Revaclier, R. (1999) Assessment of sewage treatment plant effluents impact on river water quality using dissolved rb/sr ratio. *Env. Sci. and Tech.* XX:A-E (ES981097G)

Nichols, K.M., Miles-Richardson, S.R., Snyder, E.M., Giesy, J.P. (1999) Effects of exposure to municipal wastewater in situ on the reproductive physiology of the fathead minnow (Pimephales promelas). So *Environmental Toxicology and Chemistry* 18:9.

Nichols, K.M., Snyder, E.M., Snyder, S.A., Pierens, S.L., Miles-Richardson, S.R., and Giesy, J.P. (2001) Effects of nonylphenol ethoxylate exposure on reproductive output and bioindicators of environmental estrogen exposure in fathead minnows, (*Pimephales promelas*). *Environ. Toxicol. Chem.* 20:3:510.

Nilsen, B.M., Berg, K., Arukwe, A., and Goksøyr, A. (1998) Monoclonal and polyclonal antibodies against fish bitellogenin for use in pollution monitoring. *Marine Environmental Research.* 46(1-5):153.

Nilsen, B.M., Eidem, J.E., Kristiansen, S.I., Nilsen, M.V., Berg, K., Ibsen, M.W., Kordes, C., Gutzeit, H.O., Denslow, N.D., and Goksoyr, A. (2001) Development of quantitative vitellogenin-ELISA assays for fish test species used in endocrine disruptor screening. *Toxicology* 164:1-3:210.

Nimrod, A.C. and Benson, W.H. (1998) "Assessment of estrogenic activity in fish", In *Chemically-Induced Alteration in the Functional Development and Reproduction of Fishes*, R. Rolland, M Gilbertson, and R.E. Peterson Eds., SETAC Press, Pensacola, FL, pp. 87-100.

Niven, S.J., Snape, J., Hetheridge, M., Evans, M., McEvoy, J., and Rowland, S.J. (2001) Investigations of cholesterol transformation during sewage treatment: relevance to estrogen formation pathways? *Sci. Total Environ.* 279:1-3:75.

Niven, S.J., Snape, J., Hetheridge, M., Evans, M., McEvoy, J., Sutton, P.G., and Rowland, S.J. (2001) Investigations of the origins of estrogenic A-ring aromatic steroids in UK sewage treatment works effluents. *Analyst* 126:3:285.

Okumura, H., Saeki, F., Matsubara, H., Adachi, S., and Yamauchi, K. (2001) Changes in serum vitellogenin levels and immunohistochemical localization of vitellogenin in hepatic cells during ovarian development in the Japanese eel. *Fish. Sci.* 67:5:880.

Organization for Economic Co-operation and Development (OECD). (1997) Draft detailed review paper: *Appraisal of test methods for sex-hormone disrupting chemicals.* Paris, France.

Orlando, E.F., Denslow, N.D., Folmar, L.C., and Guillette, L.J. (1999) A comparison of the reproductive physiology of largemouth bass, (*Micropterus salmoides*), collected from the Escambia and Blackwater Rivers in Florida. *Environ. Health Perspect.* 107:3:199.

Ostlund Farrants, A.K., Meyer, D.J., Coles, B., Southan, C., Aitken, A., and Ketterer, B. (1987) The separation of glutathione *S*-transferase subunits using reversed phase high-performance liquid chromatography. *Biochem. J.* 245:423.

Ostrander, G.K., Blair, J.B., Stark, B.A., Marley, G.M., Bales, W.D., Veltri, R.W., Hinton, D.E., Okihiro, M., Ortego, L.S., Hawkins, W.E. (1995) Long term primary culture of epithelial cells from rainbow trout (*O. mykiss*) liver; in vitro. *Cell Dev. Biol.* 31:367.

Ozers, M.S., Hill, J.J., Ervin, K., Royer, C.A., and Gorski, J. (2001) The dissociation rate of estrogen receptor alpha from the consensus estrogen response element. *Mol. Cell. Endocrinol.* 175:1-2:101.

Pace, P., Talow, J., Suntharalingam, S., Coombes, R.C., Ali, S. (1997) Human estrogen receptor binds DNA in a manner similar to and dimerizes with estrogen receptor. *J. Biol. Chem.* 272:25832.

Paech, K., Webb, P., Kuiper, G.G.J.M., Nilsson, S., Gustafsson, J.-A., Kushner, P.J., and Scanlan, T.S. (1997) Differential ligand activation of estrogen receptors ER and ER at API sites. *Science.* 277:1508.

Page, L.M. and Burr, B.M. 1991. *Freshwater Fishes.* Houghton Mifflin Company, Boston.

Pakdel, F., Feon, S., LeGac, F., LeMenn, F., and Valotair, Y. (1991) In vivo estrogen induction of hepatic estrogen receptor mRNAs and correlation with vitellogenin mRNA in rainbow trout. *Mol. Cell. Endocrinology.* 75:205.

Pakdel, F., LeGac, F., LeGoff, P., and Valotaire, Y. (1990) Full-length sequence and *in vitro* expression of rainbow trout estrogen receptor cDNA. *Molecular and Cellular Endocrinol.* 71:195.

Pakdel, F., Petit, F., Anglade, I., Kah, O., Delaunay, F., Bailhache, T., and Valotair, Y. (1994) Overexpression of rainbow trout estrogen receptor domains in *Escherichia coli*: characterization and utilization in the production of antibodies for immunoblotting and immunocytochemistry. *Molecular and Cellular Endocrinol.* 104:81.

Pakdel, F.L., Le Guellec, C., Vaillant, C., Le Roux, M.G., and Valotaire, Y. (1989) Identification and estrogen induction of two estrogen receptors (ER) messenger ribonucleic acids in the rainbow trout liver: Sequence homology with other ERs. *Mol Endocrinol.* 3:44.

Palmer, B.D. and Selcer, K.W. (1996) Vitellogenin as a biomarker for xenobiotic estrogens: A review. environmental toxicology and risk assessment: biomarkers and risk assessment. *ASTM STP* 1306 Vol 5:3.

Palmer, B.D., Huth, L.K., Pieto, D.L., and Selcer, K.W. (1998) Vitellogenin as a Biomarker for Xenobiotic Estrogens in an Amphibian Model System. *J. Env. Tox. and Chem.* 17:30.

Pandey, A.K. (1992) Endocrinology of calcium metabolism in amphibians, with emphasis on the evolution of hypercalcemic regulation in tetrapods. *Biol Struct Morphog.* 4(3):102.

Pankhurst, N.W., (1998) Further evidence of the equivocal effects of cortisol on *in vitro* steroidogenesis by ovarian follicles of rainbow trout Oncorhynchus mykiss. *Fish Physiol. Biochem.* 19:4:315.

Pankhurst, N.W., (2001) Stress inhibition of reproductive endocrine processes in a natural population of the spiny damselfish (*Acanthochromis polyacanthus*). *Mar. Freshw. Res.* 52:5:753.

Pankhurst, N.W., and Van der Kraak, G., (2000) Evidence that acute stress inhibits ovarian steroidogenesis in rainbow trout *in vivo*, through the action of cortisol. *Gen. Comp. Endocrinol* 117:2:225 (C) Academic Press.

Panter, G.H., Thompson, R.S., and Sumpter, J.P. (1998) Adverse reproductive effects in male fathead minnows (*Pimephales promelas*) exposed to environmentally relevant concentrations of the natural oestrogens, oestradiol and oestrone. *Aquatic Toxicology.* 42:243.

Panter, G.H., Thompson, R.S., and Sumpter, J.P. (2000) Intermittent exposure of fish to estradiol. *Environ. Sci. Technol.* 34:13:2756.

Panter, G.H., Tyler, C.R., Maddix, S., Campbell, P.M., Länge, R., Lye, C.M., and Sumpter, J.P. (1999b) *Application of an enzyme-linked immunosorbent assay (elisa) to quantify vitellogenin concentrations in fathead minnows (Pimephales promelas) exposed to endocrine disrupting chemicals.* CEFIC-EMSG Aquatic Research Programme: Progress Report Project A: Fish Screening Assay for Endocrine Active Substances.

Papoulias, D.M., Noltie, D. B., and Tillitt, D.E., (2000) Effects of methyl testosterone exposure on sexual differentiation in medaka, Oryzias latipes. *Marine Environmental Research* 50:1-5:181.

Parks, L.G., Cheek, A.O., Denslow, N.D., Heppell, S.A., McLachlan, J.A., LeBlanc, G.A., and Sullivan, C.V. (1999) *Fathead minnow (Pimephales promelas) vitellogenin: purification, characterization and quantitative immunoassay for the detection of estrogenic compounds.* Comparative Biochemistry and Physiology, accepted.

Pawlowski, S., Islinger, M., Volkl, A., and Braunbeck, T. (2000) Temperature-dependent vitellogenin-mRNA expression in primary cultures of rainbow trout (*Oncorhynchus mykiss*) hepatocytes at 14 and 18 degrees C. *Toxicol. Vitro* 14:6:531.

Pelissero, C., Bennetau, B., Babin, P., Le Menn F., and Dunogues, J. (1991) The estrogenic activity of certain phytoestrogens in the siberian sturgeon (*Acipenser baeri). J. Steroid Biochem Mol. Biol.* 38:293.

Pelissero, C., Flouriot, G., Foucher, J.L., Bennetal, B., Dunogues, J., LeGac, F., and Sumpter, J.P (1993) Vitellogenin synthesis in cultured hepatocytes: An in vitro test for the estrogenic potency of chemicals. *J. Steroid Biochem. Molec. Bio.* 44(3):263.

Pennisi, E. (1997) Differing roles found for estrogen's two receptors. *Science.* 277:1439.

Peters, L.D., Doyotte, A., Mitchelmore, C.L., McEvoy, J., and Livingstone, D.R. (2001) Seasonal variation and estradiol-dependent elevation of Thames estuary eel (*Anguilla anguilla*) plasma vitellogenin levels and comparisons with other United Kingdom estuaries. *Sci. Total Environ.* 279:1-3:137.

Peterson, D.N., Tkalcevic, G.T., Koza-Taylor, P.H., Turi, T.G., and Brown, T.A. (1998) Identification of estrogen receptor, a functional variant of estrogen receptor expressed in normal rat tissues. *Endocrinology*. 139:1082.

Petit, G., Beauchaud, M., and Buisson, B. (2001) Density effects on food intake and growth of largemouth bass (*Micropterus salmoides*). *Aquac. Res*. 32:6:495.

Pflieger, W.L. (1966) Reproduction of smallmouth bass (*Micropterus dolomieui*) in a small Ozark stream. *Am. Midl. Nat*. 76:2:410.

Pundir, R., and Saxena, A.B. (1990) Seasonal changes in the testes of fish *Puntius ticto,* and their relation to heavy metal toxicity. *Bull. Environ. Contam. Toxicol.* 45:288-293.

Purdom, C.E., Hardiman, P.A., Bye, V.J., Eno, N.C., Tyler, E.R., and Sumpter J.P. (1994) Estrogenic effects of effluents from sewage-treatment works. *Chem. Ecol.* 8:275.

Ratnasabapathy, R. (1995) *In vitro* characterization of an estrogen-regulated mRNA stabilizing activity in the avian liver. *Cell Mol. Biol. Res.* 41:583.

Ratnasabapathy, R., Tom, M., and Post, C. (1997) Modulation of the hepatic expression of the estrogen-regulated mRNA stabilizing factor by estrogenic and antiestrogenic nonsteroidal xenobiotics. *Biochem Pharmacol.* 53:1425.

Razandi, M., Pedram, A., Greene, G.L., and Levin, E.R. (1999) Cell membrane and nuclear estrogen receptors(ers) originate from a single transcript: studies of ER and ER expressed in Chinese hamster ovary cells. *Molecular Endocrinol.* 13:307.

Rice, C.D., and Xiang, Y. (2000) Immune function, hepatic CYP1A, and reproductive biomarker responses in the gulf killifish (*Fundulus grandis*), during dietary exposures to endocrine disrupters. *Marine Environmental Research* 50:1-5:163.

Rice, C.D., and Xiang, Y. (2000) Immune function, hepatic CYP1A, and reproductive biomarker responses in the gulf killifish (*Fundulus grandis*), during dietary exposures to endocrine disrupters. *Mar. Environ. Res*. 50:5:163.

Ridgway, M.S. and Shuter, B.J. (1994) The effects of supplemental food on reproduction in parental male smallmouth bass. *Environ. Biol. Fishes* 39:2:201.

Ridgway, M.S., Shuter, B.J., and Post, E.E. (1991) The relative influence of body size and territorial behavior on nesting asynchrony in male smallmouth bass, (*Micropterus-dolomieui*) (Pisces, Centrarchidae). *J. Anim. Ecol*. 60:2:665.

Roberts, S.B., Jackson, L.F., King, W., Taylor, R.G., Grier, W.J., and Sullivan, C.V. (1999) Annual reproductive cycle of the common snook: Endocrine correlates of maturation. *Trans. Am. Fish. Soc*. 128:3:436.

Rodgers-Gray, T.P., Jobling, S., Kelly, C., Morris, S., Brighty, G., Waldock, M.J., Sumpter, J.P., and Tyler, C.R. (2001) Exposure of juvenile roach (*Rutilus rutilus*) to treated sewage effluent induces dose-dependent and persistent disruption in gonadal duct development. *Environ. Sci. Technol*. 35:3:462.

Rodgers-Gray, T.P., Jobling, S., Morris, S., Kelly, C., Kirby, S., Janbakhsh, A., Harris, J.E., Waldock, M.J., Sumpter, J.P., Tyler, C.R., (2000) *Environ. Sci. Technol*. 34:1521-1528

Rosenblum, P.M., Brandt, T.M., Mayes, K.B., and Hutson, P. (1994) Annual cycles of growth and reproduction in hatchery-reared Florida largemouth bass, (*Micropterus-salmoides floridanus*), raised on forage or pelleted diets. *J. Fish Biol.* 44:6:1045.

Routledge, E.J., and Sumpter, J.P. (1996). Estrogenic activity of surfactants and some of their degradation products assessed using a recombinant yeast strain. *Environ. Toxicol. Chem.* 15(3):241.

Routledge, E.J., Sheahan, D., Desbrow, C., Brighty, G.C., Waldock, M., and Sumpter, J.P. (1998) Identification of estrogenic chemicals in STW effluent: in vivo responses in trout and roach. *Environ. Sci. Technol.* 32:1559.

Routledge, E.J., Waldock, M., and Sumpter, J.P. (1999) Response to comment on "Identification of estrogenic chemicals in STW effluent: Chemical fractionation and *in vitro* biological screening." *Environ. Sci. Technol.* 33:2:371.

Routledge, E.J., and Sumpter, J.P. (1996) Estrogenic activity of surfactants and some of their degradation products assessed using a recombinant yeast screen. *Env. Tox. and Chem.* 15:3:241.

Ruby, S.M., Hull, R., and Anderson, P. (2000) Sublethal lead affects pituitary function of rainbow trout during exogenous vitellogenesis. *Arch Environ Contam Toxicol.* 38:46.

Ryffel, G.U. (1978) Synthesis of vitellogenin, an attractive model for investigating hormone-induced gene activation. *Mol. Cell. Endocrinol.* 12:237.

Sambrook, J., Fritsch, E.F., and Maniatis, T. (1989) *Molecular Cloning: A Laboratory Manual, 2nd Ed.* Cold Spring Harbor, New York, Cold Spring Harbor Press.

Schibler, U., Hagenbuchle, O., Wellauer, P.K., and Pitted A.C. (1983) Two promoters of different strengths control the transcription of the mouse alpha-amylase gene amy-1 in the paratid gland and the liver. *Cell.* 33:501.

Schlenk, D., Stresser, D.M., Rimoldi, J., Arcan, L., McCants, J., Nimrod, A.C., and Benson, W.H. (1998) Biotransformation and estrogenic activity of methoxychlor and its metabolites in channel catfish (*Ictalurus punctatus*). *Marine Environmental Research.* 46:159.

Scholz, S. and Gutzeit, H.O. (2000) 17-a-ethinylestradiol affects reproduction, sexual differentiation and aromatase gene expression of the medaka (*Oryzias latipes*). *Aquatic Toxicology* 50:363. © Elsevier Science B.V. All Rights Reserved.

Schreck, C.B., Contreras-Sanchez, W., and Fitzpatrick, M.S. (2001) Effects of stress on fish reproduction, gamete quality, and progeny. *Aquaculture* 197:1-4:3. Published by Elsevier Science B.V.

Schreihofer, D.A., Resnick, E.M., Soh, A.Y., and Shupnik, M.A. (1999) Transcriptional regulation by a naturally occurring truncated rat estrogen receptor(er), Truncated ER Product-1 (TERP-1). *Molecular Endocrinol.* 13:320.

Schultz, I.R., and Drum, A.S. (2001) Pharmacokinetics of vitellogenin and 17 alpha-ethynylestradiol in rainbow trout. *Toxicology* 164:1-3:207.

Schultz, I.R., Orner, G., Merdink, J.L., and Skillman, A. (2001) Dose-response relationships and pharmacokinetics of vitellogenin in rainbow trout after intravascular administration of 17 alpha-ethynylestradiol. *Aquat. Toxicol.* 51:3:305.

Schwaiger, J., Spieser, O.H., Bauer, C., Ferling, H., Mallow, U., Kalbfus, W., and Negel, R.D. (2000) Chronic toxicity of nonylphenol and ethinylestradiol: haematological and histopathological effects in juvenile Common carp (*Cyprinus carpio*). *Aquatic Toxicology.* 51:69. © 2000 Elsevier Science B.V. All Rights Reserved.

Scott, A.P., and Sumpter, J.P. (1983) A comparison of the female reproductive cycles of autumn-spawning and winter-spawning strains of rainbow trout (*Salmo gairdneri* Richardson). *Gen. and Comparative Endocrinology.* 52:79.

Sedlak, D.L., Gray, J.L., Pinkston, K.E. (2000) Understanding microcontaminants in recycled water. *Environmental Science and Technology* 34:23:508

Sehgal, N., and Goswami, S.V. (2001) Biochemical changes in the liver of the Indian freshwater mussel, (*Channa punctatus*) (Bloch) during estradiol-induced vitellogenin synthesis. *Fish Physiol. Biochem.* 24:2:149.

Selman, K. and Wallace, R. (1986) Gametogenesis in *Fundulus heteroclitus. American Zoology* 26:173.

Selman, K., Wallace, R., Sarka, A., and Qi, X. (1993) Stages of Oocyte Development in the Zebrafish, *Brachydanio rerio. Journal of Morphology* 218:203.

Selman. K. and Wallace, R. (1986). Gametogenesis in Fundulus heteroclitus. *Amer. Zool.* 26:173.

Sepulveda, M.S., Johnson, W.E., Higman, J.C., Denslow, N.D., Schoeb, T.R., and Gross, T.S. (2001) An evaluation of biomarkers of reproductive function and potential contaminant effects in Florida largemouth bass (*Micropterus salmoides floridanus*) sampled from the St. Johns River. *The Science of the Total Environment.* Article in Press.

Sepulveda, M.S., Ruessler, D.S., Denslow, N.D., Holm, S.E., Schoeb, T.R., Gross, T.S. (2001) Assessment of reproductive effects in largemouth bass (*Micropterus salmoides*) exposed to bleached/unbleached kraft mill effluents. *Archives of Environmental Contamination and Toxicology* 41:475.

Sepulveda, M.S., Ruessler, D.S., Denslow, N.D., Holm, S.E., Schoeb, T.R., and Gross, T.S. (2001) Assessment of reproductive effects in largemouth bass (*Micropterus salmoides*) exposed to bleached/unbleached kraft mill effluents. *Arch. Environ. Contam. Toxicol.* 41:4:475.

Shapiro, D. (1982) Steroid hormone regulation of vitellogenin gene expression. *Crit. Rev. Biochem.* 12:187.

Shelby, M.D., Newbold, R.R., Tully, D.B., Chae, K., and Davis, V.L. (1996) Assessing environmental chemicals for estrogenicity using a combination of *in vitro* and in vivo assays. *Environ. Health Perspect.* 104:1296.

Simpson, M.G., Parry, M., Kleinkauf, A., Swarbreck, D., Walker, P., and Leah, R.T. (2000) Pathology of the liver, kidney and gonad of flounder (*Platichthys flesus*) from a UK estuary impacted by endocrine disrupting chemicals. *Mar. Environ. Res.* 50:5:283.

Sinks, G.D., and Schultz, T.W. (2001) Correlation of tetrahymena and pimephales toxicity: Evaluation of 100 additional compounds. *Environ. Toxicol. Chem.* 20:4:917.

Smith, F.J.F. (1978) Seasonal changes in the histology of the gonads and dorsal skin of the fathead minnow, (*Pimephales promelas*). *Can. J. Zool.* 56:2103.

Snyder, S.A., Keith, T.L., Verbrugge, D.A., Snyder, E.M., Gross, T.S., Kannan, L., and Giesy, J.P. (1999) Analytical methods for detection of selected estrogenic compounds in aqueous mixtures. *SO Environmental Science & Technology* 33:16.

Sohoni, P. and Sumpter, J.P. (1998) Several environmental oestrogens are also anti-androgens. *J. Endocrinol* 158:327.

Sohoni, P., Tyler, C.R., Hurd, K., Caunter, J., Hetheridge, M., Williams, T., Woods, C., Evans, M., Toy, R., Gargas, M., and Sumpter, J.P. (2001) Reproductive effects of long-term exposure to bisphenol a in the fathead minnow (*Pimephales promelas*). *Environ. Sci. Technol.* 20:6:1374.

Sohoni, P., Tyler, C.R., Hurd, K., Caunter, J., Hetheridge, M., Williams, T., Woods, C., Evans, M., Toy, R., Gargas, M., and Sumpter, J.P. (2001) Reproductive effects of long-term exposure to bisphenol a in the fathead minnow (*Pimephales promelas*). *Environ. Sci. Technol.* 35:14:2917.

Sole, M., de Alda, M.J.L., Castillo, M., Porte, C., Ladegaard-Pedersen, K., and Barcelo, D. (2000) Estrogenicity determination in sewage treatment plants and surface waters from the Catalonian area (NE Spain). *Environ. Sci. Technol.* 34:24:8.

Sole, M., Porte, C., and Barcelo, D. (2001) Analysis of the estrogenic activity of sewage treatment works and receiving waters using vitellogenin induction in fish as a biomarker. *Trac-Trends Anal. Chem.* 20:9:518.

Specker, J.L. and Sullivan, C V. (1994) "Vitellogenesis in fishes; status and perspectives." In *Perspectives in Comparative Endocrinology*, K.G. Davey, R.E. Peter and S.S. Tobe, Eds., National Research Council of Canada, pp. 304.

Springer, C.L. 2001 Redeye Bass: The Deep South's brook trout. Retrieved from the web on May 22, 2001 from http://www.fedflyfishers.org/flyfisher/Autumn99/redeyebass.htm

Stevens, J.T., Gfeller, W., Machemer, L., and Leist, K.H. (1998) Adequacy of required regulatory hazard testing for the detection of potential hormonal activity of crop protection chemicals. *J. of Tox and Env. Health.* 1:59-79.

Sumpter, J.P. and Jobling, S. (1995) Vitellogenesis as a biomarker of estrogenic contamination of the aquatic environment. *Environmental Health Perspectives.* 103:173.

Sun, G., Porter, W., and Safe, S. (1998) Estrogen-induced retinoic acid receptor alpha 1 gene expression: role of estrogen receptor-SP1 complex. *Mol. Endocrinol.* 12:882-890.

Sweet, L.I., and Zelikoff, J.T., (2001) Toxicology and immunotoxicology of mercury: A comparative review in fish and humans. *J. Toxicol. Env. Health-Pt b-Crit. Rev.* 4:2:161.

Takemura, A., and Kim, B.H. (2001) Effects of estradiol-17 beta treatment on in vitro and *in vivo* synthesis of two distinct vitellogenins in tilapia. *Comp. Biochem. Physiol. A-Mol. Integr. Physiol.* 129:2-3:641.

Tan, N.S., Frecer, V., Lam, T.J., and Ding, J.L. (1999) Temperature dependence of estrogen binding: importance of a subzone in the ligand binding domain of a novel piscine estrogen receptor. *Biochim Biophys Acta.* 1452:103.

Tang, M.X., Redemann, C.T., and Szoka, Jr., F.C. (1996) In vitro gene delivery by degraded polyamidoamine dendrimers. *Bioconjugate Chem.* 7:703.

Tchoudakova, A., Pathak, S., and Callard, G.V. (1999) Molecular cloning of an estrogen receptor subtype from the goldfish, (*Carassius auratus*). *Gen. and Compar. Endocrinol.* 113:388-400.

Ternes, T.A., Kreckel, P., Mueller, J. (1999) Behavior and occurrence of estrogens in municipal sewage treatment plants- II. Aerobic batch experiments with activated sludge. *Sci Total Environ.* 225:91-99.

Texas Parks and Wildlife Department, 2001. Freshwater Fishing. Retrieved from the web on May 22, 2001 from http://tpwd.state.tx.us/fish/infish/species/spb/spb.htm.

Thomas, P. and Smith J. (1993) Binding of xenobiotics to the estrogen receptor of spotted seatrout: A screening assay for potential estrogenic effects. *Marine Environmental Research.* 35:147-151.

Thomas, P., and Smith, J. (1992) Binding of Xenobiotics to the Estrogen Receptor of Spotted Seatrout: A Screening Assay for Potential Estrogenic Effects. *Marine Environmental Research.* 33:147-151.

Thompson, J.D., Higgins, D.G. and Gibson, T.J. (1994) CLUSTAL W.: Improving the sensitivity of progressive multiple sequence alignment through sequence weighting, positions-specific gap penalties and weight matrix choice. *Nuc. Acid. Res.* 22: 4673.

Thompson, S., Tilton, F., Schlenk, D., and Benson, W.H. (2000) Comparative vitellogenic responses in three teleost species: extrapolation to in situ field studies. *Marine Environmental Research* 50:1-5:185.

Thorpe, K.L., Hutchinson, T.H., Hetheridge, M.J. Scholze, M., Sumpter, J.P., and Tyler, C.R. (2001) Assessing the biological potency of binary mixtures of environmental estrogens using vitellogenin induction in juvenile rainbow trout (*Oncorhynchus mykiss*). *Environ. Sci. Technol.* 35:12:2476.

Thorpe, K.L., Hutchinson, T.H., Hetheridge, M.J., Sumpter, J.P., and Tyler, C.R. (2000) Development of an *in vivo* screening assay for estrogenic chemicals using juvenile rainbow trout (*Oncorhynchus mykiss*). *Environ. Toxicol. Chem.* 19:11:2812.

Tillit, D.E., Ankley, G.T., Giesy, J.P., Ludwig, J.P., Kurita-Matsuba, H., Wesehloh, D.V., Ross, P.S., Bishop, C.A., Sileo, L., Stromborg, K.L., Larson, J., and Kubiak, T.J. (1992) Polychlorinated biphenyl residues and egg mortality in double crested cormorants from the great lakes. *Environ. Toxicol. Chem.* 11:1281.

Todo, T., Adachi, S., and Yamauchi, K. (1996) Molecular cloning and characterization of Japanese eel estrogen receptor cDNA. *Mol. Cell. Endocrinol.* 119:37.

Tolar, J.F., Mehollin, A.R., Watson, R.D., and Angus, R.A. (2001) Mosquitofish (*Gambusia affinis*) vitellogenin: identification, purification, and immunoassay. *Comp. Biochem. Physiol. C-Pharmacol. Toxicol. Endocrinol.* 128:2:237.

Trebitz, A.S. (1991) Timing of spawning in largemouth bass - implications of an individual-based model. *Ecol. Model.* 59:3-4:203.

Tremblay, G.B., Tremblay, A., Copeland, N.G., Gilbert, D.J., Jenkins, N.A., Labrie, F., and Giguere, V. (1997) Cloning, chromosomal localization, and functional analysis of the murine estrogen receptor beta. *Mol. Endocrinol.* 11:353.

Tremblay, L., and Van Der Kraak, G. (1999) Comparison between the effects of the phytoserol β-sitosterol and pulp and paper mill effluents on sexually immature rainbow trout. *Env. Tox. and Chem.* 18:2:329.

Trichet, V., Buisine, N., Mouchel, N., Moran, P., Pendas, A.M., LePennex, J.P. and Wolff, J. (2000) Genomic analysis of the vitellogenin locus in rainbow trout (*Oncorhynchus mykiss*) reveals a complex history of gene amplification and retroposon activity. *Mol. Gen. Genet.* 263:5:828.

Tyler, C.R. and Routlege, E.J. (1998) Oestrogenic effects in fish in English rivers with evidence of their causation. *Pure and Applied Chemistry* 70:1795.

Tyler, C.R., and Sumpter, J.P. (1990) The development of a radioimmunoassay for carp (*Cyprinus carpio)* vitellogenin. *Fish Physio. Biochem.* 8:129.

Tyler, C.R., van Aerle, R., Hutchinson, T.H., Maddix, S., and Trip, H. (1999) An *in vivo* testing system for endocrine disruptors in fish early life stages using induction of vitellogenin. E*nvironmental Toxicology and Chemistry* 18:2:337.

Tyler, C.R., Vandereerden, B., Jobling, S., Panter, G., and Sumpter, J.P. (1996) Measurement of vitellogenin, a biomarker for exposure to oestrogenic chemicals, in a wide variety of cyprinid fish. J. Comp. Physio. 166B:418.

Umayahara, Y., Kawamori, R., Watada, H., Imano, E., Iwama, N., Morishima, T., Yamasaki, Y., Kajimoto, Y., and Kamada, T. (1994) Estrogen regulation of the insulin-like growth factor I gene transcription involves an AP-1 enhancer. *J. Biol. Chem.* 269:16433.

United States Geological Survey (2000) *Apalachicola-Chattahoochee-Flint River Basin NAWQA Study. Description of the ACF River Basin Study Area.* Retrieved December 12, 2001 from http://ga.water.usgs.gov/nawqa/main.description.html.

United States Geological Survey (2001) "National Reconnaissance of Emerging Contaminants in the Nation's Water Resources." Retrieved December 17, 2001 from http://toxics.usgs.gov/regional/emc.html.

United States Geological Survey (2002a) *Real time data for USGS 02335450 Chattahoochee River above Roswell, Georgia.* Retrieved January 8, 2002 from http://water.usgs.gov/ga/nwis/uv/?site_no=02335450&PARAmeter_cd=00065,00060.

United States Geological Survey (2002b) *Real time data for USGS 02337170 Chattahoochee River near Fairburn, Georgia.* Retrieved January 8, 2002 from http://water.usgs.gov/ga/nwis/uv/?site_no=02337170&PARAmeter_cd=00065,00060.

United States Geological Survey (2002c) *Real time data for USGS 02384500 Conasauga River near Eton, Georgia.* Retrieved January 8, 2002 from http://water.usgs.gov/ga/nwis/uv/?site_no=02384500&PARAmeter_cd=00065,00060.

United States Geological Survey (2002d) *Real time data for USGS 02387000 Conasauga River at Tilton, Georgia*. Retrieved January 8, 2002 from http://water.usgs.gov/ga/nwis/discharge/?site_no=02387000&agency_cd=USGS.

United States Geological Survey. 2001. Nonindigenous Aquatic Species. Retrieved from the web on May 22, 2001 from http://nas.er.usgs.gov/fishes/accounts/centrarc/.

US Environmental Protection Agency. (1999) Preliminary EPA List of 87 compounds for high throughput pre-screening (HTPS) pilot project. EDSTAC 1998 draft report. not included in final report.

Van den Belt, K., Verheyen, R., and Witters, H. (2001) Reproductive effects of ethynylestradiol and 4t-octylphenol on the zebrafish (*Danio rerio*). *Arch. Environ. Contam. Toxicol.* 41:4:458.

Van Der Kraak, G., Hewitt, M., Lister, A., McMaster, M.E., and Munkittrick, K.R. (2001) Endocrine toxicants and reproductive success in fish. *Hum. Ecol. Risk Assess.* 7:5:1017.

Van Der Kraak, G.J., Munkittrick, K.R., McMaster, M.E., Port, C.B., and Chang, J.P. (1992) Exposure to bleached kraft pulp mill effluent disrupts the pituitary-gonadal axis of white sucker at multiple sites. *Toxicol Appl. Pharmacol.* 115:224.

Vanwinkle, W., Rose, K.A., Winemiller, K.O., Deangelis, D.L., Christensen, S.W., Otto, R.G., and Shuter, B.J. (1993) Linking life-history theory, environmental setting, and individual-based modeling to compare responses of different fish species to environmental-change. *Trans. Am. Fish. Soc.* 122:3:459.

Varanasi, U., Nishimoto, M., Reichert, W. L., and Eberhart, B.T. (1986) Comparative metabolism of benzo(a)pyrene and covalent binding to hepatic dNA in English sole, starry flounder, and rat. *Cancer Res.* 46:3817.

Vijayan, M.M., Takemura, A., and Mommsen, T.P. (2001) Estradiol impairs hyposmoregulatory capacity in the euryhaline tilapia, Oreochromis mossambicus. *Am. J. Physiol.-Regul. Integr. Comp. Physiol.* 281:4:R1161.

Wahli, W., Dawid, I.B., Ryffel, G.U. and Weber, R. (1981) Vitellogenesis and the vitellogenin gene family. *Science* 212: 298.

Walker, D. (2000) Oestrogenicity and wastewater recycling: Experience from Essex and Suffolk water. *J. Chart. Inst. Water. Environ. Manage.* 14:6:427.

Wallace, R.A. (1985) "Vitellogenesis and oocyte growth in non-mammalian vertebrates." pp. 127. In *Developmental Biology, Vol. 1* (Browder, L.W. ed) Plenum Press, New York.

Waller, C.L., Juma, B.W., Gray, Jr., E.L., and Kelce, W.R. (1996) Three-Dimensional Quantitative Structure-Activity Relationships for Androgen Receptor Ligands. *Toxicol. Appl. Pharmacol.* 137:219-227.

Walsh, G.E. and Bahner, L.H. (1980) Toxicity of textile mill effluents to freshwater and estuarine algae, crustaceans and fishes. *Environmental Pollution.* 21:169.

Wang, H., Yan, T., Tan, J.T.T., and Gong, Z.Y. (2000) A zebrafish vitellogenin gene (vg3) encodes a novel vitellogenin without a phosvitin domain and may represent a primitive vertebrate vitellogenin gene. *Gene* 256:1-2:303.

Warrier, S.R., Tirumalai, R., and Subramoniam, T. (2001) Occurrence of vertebrate steroids, estradiol 17 beta and progesterone in the reproducing females of the mud crab (*Scylla serrata*). *Comp. Biochem. Physiol. A-Mol. Integr. Physiol.* 130:2:283.

Weatherman, R.V., Clegg, N.J., and Scanlon, T.S. (2001) Differential SERM activation of the estrogen receptors (ER alpha and ER beta) at AP-1 sites. *Chem. Biol.* 8:5:427.

Webb, P., Lopez, G.N., Uht, R.M., and Kushner, P.J. (1995) Tamoxifen activation of the estrogen receptor/AP-1 pathway: potential origin for the cell-specific estrogen-like effects of antiestrogens. *Mol. Endocrinol.* 9:443.

Wester, P.W. (1991) Histopathological effects of environmental pollutants -hch and methyl mercury on reproductive organs in freshwater fish. *Comp. Biochem. Physiol.* 100C:237.

White, R., Jobling, S., Hoiare, S.A., Sumpter, J.P., Parker, M.G. Iwama, N., Morishima, T., Yamasaki, Y., Kajimoto, Y., and Kamada, T. (1994) Environmentally persistent alkylphenolic compounds are estrogenic. *Endocrinology.* 135:175.

Wiegand, C., Krause, E., Steinberg, C., and Pflugmacher, S. (2000) Metabolism of the S-triazine herbizide atrazine in early life stages of the zebrafish (*Danio rerio*). *Marine Environmental Research* 50:1-5:73.

Wiegmann, D.D., Baylis, J.R., and Hoff, M.H. (1997) Male fitness, body size and timing of reproduction in smallmouth bass, (*Micropterus dolomieui*). *Ecology* 78:1:128.

Williams, P.J., Courtenay, S.C., and Wilson, C.E. (1998) Annual sex steroid profiles and effects of gender and season on cytochrome P450 mRNA induction in Atlantic tomcod (*Microgadus tomcod*). *ET&C* 17:1582-1588.

Wood, J.R., Likhite, V.S., Loven, M.A., and Nardulli, A.M. (2001) Allosteric modulation of estrogen receptor conformation by different estrogen response elements. *Mol. Endocrinol.* 15:7:1114.

Woodward, A.R., Jenning, M.L., and Percival, H.F. (1989) Egg collecting and hatch rates of American alligator eggs in Florida. *Wildlife Soc. Bull.* 17:124.

Woodward, A.R., Jennings, M.L., Percival, H.F., and Moore, C.T. (1993) Low clutch viability of American alligators on Lake Apopka. *Fl. Sci.* 56:52.

Xie, W., Duan, R., and Safe, S. (1999) Estrogen induces adenosine deaminase gene expression in MCF-7 human breast cancer cells: Role of estrogen receptor-SP1 interactions. *Endocrinol.* 140:219.

Yoon, J.M., Chung, E.Y., and Kim, G.W. (2001) Ultrastructures of oocyte development and electrophoretic patterns of the yolk protein following HCG treatment in Korean native catfish (*Silurus asotus*). *Asian Australas. J. Anim. Sci.* 14:2:174.

Zacharewski, T. (1997) *In vitro* bioassays for assessing estrogenic substances. *American Chemical Society* 31(3):613.

Zacharewski, T. (1997) *In vitro* bioassays for assessing estrogenic substances. *Environ Sci Technol.* 31:613.

Zaroogian, G., Gardner, G., Horowitz, D.B., Gutjahr-Gobell, R., Haebler, R., and Mills, L. (2001) Effect of 17 beta-estradiol, o,p '-DDT, octylphenol and p,p '-DDE on gonadal development and liver and kidney pathology in juvenile male summer flounder (*Paralichthys dentatus*). *Aquat. Toxicol.* 54:1-2:101.

Zhou, T., Weis, J.S., John-Alder, H.B., and Weis, P. (1999). Thyroidal status of mummichog (*Fundulus heteroclitus*) from a polluted versus a reference habitat. *ET&C* 18:2817.

Zillioux, E.J.; Johnson, I.C.; Kiparissis, Y.; Mecalfe, C.D.; Wheat, J.V.; Ward, S.G.; and Liu, H. (2001). The sheepshead minnow as an in vivo model for endocrine disruption in marine teleosts: A partial life-cycle test with 17β-ethynlestradiol-33. *Environmental Toxicology and Chemistry* 20(9):1968. © SETAC.

GLOSSARY

17β-Estradiol
Natural estrogen; a female steroid hormone.

Aeration
The process of charging or treating with air or a gas.

Aliquot
In mathematics, forming an exact proper divisor (an aliquot part of 15 is 5). In chemistry, comprising a definite part of a whole (an aliquot quantity of acid for analysis).

Androgen
The natural sex hormones of male animals, responsible for the development of an animal in the male form and for the development of secondary sexual characteristics at puberty (e.g., hair and muscle growth and deepening of the voice). The major androgen is testosterone.

Anti-androgen
Substances that block the function of normal androgens. Like androgens, they bind to androgen receptors, but unlike androgens, this binding does not result in activation of the receptors. Thus anti-androgens block the receptors from stimulation by androgens.

Antagonistic
The situation in which the combined effect of two chemicals is much less than the sum of the effect of each agent given alone.

Aseptic
A sterile technique.

Assay
The qualitative or quantitative analysis of a chemical substance.

Asynchronous spawning
Multiple and daily spawning with all oocyte development stages present.

Bioaccumulation The process through which chemicals concentrate in an organism.

Bioassay
A determination of the biological activity or potency of a substance, such as a vitamin or hormone, by testing its effect on the growth of an organism.

Biomarker
A physiological change (biochemical, cellular, or structural) that is measured in a living organism or biological sample that results from exposure to a chemical or condition.

Biomass
That part of a given habitat consisting of living matter, expressed either as the weight of organisms per unit area or as the volume of organisms per unit volume of habitat.

Cytotoxicity
The adverse effect on cell survival proliferation and/or function.

Degradation
The breakdown of a compound, especially an organic hydrocarbon.

Denature
To treat (a protein or the like) by chemical or physical means so as to alter the original state.

Disinfection
The cleansing of all infections and disease-bearing germs.

Dose–response relationship
In an experimental context, this is a relationship between the applied dose (chemical or condition) and the measured response (biochemical, physiological, behavioral, reproductive, genetic, etc.). The response is generally proportional to the dose.

EDSTAC
Endocrine Disruptor Screening and Testing Advisory Group.

ELISA
Enzyme Linked Immunosorbant Assay.

Endocrine
The hormonal pathway characterized by the production of a biologically active substance by a ductless gland which travels via the blood stream to exert effects on a distant target cell.

Endocrine disrupting compound/endocrine disruptor
As defined by the EDSTAC, this is an exogenous substance that changes endocrine function and causes adverse effects at the level of the organism, its progeny and/or (sub) population of organisms.

Endocrine disruptor
An exogenous substance or mixture that alters the function(s) of the endocrine system and consequently causes adverse health effects in an intact organism, or its progeny, or (sub) populations.

Endocrine glands
Glands (hypothalamus, pituitary, thyroid, parathyroid, pancreas, adrenal, ovary, and testis) that produce hormones that regulate metabolism, reproduction, and other bodily

functions. Some tissues that also produce hormones include the placenta, kidney, and cells in the gastrointestinal tract, mammary glands, bone, muscle, and the nervous system.

Endocrine modulator
Any chemical that affects the normal operation of the endocrine system.

Endocrine system
A network of glands that monitors and releases hormones that control things such as growth and sexual development.

Endogenous
A chemical substance formed inside the body; for instance a hormone.

Estrogen
A steroid hormone produced by the ovary that regulates certain female reproductive functions and the development and maintenance of female secondary sexual characteristics.

Exogenous
A chemical substance formed outside of the body.

Exogenous estrogen
A chemical originating outside the body that can produce estrogenic effects in the body.

Fate and exposure modeling
A process for simulating the distribution of chemicals released to the environment.

Gonad
A sex organ like an ovary or testes.

Half-life
The amount of time it takes for 50% reduction in concentration.

Hematocrit
The relative volume of blood occupied by red blood cells.

HER
Human estrogen receptor.

Histology
The study of cells and tissues on the microscopic level.

Histopathology
Pathologic histology; dealing with the minute structure, composition and function of diseased tissues.

Hormones
Tiny chemical messengers that help our body perform different tasks. Hormones are produced by the endocrine glands and sent all over the body to stimulate certain activities. For example, insulin is a well-known hormone that helps a body digest food. Growth, digestion, reproduction and sexual functions are all triggered by hormones.

Hypothalamus
The hypothalamus contains vital centers for controlling the autonomic nervous system, body temperature and water and food intake, and is the center for primitive physical and emotional behavior. It also produces hormones for regulating pituitary secretion and two systemic hormones (e.g., vasopressin).

Immunotoxicity
An adverse effect on the immune system.

In vitro
(As defined by the ESDTAC when discussing screening and testing.) Within an artificial environment outside a living organism (i.e., a glass dish).

In vivo
(As defined by the ESDTAC when discussing screening and testing.) Within a living thing.

In situ
In the natural or normal place; confined to the site of origin without invasion of neighboring tissues.

Ligand
A molecule that binds to a receptor-binding site.

Non-aromatizable
An androgen that is not converted by the aromatase enzyme to estrogen.

Oogenesis
The origin and development of ovum (female reproductive cell).

Ovaries
These double organs are the sex glands. The ovaries in females produce the egg cells (ova) as well as sex hormones that flow through the blood and give rise to such secondary traits as breasts.

Pituitary
The hypophysis or pituitary gland is the master gland of the body. Compared with other endocrine glands, it produces the largest number of hormones, including some that control the other endocrine glands.

Plasma
The fluid in which blood cells are suspended.

Plasmids
Small extra-chromosomal circular DNAs that have an origin of replication and grow and replicate within E. coli.

Potential endocrine disrupter
An exogenous substance or mixture that possesses properties that might be expected to lead to endocrine disruption in an intact organism, its progeny, or its (sub) populations.

Receptor
The site at which a hormone or an agent acting as a hormone binds.

Reproductive toxicity
An adverse effect on the reproductive system (behavior, fertility, or development).

Risk
The probability that damage to life, health, and/or the environment will occur as a result of a given hazard (such as exposure to a toxic chemical).

Secretion
Any substance produced naturally by the body or more specifically, the endocrine glands; i.e., estrogen or testosterone.

Steroids
Hormones that are derivatives of cholesterol including testosterone estradiol and progesterone.

Synergistic
When the combined effect of two substances or conditions exceeds the sum of their individual effects.

Testes
These double organs are the sex glands. The testes in males produce the germinating cells, the sperm cells, as well as sex hormones that flow through the blood and give rise to such secondary traits as the beard.

Toxin
A poisonous chemical substance that is toxic to an organism.

Vitellogen (Vtg.)
The protein precursor to yolk.

YES
The yeast estrogen screening assay with *Saccharomyces cerevisiae* hER.

REFERENCES

Allen, Y.; Matthiessen, P.; Haworth, S.; and Thain, J.E. 1999. Survey of estrogenic activity in United Kingdom estuarine and coastal waters and its effects on gonadal development of the flounder *Platichthys flesus*. *Environ. Toxicol. Chem.* 18:1791.

Ankley, G.; Mihaich, E.; Stahl, R.; Tillitt, D.; Colborn, T.; McMaster, S.; Miller, R.; Bantle, J.; Campbell, P.; Denslow, N.; Dickerson, R.; Folmar, L.; Fry, M.; Giesey, J.; Grey, L.E.; Guiney, P.; Hutchinson, T.; Kennedy, S.; Peterson, R.; Purdy, R.; Ringer, R.; Thomas, P.; Touart, L.; Van der Kraak, G.; Zacharewski, T. 1998. Overview of a workshop on screening methods for detecting potential (anti-) estrogenic/androgenic chemicals in wildlife. *Environ. Toxicol. Chem.* 17:68-87.

Ankley, G.T.; Jensen, K.M.; Kahl, M.D.; Korte, J.J.; and Makynen, E.A. 2001. Description and evaluation of a short-term reproduction test with the fathead minnow (*Pimephales promelas*). *Environ. Toxicol. Chem.* 20:6:1276.

Arukwe, A.; Forlin, L.; and Goksøyr, A. 1997. Xenobiotic and steroid biotransformation enzymes in Atlantic salmon (*Salmo salar*) liver treated with an estrogenic compound 4-nonylphenol. *Environ. Toxicol. Chem.* 16:12:2576.

Baronti, C.; Curini, R.; D'Ascenzo, G.; Di Corcia, A.; Gentili, A.; and Samperi, R. 2000. Monitoring Natural and Synthetic Estrogens at Activated Sludge Sewage Treatment Plants and in a Receiving River Water. *Environ. Sci. Technol.* 34:24:5059.

Batty, J.L.R. 1999. Morphological and reproductive characteristics of male mosquitofish (*Gambusia affinis holbrook*) inhabiting sewage-contaminated waters in New South Wales, Australia. *Arch. Environ. Contam. Tox.* 36:301.

Belfroid, A.C.; Van der Horst, A.; Vethhak, A.D.; Schäfer, A.J.; Rijs, G.B.J.; Wegener, J.; Cofino, W.P. 1999. Analysis and occurrence of estrogenic hormones and their glucuronides in surface water and wastewater in The Netherlands. *Sci. Total Environ.* 225:101.

Bowman, C.J. and Denslow, N.D. 1999. Development and validation of a species- and gene-specific molecular biomarker: Vitellogenin mRNA in largemouth bass (*Micropterus salmoides*). *Ecotoxicology* 8:5:399.

Bowman, C.J.; Kroll, K.J.; Hemmer, M.J.; Folmar, L.C.; and Denslow, N.D. 2000. Estrogen-induced vitellogenin mRNA and protein in sheepshead minnow (*Cyprinodon variegatus*). *Gen. Comp. Endocrinol.* 120:3:300.

Burse, J.R.; Cheek, A.O.; and Bart, H.L. Jr. 2000. Effects of exposure to paper mill effluents in *Lepomis macrochirus* from Elevenmile Creek, Florida. Poster #12 Abstract presented at *e .hormone 2000 Conference*, October 15-18, 2000, New Orleans, La.

Colborn, T. and Clement, C. 1992. Chemically-induced alterations in sexual and functional development: the wildlife/human connection. *Advances in Modern Environmental Toxicology*. Vol.21. Princeton Scientific, Princeton, N.J.

Colucci, M.S.; Bork, H.; and Topp, E. 2001. Persistence of estrogenic hormones in agricultural soils: I. 17β-estradiol and estrone. *J. Environ. Qual.* Vol. 30.

Colucci, M.S. and Topp, E. 2001. Persistence of estrogenic hormones in agricultural soils: II 17α-ethynylestradiol. *J. Environ. Qual.* Vol. 30.

Danylchuk, A.J. and Tonn, W.M. 2001. Effects of social structure on reproductive activity in male fathead minnows (*Pimephales promelas*). *Behav. Ecol.* 12:4:482.

Denslow, N.D.; Bowman, C.J.; Ferguson, R.J.; Lee, H.S.; Hemmer, M.J.; and Folmar, L.C. 2001. Induction of gene expression in sheepshead minnows (*Cyprinodon variegatus*) treated with 17beta-estradiol, diethylstilbestrol, or ethinylestradiol: The use of mRNA fingerprints as an indicator of gene regulation. *Gen. Comp. Endocrinol.* 121:3:250.

Desbrow, C.; Routledge, E.J.; Brighty, G.C.; Sumpter, J.P.; and Waldock, M. 1998. Identification of estrogenic chemicals in STW effluent: Chemical fractionation and in vitro biological screening. *Environ. Sci. Technol.* 32:11:1549.

Endocrine Disruptor Screening and Testing Advisory Committee (EDSTAC). Final Report. August 1998. Available at http://www.epa.gov/scipoly/oscpendo/history/finalrpt.htm.

Englebienne, P. 2000. Immune and Receptor Assay in Theory and Practice. CRC Press LLC, Boca Raton, Fla.

Fenske, M.; van Aerle, R.; Brack, S.; Tyler, C.R.; and Segner, H. 2001. Development and validation of a homologous zebrafish (*Danio rerio Hamilton-Buchanan*) vitellogenin enzyme-linked immunosorbent assay (ELISA) and its application for studies on estrogenic chemicals. *Comp. Biochem. Physiol. C-Toxicol. Pharmacol.* 129:3:217.

Folmar, L.; Hemmer, M.; Webb, R.; Robinson, G.; Chow, M.; Kroll, K.; and Denslow, N. 1997. Development of an in vivo estrogen responsive bioassay in the sheepshead minnow. *Abstract to SETAC 18th Annual Meeting.* San Francisco, Calif.

Folmar, L.C. 1992. *Effects of chemical contaminants on blood chemistry of teleost fish: A bibliography and synopsis of selected effects.* U.S. Environmental Protection Agency, Environmental Research Laboratory, Sabine Island, Gulf Breeze, Fla. 32561.

Folmar, L.C.; Denslow, N.D.; Kroll, K.; Orlando, E.F.; Enblom, J.; Marcino, J.; Metcalfe, C.; and Guillette, L.J. 2001. Altered serum sex steroids and vitellogenin induction in walleye (*Stizostedion vitreum*) collected near a metropolitan sewage treatment plant. *Arch. Environ. Contam. Toxicol.* 40:3:392.

Folmar, L.C.; Denslow, N.D.; Rao, V.; Chow, M.; Crain, A.D.; Enblom, J.; Marcino, J.; and Guillette Jr., L.J. 1996. Vitellogenin induction and reduced serum testosterone concentrations in feral male carp (*Cyprinus carpio*) captured near a major metropolitan sewage treatment plant. *Environ. Health Perspect.* 104:1096.

Folmar, L.C.; Denslow, N.D.; Wallace, R.A.; LaFleur, G.; Gross, T.S.; Bonomelli, S.; and Sullivan, C.V. 1995. A highly conserved N-terminal sequence for teleost vitellogenin with potential value to the biochemistry, molecular biology and pathology of vitellogenesis. *J. Fish Biol.* 46:255.

Folmar, L.C.; Gardner, G.R.; Schreibman, M.P.; Magliulo–Cepriano, L.; Mills, L.J.; Zaroogian, G.; Gutjahr–Gobell, R.; Haebler, R.; Horowitz, D.B.; and Denslow, N.D. 2001. Vitellogenin-induced pathology in male summer flounder (*Paralichthys dentatus*). *Aquat. Toxicol.* 51:4:431.

Folmar, L.C.; Hemmer, M.; Webb, R.; Bowman, C.; Kroll, K.; and Denslow, N.D. 1999. Relative estrogenicity of estradiol, ethynyl estradiol and diethylstilbestrol in an in vivo, male sheepshead minnow (*Cyprinodon variegatus*) vitellogenin bioassay, submitted.

Folmar, L.D.; Denslow, N.D.; Wallace, R.A.; LaFleur, G.; Gross, T.S.; Bonomelli, S.; and Sullivan, C.V. 1995. A highly conserved N-terminal sequence for teleost vitellogenin with potential value to the biochemistry, molecular biology and pathology of vitellogenesis. *J. Fish Biol.* 46:255.

Foran, C.M.; Bennett, E.R.; and Benson, W.H. 2000. Exposure to environmentally relevant concentrations of different nonylphenol formulations in Japanese medaka. *Marine Environmental Research* 50:1-5:135.

Garcia–Reyero, N.; Grau, E.; Castillo, Jose Lopez De; Alda, M.; Barcelo, D.; and Pina, B. 2001 Monitoring of endocrine disruptors in surface waters by the yeast recominant assay. *Environ. Toxicol. Chem.* 20:6:1152

Giesy, J.P.; Tillitt, D.E.; Verbrugge, D.A.; Sanderson, J.T.; Jude, D.J.; Gale, R.W.; Meadows, J.C.; Zajieck, J.L.; Peterman, P.H.; Schwartz, T.R.; and Tuchman, M.L. 1997. Polychlorinated dibenzo-*p*-dioxins, dibenzofurans, biphenyls and 2,3,7,8-tetrachlorodibenzo-*p*-dioxin equivalents in fishes from Saginaw Bay, Michigan. *Environ. Toxicol. Chem.* 16:713.

Goodbred, S.L.; Gilliom, R.J.; Gross, T.S.; Denslow, N.P.; Bryant, W.L.; and Schoeb, T.R. 1996. *Reconnaissance of 17β-estradiol, 11-ketotestosterone, vitellogenin, and gonad histopathology in common carp of United States streams: Potential for contaminant-induced endocrine disruption.* U.S. Geological Survey Open-File Report 96-627. National Water Quality Assessment, Pesticide National Synthesis Project, USGS.

Gray, M.A. and Metcalf, C.D. 1997. Induction of the testis–ova in Japanese medaka (*Oryzias latipes*) exposed to *p*-nonylphenol. *Env. Tox. and Chem.* 18(5):1082-1086.

Greenberg, M. E. and Bender, T. P. 1997. *Identification of newly transcribed RNA*. Current protocols in molecular biology. (V.B. Chanda, Ed.), pp. 4.10.1. John Wiley and Sons.

Gronen, S.; Denslow, N.; Manning, S.; Barnes, S.; Barnes, D.; and Brouwer, M. 1999. Serum vitellogenin levels and reproductive impairment of male Japanese medaka (*Oryzias latipes*) exposed to 4-*tert*-octylphenol. *Environmental and Health Perspectives*, accepted.

Gross, T.S.; Sepulveda, M.S.; Wieser, C.M.; Wiebe, J.J.; Schoeb, T.R.; Denslow, N.D.; and Johnson, W.E. 2001. "Seasonal reproductive cycles of Florida largemouth bass (*Micropterus salmoides floridanus*)." Submitted to the *Proceedings of the 2000 Black Bass Symposium*, American Fisheries Society.

Harries, J.E.; Sheahan, D.A.; Jobling, S.; Matthiessen, P.; Neall, P.; Sumpter, J.P.; Tylor, T.; and Zaman, N. 1997. Estrogenic activity in five United Kingdom rivers detected by measurement of vitellogenesis in caged male trout. *Environ. Toxicol. Chem.* 16(3):534-542.

Harries, J.E.; Sheahan, D.A.; Jobling, S.; Matthiessen, P.; Neall, P.; Routledge, E.J.; Rycroft, R.; Sumpter, J.P.; and Taylor, T. 1996. A survey of estrogenic activity in United Kingdom inland waters. *Environ. Toxicol. Chem.* 15:11:1993.

Hashimoto, S.; Bessho, H.; Hara, A.; Nakamura, M.; Iguchi, T.; and Fujita, K. 2000. Elevated serum vitellogenin levels and gonadal abnormalities in wild male flounder (*Pleuronectes yokohamae*) from Tokyo Bay, Japan. *Mar. Environ. Res.* 49:37-53.

Hemmer, M.J.; Hemmer, B.L.; Bowman, C.J.; Kroll, K.J.; Folmar, L.C.; Marcovich, D.; Hoglund, M.D.; and Denslow, N.D. 2001. Effects of p-nonylphenol, methoxychlor, and endosulfan on vitellogenin induction and expression in sheepshead minnow (*Cyprinodon variegatus*). *Environ. Toxicol. Chem.* 20:2:336.

Hemming, J.M.; Waller, W.T.; Chow, M.C.; Denslow, N.D.; and Venables, B. 2001. Assessment of the estrogenicity and toxicity of a domestic wastewater effluent flowing through a constructed wetland system using biomarkers in male fathead minnows (*Pimephales promelas* rafinesque, 1820). *Environ. Toxicol. Chem.* 20(10):2268. © SETAC.

Holbech, H.; Andersen, L.; Petersen, G.I.; Korsgaard, B.; Pedersen, K.L.; and Bjerregaard, P. 2001. Development of an ELISA for vitellogenin in whole body homogenate of zebrafish (*Danio rerio*). *Comp. Biochem. Physiol. C-Toxicol. Pharmacol.* 130:1:119.

Jardine, J.J.; Van der Kraak, G.J.; and Munkittrick, K.R. 1996. Capture and confinement stress in white sucker exposed to bleached Kraft pulp mill effluent. *Ecotoxicol Environ Saf.* 33:3:287.

Jensen, K.M.; Korte, J.J.; Kahl, M.D.; Pasha, M.S.; and Ankley, G.T. 2001. Aspects of basic reproductive biology and endocrinology in the fathead minnow (*Pimephales promelas*). *Comp. Biochem. Physiol. C-Pharmacol. Toxicol. Endocrinol.* 128:1:127. Published by Elsevier Science Inc.

Jobling, S.; Nolan, M.; Tyler, C.R.; Brighty, G.; and Sumpter, J.P. 1998. Widespread sexual disruption in wildfish. *Environ. Sci. Technol.* 32:2498.

Johnson, A.C. and Sumpter, J.P. 2001. Removal of endocrine-disrupting chemicals in activated sludge treatment works. *Centre for Ecology and Hydrology,* Wallingford, Oxfordshire OX10 8BB, UK, and Dept. of Biology and Biochem., Brunel Univ., Uxbridge, Middlesex UB8 3PH, UK.

Kahl, M.D.; Jensen, J.J.; Korte, G.T.; and Ankley, G.T. 2001. Effects of handling on endocrinology and reproductive performance of the fathead minnow. *J. Fish Biol.* 59:3:515.

Kolpin, D.W.; Furlong, E.T.; Meyer, M.T.; Thurman, E.M.; Zaugg, S.D.; Barber, L.B.; and Buxton, H.T. 2002. Pharmaceuticals, hormones, and other organic wastewater contaminants in U.S. streams, 1999-2000: a national reconnaissance. *Environ. Sci. Technol.* 36-1202.

Kramer, V.J.; Miles–Richardson, S.; Pierens, S.L.; and Giesy, J.P. 1998. Reproductive impairment and induction of alkaline-labile phosphate, a biomarker of estrogen exposure, in fathead minnows (*Pimephales promelas*) exposed to waterborne 17 beta-estradiol. *Aquatic Toxicology* 40:4. (C) 1998 Elsevier Science B.V. All rights reserved.

Lange, R.; Hutchinson, T.H.; Croudace, C.P.; and Siegmund, F. 2001. Effects of the synthetic estrogen 17 alpha-ethinylestradiol on the life-cycle of the fathead minnow (*Pimephales promelas*). *Environ. Toxicol. Chem.* 20:6:1216.

Larsson, D.G.J.; Adolfsson–Erici, M.; Parkkonen, J.; Pettersson, M.; Berg, A.H.; Olsson, P.E.; and Förlin, L. 1999. Ethinyloestradiol — an undesired fish contraceptive? *Aquatic Toxicology* 45:91.

Latonnelle, K.; Le Menn, F.; and Bennetau–Pelissero, C. 2000. In vitro estrogenic effects of phytoestrogens in rainbow trout and Siberian sturgeon. *Ecotoxicology* 9:1-2:115.

Laws, S.C.; Carey, S.A.; Ferrell, J.M.; Bodman, G.J.; and Cooper, R.L. 2000. Estrogenic activity of octylphenol, nonylphenol, bisphenol A, and methoxychlor in rats. *Toxicological Sciences* 54:154.

Legler, J.; van den Brink, C.E.; Brouwer, A.; Murk, A.J.; van der Saag, P.T.; Vethaak, A.D.; and van der Burg, B. 1999. Development of a stably transfected estrogen receptor-mediated luciferase reporter gene assay in the human t47d breast cancer cell line. *Toxicological Sciences* 48:55.

Lye, C.M.; Frid, L.J.; Gill, M.E.; Cooper, D.W.; and Jones, D.M. 1999. Estrogenic alkylphenols in fish tissues, sediments, and waters from the UK Tyne and Tees estuaries. *Environ. Sci. Technol.* 33:1009.

McNatt, H.B.; Angus, R.A.; and Howell, W.M. 2000. Effects of paper mill effluent on a population of eastern mosquitofish, *Gambusia holbrooki*. Poster #13. Abstract presented at *e.hormone 2000 Conference*, October 15-18, 2000, New Orleans, La.

Metcalfe, C.D.; Metcalfe, T.L.; Kiparissis, Y.; Koenig, B.G.; Khan, C.; Hughes, R.J.; Croley, T.R.; March, R.E.; and Potter, T. 2001. Estrogenic potency of chemicals detected in sewage treatment plant effluents as determined by *in vivo* assays with Japanese medaka (*Oryzias latipes*). *Environ. Toxicol. Chem.* 20:2:297.

Miles–Richardson, S.R.; Kramer, V.J.; Fitzgerald, S.D.; Render, J.A.; Yamini, B.; Barbee, S.J.; and Giesy, J.P. 1999. Effects of waterborne exposure of 17 beta-estradiol on secondary sex characteristics and gonads of fathead minnows (*Pimephales promelas*). *Aquatic Toxicology* (C) 1999 Elsevier Science B.V. All rights reserved. 1999 NOV.

Munkittrick, D.R. and Van Der Kraak, G. 1999. "Appropriate uses of physiological techniques for endocrine studies" environmental toxicology and risk assessment: standardization of biomarkers for endocrine disruption and environmental assessment," ASTM STP Vol. 8:1364, D. S. Henshel Ed. M. C. Black, M. C. Harrass, American Society of Testing and Materials, West Conshohocken, Pa.

Munkittrick, K.R.; Servos, M.R.; Carey, J.H.; and Van Der Kraak, G.J. 1997. Environmental impacts of pulp and paper wastewater: Evidence for a reduction in environmental effects at North American pulp mills since 1992. *Water Sci. Technol.* 35:329.

Munkittrick, K.R.; Van Der Kraak, G.J.; McMaster, M.E.; Van Den Heuvel, M.R.; and Servos, M.R. 1994. Survey of receiving-water environmental impacts associated with discharges from pulp mills. 2. Gonad size, liver size, hepatic EROD activity and plasma sex steroid levels in white sucker. *Environ. Toxicol. Chem.*13:1089.

Nagler, J.J.; Bourma, J.; Thorgaard, G.H.; and Dauble, D.D. 2001. High incidence of a male-specific genetic marker in phenolic female Chinook salmon from the Columbia River. *Environmental Health Perspectives* 109:67.

National Research Council. 1999. *Hormonally active agents in the environment*. National Academy Press, Washington, DC.

National Research Council, Committee on Biological Markers. 1987. Biological markers in environmental health research. *Environ Health Perspect.* 74:3-9.

National Research Council. 1989. *Biological markers in reproductive toxicology*. National Academy Press, Washington, DC.

Nichols, K.M.; Miles–Richardson, S.R.; Snyder, E.M.; Giesy, J.P. 1999. Effects of exposure to municipal wastewater in situ on the reproductive physiology of the fathead minnow (Pimephales promelas). *Environ. Toxicol. Chem.* 18:9.

Orlando, E.F.; Denslow, N.D.; Folmar, L.C.; and Guillette, L.J. 1999. A comparison of the reproductive physiology of largemouth bass (*Micropterus salmoides*) collected from the Escambia and Blackwater Rivers in Florida. *Environ. Health Perspect.* 107:3:199.

Panter, G.H.; Thompson, R.S.; and Sumpter, J.P. 1998. Adverse reproductive effects in male fathead minnows (*Pimephales promelas*) exposed to environmentally relevant concentrations of the natural oestrogens, oestradiol, and oestrone. *Aquatic Toxicol..* 42:243.

Panter, G.H.; Thompson, R.S.; and Sumpter, J.P. 2000. Intermittent exposure of fish to estradiol. *Environ. Sci. Technol.* 34:13:2756.

Rodgers–Gray, T.P.; Jobling, S.; Kelly, C.; Morris, S.; Brighty, G.; Waldock, M.J.; Sumpter, J.P.; and Tyler, C.R. 2001. Exposure of juvenile roach (*Rutilus rutilus*) to treated sewage effluent induces dose-dependent and persistent disruption in gonadal duct development. *Environ. Sci. Technol.* 35:3:462.

Routledge, E.J. and Sumpter, J.P. 1996. Estrogenic activity of surfactants and some of their degradation products assessed using a recombinant yeast strain. *Environ. Toxicol. Chem.* 15(3):241.

Scholz, S. and Gutzeit, H.O. 2000. 17-a-ethinylestradiol affects reproduction, sexual differentiation and aromatase gene expression of the medaka (*Oryzias latipes*). *Aquatic Toxicology* 50:363. © Elsevier Science B.V. All Rights Reserved.

Sedlak, D.L.; Gray, J.L.; and Pinkston, K.E. 2000. Understanding microcontaminants in recycled water. *Environ. Sci. Technol.* 34:23:508.

Sepulveda, M.S.; Johnson, W.E.; Higman, J.C.; Denslow, N.D.; Schoeb, T.R.; and Gross, T.S. 2001. An evaluation of biomarkers of reproductive function and potential contaminant effects in Florida largemouth bass (*Micropterus salmoides floridanus*) sampled from the St. Johns River. *The Science of the Total Environment.* Article in Press.

Sepulveda, M.S.l Ruessler, D.S.; Denslow, N.D.; Holm, S.E.; Schoeb, T.R.; and Gross, T.S. 2001. Assessment of reproductive effects in largemouth bass (*Micropterus salmoides*) exposed to bleached/unbleached Kraft mill effluents. *Arch. Environ. Contam. Toxicol.* 41:4:475.

Smith, F.J.F. 1978. Seasonal changes in the histology of the gonads and dorsal skin of the fathead minnow, *(Pimephales promelas)*. Can. J. Zool. 56:2103.

Sohoni, P. and Sumpter, J.P. 1998. Several environmental oestrogens are also anti-androgens. *J. Endocrinol* 158:327.

Suter, W. II. 1993. *Ecological risk assessment.* Lewis Publishers, Chelsea, Mich.

Ternes, T.A.; Kreckel, P.; and Mueller, J. 1999. Behavior and occurrence of estrogens in municipal sewage treatment plants II. Aerobic batch experiments with activated sludge. *Sci Total Environ.* 225:91-99.

Thompson, S.; Tilton, F.; Schlenk, D.; and Benson, W.H. 2000. Comparative vitellogenic responses in three teleost species: extrapolation to in situ field studies. *Marine Environmental Research* 50:1-5:185.

Thorpe, K.L.; Hutchinson, T.H.; Hetheridge, M.J.; Sumpter, J.P.; and Tyler, C.R. 2000. Development of an in vivo screening assay for estrogenic chemicals using juvenile rainbow trout (*Oncorhynchus mykiss*). *Environ. Toxicol. Chem.* 19:11:2812.

Tyler, C.R. and Routlege, E.J. 1998. Oestrogenic effects in fish in English rivers with evidence of their causation. *Pure and Applied Chemistry* 70:1795.

Tyler, C.R.; van Aerle, R.; Hutchinson, T.H.; Maddix, S.; and Trip, H. 1999. An in vivo testing system for endocrine disruptors in fish early life stages using induction of vitellogenin. *Environ. Toxicol. Chem.* 18:2:337.

Van den Belt, K.; Verheyen, R.; and Witters, H. 2001. Reproductive effects of ethynylestradiol and 4t-octylphenol on the zebrafish (*Danio rerio*). *Arch. Environ. Contam. Toxicol.* 41:4:458.

Van Der Kraak, G.; Hewitt, M.; Lister, A.; McMaster, M.E.; and Munkittrick, K.R. 2001. Endocrine toxicants and reproductive success in fish. *Hum. Ecol. Risk Assess.* 7:5:1017.

Wang, H.; Yan, T.; Tan, J.T.T.; and Gong, Z.Y. 2000. A zebrafish vitellogenin gene (vg3) encodes a novel vitellogenin without a phosvitin domain and may represent a primitive vertebrate vitellogenin gene. *Gene* 256:1-2:303.

Williams, P.J.; Courtenay, S.C.; and Wilson, C.E. 1998. Annual sex steroid profiles and effects of gender and season on cytochrome P450 mRNA induction in Atlantic tomcod (*Microgadus tomcod*). *Environ. Toxicol. Chem.* 17:1582-1588.

Zacharewski, T. 1997. In vitro bioassays for assessing estrogenic substances. *American Chemical Society* 31(3):613.

Zacharewski, T. 1997. In vitro bioassays for assessing estrogenic substances. *Environ. Sci. Technol.* 31:613.

Zhou, T.; Weis, J.S.; John–Alder, H.B.; and Weis, P. 1999. Thyroidal status of mummichog (*Fundulus heteroclitus*) from a polluted versus a reference habitat. *Environ. Toxicol. Chem.*18:2817.

Zillioux, E.J.; Johnson, I.C.; Kiparissis, Y.; Metcalfe, C.D.; Wheat, J.V.; Ward, S.G.; and Liu, H. 2001. The sheepshead minnow as an in vivo model for endocrine disruption in marine teleosts: A partial life-cycle test with 17β-ethynlestradiol-33. *Environ. Toxicol. Chem.* 20(9):1968. © SETAC.

WERF Subscribers

The following agencies and corporations are WERF's charter stormwater research supporters:

Stormwater Agency

California
Monterey, City of
Santa Rosa, City of

Colorado
Boulder, City of

Georgia
Griffin, City of

Kansas
Overland Park, City of

Kentucky
Louisville & Jefferson County Metropolitan Sewer District

North Carolina
Cary, Town of
Charlotte, City of/Mecklenburg County

Pennsylvania
Philadelphia, City of

Corporate

BaySaver Inc.
Vortechnics Inc.
WWETCO, LLC

State

Fresno Metropolitan Flood Control District, Calif.
Urban Drainage & Flood Control District, Colo.

Wastewater Utility

Alabama
Montgomery Water Works & Sanitary Sewer Board

Alaska
Anchorage Water & Wastewater Utility

Arizona
Gila Resources
Glendale, City of, Utilities Department
Mesa, City of
Phoenix Water Services Department
Pima County Wastewater Management

Arkansas
Little Rock Wastewater Utility

California
Central Contra Costa Sanitary District
Contra Costa Water District
Crestline Sanitation District
Delta Diablo Sanitation District
Dublin San Ramon Services District
East Bay Municipal Utility District
El Dorado Irrigation District
Escondido, City of
Fairfield-Suisun Sewer District
Irvine Ranch Water District
Las Virgenes Municipal Water District
Lodi, City of
Los Angeles, City of
Los Angeles County, Sanitation Districts of
Orange County Sanitation District
Palo Alto, City of
Riverside, City of
Sacramento Regional County Sanitation District
San Diego Metropolitan Wastewater Department, City of
San Francisco, City & County of
San Jose, City of
Santa Barbara, City of
Santa Rosa, City of
South Bayside System Authority
South Orange County Wastewater Authority
Union Sanitary District

Colorado
Boulder, City of
Colorado Springs, City of
Littleton/Englewood Water Pollution Control Plant
Metro Wastewater Reclamation District, Denver

Connecticut
New Haven, City of, WPCA

District of Columbia
District of Columbia Water & Sewer Authority

Florida
Broward, County of
Fort Lauderdale, City of
Gainesville Regional Utilities
JEA
Kissimmee, City of, Department of Water Resources
Miami-Dade Water & Sewer Authority
Orange County Utilities Department
Orlando, City of
Reedy Creek Improvement District
Seminole County Environmental Services
St. Petersburg, City of
Stuart Public Utilities
Tallahassee, City of
Tampa, City of
West Palm Beach, City of

Georgia
Atlanta Wastewater Services
Augusta, City of
Clayton County Water Authority
Cobb County Water System
Columbus Water Works
Fulton County
Gwinnett County Department of Public Utilities
Macon Water Authority
Savannah, City of

Hawaii
Honolulu, City and County of

Illinois
American Bottoms Wastewater Treatment Plant
Dupage, County of
Greater Peoria Sanitary District
Metropolitan Water Reclamation District of Greater Chicago
Wheaton Sanitary District

Indiana
Fort Wayne, City of
Indianapolis, City of

Iowa
Ames, City of
Des Moines Metro Wastewater Reclamation Authority

Kansas
Johnson County Unified Wastewater Districts
Unified Government of Wyandotte County/Kansas City, City of

Kentucky
Louisville & Jefferson County Metropolitan Sewer District

Louisiana
Sewerage & Water Board of New Orleans

Maine
Bangor, City of

Maryland
Anne Arundel County Bureau of Utility Operations
Howard County Department of Public Works
Washington Suburban Sanitary Commission

Massachusetts
Boston Water & Sewer Commission
Massachusetts Water Resources Authority
Upper Blackstone Water Pollution Abatement District

Michigan
Ann Arbor, City of
Detroit, City of

Grand Rapids, City of
Holland Board of Public Works
Lansing, City of
Owosso Mid-Shiawassee County WWTP
Saginaw, City of
Wayne County Department of Environment
Wyoming, City of

Missouri
Independence, City of
Kansas City Missouri Water Services Department
Little Blue Valley Sewer District
Metropolitan St. Louis Sewer District

Nebraska
Lincoln Wastewater System

Nevada
Henderson, City of

New Jersey
Passaic Valley Sewerage Commissioners

New York
New York City Department of Environmental Protection

North Carolina
Cary, Town of
Charlotte/Mecklenburg Utility Department
Durham, City of
Metropolitan Sewerage District of Buncombe County
Orange Water & Sewer Authority

Ohio
Akron, City of
Butler County Department of Environmental Services
Columbus, City of
Metropolitan Sewer District of Greater Cincinnati
Northeast Ohio Regional Sewer District

Oklahoma
Tulsa, City of

Oregon
Clean Water Services
Eugene/Springfield Water Pollution Control
Water Environment Services

Pennsylvania
Philadelphia, City of
University Area Joint Authority, State College

South Carolina
Charleston Commissioners of Public Works
Mount Pleasant Waterworks & Sewer Commission
Spartanburg Sanitary Sewer District

Tennessee
Cleveland, City of
Knoxville Utilities Board
Murfreesboro Water & Sewer Department
Nashville Metro Water Services

Texas
Amarillo, City of
Austin, City of
Dallas Water Utilities
Denton, City of
El Paso Water Utilities
Fort Worth, City of
Gulf Coast Waste Disposal Authority
Houston, City of
San Antonio Water System
Trinity River Authority

Utah
Salt Lake City Corporation

Virginia
Alexandria Sanitation Authority
Arlington, County of
Fairfax County Virginia
Hampton Roads Sanitation District
Henrico, County of
Hopewell Regional Wastewater Treatment Facility
Loudoun County Sanitation Authority
Prince William County Service Authority
Richmond, City of

Washington
Edmonds, City of
Everett, City of
King County Department of Natural Resources
Seattle Public Utilities

Wisconsin
Green Bay Metro Sewerage District
Madison Metropolitan Sewerage District
Milwaukee Metropolitan Sewerage District
Racine, City of
Sheboygan Regional Wastewater Treatment
Wausau Water Works

Australia
Sydney Water Corporation

Canada
Toronto, City of, Ontario

Mexico
Servicios de Agua y Drenaje de Monterrey, I.P.D.

United Kingdom
Yorkshire Water Services Limited

Corporate

The ADVENT Group Inc.
Alan Plummer & Associates
Alpine Technology Inc.
Aquateam–Norwegian Water Technology Centre A/S
Black & Veatch
Boyle Engineering Corporation
BPR CSO
Brown & Caldwell
Burns & McDonnell
The Cadmus Group
Camp Dresser & McKee Inc.
Carollo Engineers Inc.
Carpenter Environmental Associates Inc.
CDS Technologies Inc.
Chemtrac Systems Inc.
CH2M HILL
Clancy Environmental Consultants Inc.
Damon S. Williams Associates, LLC
Earth Tech Inc.
Ecolab Water Care Services
EMA Inc.
The Eshelman Company Inc.
Finkbeiner, Pettis, & Strout Inc.
Frontier Geosciences Inc.
ftn Associates Inc.
Gannett Fleming Inc.
GE Betz
Golder Associates Inc.
Greeley and Hansen LLC
The HACH Company
Hazen & Sawyer, P.C.
HDR Engineering Inc.
HNTB Corporation
HydroQual Inc.
Insituform Technologies Inc.
Institute for Environmental Technology & Industry, Korea
Jacobson Helgoth Consultants Inc.
Jason Consultants Inc.
Jordan, Jones, & Goulding Inc.
KCI Technologies Inc.
Kelly & Weaver, P.C.
Kennedy/Jenks Consultants
Komline Sanderson Engineering Corporation
Lawler, Matusky & Skelly Engineers, LLP
Limno-Tech Inc.
Lombardo Associates Inc.
Malcolm Pirnie Inc.
McKim & Creed
MWH
Odor & Corrosion Technology Consultants Inc. (OCTC)
ONDEO Degremont Inc.
PA Government Services Inc.
Parametrix Inc.
Parsons
Post, Buckley, Schuh & Jernigan
The RETEC Group
R.M. Towill Corporation
Royce Technologies
Sear-Brown
SYNAGRO
Tetra Tech Inc.
Trojan Technologies Inc.
URS Corporation
USFilter NATC
Wade-Trim Inc.
Weston Solutions Inc.
Woodard & Curran
Woodruff & Howe Environmental Engineering Inc. (WHEE)
WRc/D&B, LLC

Industry

American Electric Power
BP-Amoco Corporation
Chevron Research & Technology Company
The Coca-Cola Company
Dow Chemical Company
DuPont Company
Eastman Chemical Company
Eastman Kodak Company
Merck & Company Inc.
ONDEO Services
Procter & Gamble Company
Reliant Energy
Severn Trent Services Inc.
Shell Global Solutions
Thames Water Plc
United Water Services LLC

WERF Product Order Form

As a benefit of joining the Water Environment Research Foundation, subscribers are entitled to receive one complimentary copy of all final reports and other products. Additional copies are available at cost (usually $10). To order your complimentary copy of a report, please write "free" in the unit price column. WERF keeps track of all orders. If the charge differs from what is shown here, we will call to confirm the total before processing.

Name ____________________ Title ____________________

Organization ____________________

Address ____________________

City ________ State ________ Zip Code ________ Country ________

Phone ________ Fax ________ Email ________

Stock #	Product	Quantity	Unit Price	Total
			Postage & Handling	
			VA Residents Add 4.5% Sales Tax	
			Canadian Residents Add 7% GST	
			TOTAL	

Method of Payment: (All orders must be prepaid.)

❑ Check or Money Order Enclosed

❑ Visa ❑ Mastercard ❑ American Express

Account No. ____________________ Exp. Date ________

Signature ____________________

Shipping & Handling:

Amount of Order	United States	Canada & Mexico	All Others
Up to but not more than:	Add:	Add:	Add:
$20.00	$5.00*	$8.00	50% of amount
30.00	5.50	8.00	40% of amount
40.00	6.00	8.00	
50.00	6.50	14.00	
60.00	7.00	14.00	
80.00	8.00	14.00	
100.00	10.00	21.00	
150.00	12.50	28.00	
200.00	15.00	35.00	
More than $200.00	Add 20% of order	Add 20% of order	

*minimum amount for all orders

Note: Please make checks payable to the Water Environment Research Foundation.

To Order (Subscribers Only):

Log on to www.werf.org and click on the "Product Catalog."

Phone: (703) 684-2470
Fax: (703) 299-0742.

WERF
Attn: Subscriber Services
601 Wythe Street
Alexandria, VA 22314-1994

To Order (Non-Subscribers):

Non-subscribers may be able to order WERF publications either through WEF (www.wef.org) or IWAP (www.iwap.co.uk). Visit WERF's website at www.werf.org for details.

IWA Publishing's authorised EU representative for General Product Safety Regulations is Diane D'Arras, 15 rue Duret, 75116 Paris, France, e-mail: safety@iwap.co.uk.

Printed and bound by CPI Group (UK) Ltd, Croydon, CR0 4YY
12/05/2026
02108865-0006